Conceiving God

DAVID LEWIS-WILLIAMS

Conceiving God

THE COGNITIVE ORIGIN AND EVOLUTION OF RELIGION

with 49 illustrations

Thames & Hudson

... throughout the world
Impious war is raging.

VIRGIL, *GEORGICS* I.510–511

And we are here as on a darkling plain;
Swept with confused alarms of struggle and flight,
Where ignorant armies clash by night.

MATTHEW ARNOLD, *DOVER BEACH*

[B]e ready always to give an answer to every man that
asketh you a reason of the hope that is in you with
meekness and fear.

1 PETER 3:15

First published in the United Kingdom in 2010 by
Thames & Hudson Ltd, 181A High Holborn, London WC1V 7QX

thamesandhudson.com

Lines quoted from T. S. Eliot's poem *The Love Song of J. Alfred Prufrock* are
reproduced by permission of the publishers, Faber and Faber Ltd.

British Library Cataloguing-in-Publication Data
A catalogue record for this book is available from the British Library

ISBN 978-0-500-05164-1

Printed and bound in Slovenia by DZS

Contents

Preface   6

Prolegomena: Windows on the Past and
Some of the Questions they Raise   11

1   A New Way of Thinking   23

2   From Skies to Species   49

3   A Tale of Two Scientists   87

4   Explaining Religion   115

5   Religious Experience   139

6   Religious Belief   161

7   Religious Practice   184

8   Stone Age Religion   207

9   Hildegard on the African Veld   232

10   God's Empire Strikes Back   257

Epilegomena: Of Babies and Bathwater   290

Notes   294

Bibliography and Guide to Further Reading   301

Acknowledgments   315

Sources of Illustrations   315

Index   316

Preface

It appears to me (whether rightly or wrongly) that direct arguments against Christianity and theism produce hardly any effect on the public; and freedom of thought is best promoted by the gradual illumination of men's minds which follows from the advance of science. It has, therefore, been always my object to avoid writing on religion, and I have confined myself to science.[1]

No evidence is powerful enough to force acceptance of a conclusion that is emotionally distasteful.[2]

In 1880, Charles Darwin wrote the first of these quotations to the British socialist Edward Aveling. In the second quotation, written some 60 years later, Theodosius Dobzhansky acknowledged the difficulty of getting people to change their minds about some immensely important things.

These epigraphs may seem a depressing start to a book such as this, but both writers were correct. Today, in the current science *versus* religion controversy, we see the publication of numerous books filled with arguments designed to show that humankind would be better off without religion. The reasoning in these publications is often rigorous and inescapable; we need such books. Yet whether they have any impact on the thinking of religious people is doubtful. Perhaps some do. And that is why I have somewhat reluctantly decided to address some key, indeed unavoidable, logical issues. Still, it seems clear to me that the gradual advance of scientific knowledge over the last three and more centuries has done better than logical arguments against supernaturalism: it has led to Darwin's 'freedom of thought'.

My own experience has tended to confirm Darwin's lack of faith in 'direct arguments against Christianity and theism'. I did not wrestle with arguments and then, after much intellectual agonizing, come to the conclusion that, logically, there is no such thing as a supernatural realm – and consequently no such thing as God, a Devil, angels, divinely inspired books, miracles, and so forth. Instead, over the years I pondered the long history of religion. In particular, I thought about the implications of the earliest archaeological evidence for religion. It seems indisputable that Upper Palaeolithic people who lived in France and Spain from about 45,000 to 10,000 years ago believed in a supernatural realm and spirit beings whom they tried to contact. They therefore clambered and crawled to a nether world deep inside caves. There they made images of supernaturally powerful spirit animals. Was the God of present-day monotheistic religions trying to get through to these ancient hunters? Or were they struggling to come to terms with something quite different, something going on in their own brains? And so it was through the ages, through the classical civilizations of Greece and Rome, through medieval times and up to the present day. People seem always to have believed in two domains: the material world in which they conduct their daily lives and a spirit realm that they try to contact.

Then I found it salutary to explore social (cultural) anthropology. People live in vastly different societies with different economies, social structures and, of course, religions. There were (and still are) those who believed that God wanted them to sacrifice other people; those who believed that God wanted them to kill animals by bleeding them to death; those who believed that stubborn infidels should be killed; those who claimed that special people could float up into the sky; those who believed that they should cause, by whatever means necessary, other people to accept their religion. For all these people, their own beliefs were divinely revealed and therefore beyond discussion or evaluation. Famine, fire and sword have been religion's dogs of war for so long that there are now many who doubt the possibility of tolerance between major religions. As we look over this sorry tapestry, we must face a fundamental question, one that many today, believers and non-believers alike, try to avoid: Is there really a spirit realm occupied by supernatural beings and forces that are concerned with human life on earth? By contemplating the history of religion and science we are able to answer that question in a way that gradually leads to 'freedom of thought'.

Half a century ago, it seemed that politics was the dangerous dividing factor in the world. Democracy, fascism and communism contended for (in an awful modern cliché) the hearts and minds of the world's people – not to mention economic wealth. Today, many speak of an impending clash between the Christian (or post-Christian) West and Islam. They realize that religion, rather than politics, lies at the very root of the matter, but they feel that they dare not say so. In trying to dodge the issue, they claim that all religions are fundamentally and intrinsically good; freed from 'fanaticism', diverse religions will eventually come to live in contented harmony with one another. 'Ecumenical', 'multi-cultural', 'dialogue' and 'inter-faith' are the key buzzwords.

If all religions are equal, why, we must ask, did the Christian Church spend so many centuries spreading its gospel throughout the world, hunting down heretics, and waging religious wars? Is it conceivable that, for most of its history, the Church was founded on an error? If each religion, not just Christianity, believes that it is a recipient of supernatural intimations, can we really expect them to cease proselytizing? Belief in supernaturally revealed knowledge is fundamentally incompatible with the equality of all organized religions. This gloomy thought is, however, best kept out of sight: to express it is to court accusations of defeatism, moral blindness, incitement to violence, and much else.

Viewing current conflicts from the perspective of an immense past means that you do not see features of present-day religion and the strife that it brings as the product of a few hundred (or even two thousand) years of history. Instead, you see continuities that seem to be independent of history, that are so deep-seated that they cannot merely be products of specific historical events and processes. The long view shows that there are underlying currents that point to something innate in human beings, something that from at least Upper Palaeolithic times has produced what we recognize as 'religion'.

I therefore begin with three Windows on the Past. They are vignettes taken from strongly contrasted societies and periods of history. Why, despite their obvious differences, do they seem to have something in common? These windows introduce themes that the following chapters explore.

The first of these sketches the history of scientific thought from ancient Greece to the Roman emperor Constantine and his official, political, recognition of Christianity. It also deals with the role of Greek philosophy in the formulation of Christian theology, especially as the new religion was conceived by St Paul. Crucially, this chapter identifies the origin and development of two contrasting kinds of knowledge: supernaturally revealed knowledge and that which comes from scientific thought and observation – two warring empires of the mind. This epistemological question lies at the root of the present conflict between religion and science.

Chapter 2 takes the story forward from the time of St Augustine of Hippo, one of the most influential early Church Fathers, to the first two great public clashes between science and religion: the Galileo affair and, four centuries later, the impact of evolution on not just religious thought but all aspects of life. Here we see the gap between the two types of knowledge widening. Science has emerged from the cocoon of religion but still finds it difficult to free itself from the shreds of its previous existence.

Scientists are ordinary people living in diverse communities, not insensate robots. The third chapter therefore outlines the lives and beliefs of Charles Darwin and Alfred Russel Wallace. Wallace also thought of natural selection as a mechanism for evolution but nevertheless pursued an interest in spiritualism. Tensions caused by the contradictions between science and religion were playing themselves out in real life.

Chapter 4 takes the previous chapters as its foundation and discusses some of the ideas that have been advanced to explain how human activities that are readily identifiable as religious have been understood. Any explanation for religion must take into account both its continuities and its diversity. Why do all people at all times seem to have had a religion?

In answering this question, Chapters 5, 6 and 7 consider, respectively, religious experience, belief and practice. Although certain mental experiences are foundational, religion cannot be reduced to them. Rather, religion is the outcome of complex interactions between human neurology, social contexts and repeated practices.

Chapter 8 takes up the story that began at, or even before, the time of the first of the Prolegomena's Windows on the Past. It begins with a brief outline of Upper Palaeolithic religious experience, belief and practice in the deep caves of France and Spain. I then ask how Upper Palaeolithic people used the varied spaces provided by the configurations of the caves. The exploitation of space in Upper Palaeolithic times can be compared with the structure and use of a Christian cathedral. The Upper Palaeolithic caves are the earliest evidence for what I argue is an inevitable relationship between religion and social discrimination.

Chapter 9 shows how religious experience, belief and practice work in vastly differing societies. The two contrasted examples are medieval Europe, with special reference to Hildegard of Bingen, and southern Africa, with its archaeologically known San rock paintings. This rather unexpected juxtaposition shows that specific religious visions are evident in the writings and manuscript illuminations of Hildegard and also in rock paintings made by the San. A coincidence? Almost certainly not.

The final chapter considers responses that religious people are today advancing as they confront the rising tide of disbelief and the secularization of Western civilization. I argue that this sceptical surge derives from the fact that many people no longer believe in a supernatural realm and divinely revealed knowledge – despite the apparently contradictory example of the United States. Increasingly, people are achieving 'freedom of thought'.

The respected Harvard philosopher A. N. Whitehead tackled the problem of religion and science in his 1925 Lowell Lectures:

The clash is a sign that there are wider truths and finer perspectives within which a reconciliation of a deeper religion and a more subtle science will be found.... Science is concerned with the general conditions which are observed to regulate physical phenomena; whereas religion is wholly wrapped up in the contemplation of moral and aesthetic values. On the one side there is the law of gravitation, and on the other the contemplation of beauty and holiness. What one side sees, the other misses; and vice versa.[3]

We may be forgiven for thinking that Whitehead was writing at the beginning of the twenty-first century, so little has changed. But the position he adopts is manifestly untrue. It is one of those apparently finely balanced and therefore rather attractive statements that do not stand up to scrutiny. As the following chapters show, religion does impinge on 'physical phenomena': Christians are asked to believe that Jesus was born of a virgin, that he rose from the dead, and that prayers for healing, rain and so forth are sometimes answered. At the same time, it is clear that 'moral and aesthetic values' are not the exclusive preserve of religion. People of no religious beliefs whatsoever are capable of moral action and human empathy.

Throughout the writing of this book I have been deeply aware that some readers will find any criticism of religion offensive. I have no wish to hurt their feelings. There are a few aspects of religion that almost have me too 'hoping it might be so'. I merely ask believers to ponder Darwin's words at the beginning of this Preface.

Prolegomena
Windows on the Past and
Some of the Questions they Raise

I Ochre

Today, the southernmost tip of Africa is a long way from anywhere important in world affairs, a continental *cul-de-sac* where nothing much ever happens. This was not always so.

Seventy-five thousand years ago, the coastline looked much the same as it still does. Cliffs rose above jagged rocks and sandy coves. A thin coastal haze diffused sharp outlines. High on one of the cliffs was a small cave, no more than a rock shelter. All those millennia ago, people sheltered in it from the wind and lived out their daily lives (Fig. 1). They were among the earliest, though not the very first, anatomically fully modern human beings: earlier species of pre-human beings had, by that time, disappeared from Africa, though some survived for longer in other parts of the world, as did the Neanderthals in Western Europe.

In their seaside rock shelter, generations of this comparatively new *Homo sapiens* species lived comfortably enough.[1] The ocean spawned a variety of fish, shellfish and, occasionally, a beached whale or dolphin. Exposed by the daily retreating tides, nutritious food was there for the taking – an alfresco larder replenished daily. But the people did more than eat. They fashioned enigmatic objects that seem to have had no practical purpose.[2]

Then, about 70,000 years ago, the ice caps at the north and south poles slowly grew in size; as a result the sea-level fell worldwide. The Pleistocene Ice Age was entering a new cold phase. Along the southern African coast there was no ice, but the retreating waves exposed a shelf of rock and sea sand between 10 and 25 km (6 and 16 miles) wide. The winds swept this sand into the cliff-hanging cave and covered the surface on which the occupants had been living. No sign of their presence remained visible, and no one lived there during this period. Probably, the people and generations of their descendants left their rock shelter and followed the slowly retreating sea with its sure food supply. Later, in the long-term rhythm of climate change, the ice caps melted and the sea returned to its present level. People again lived in the cave and sea foods awaited them on the shore below. As their own debris accumulated on top of the wind-blown sand, they were unaware of what lay hidden beneath their floor.

Finally, at the end of the twentieth century, archaeologists, working in what is now known as Blombos Shelter, dug down through the recent occupation layers and on through the sterile, wind-blown sand to the levels sealed and preserved below. Their labours were richly rewarded. They discovered that the early *Homo sapiens* people living in the Blombos rock shelter 75,000 years ago had sought, found and brought home lumps of ochre, as had their even more ancient forebears throughout Africa. They ground this type of rock to a fine red, often sparkling, powder that they may have used for curing animal skins but, more significantly, for body decoration – that is, for marking themselves according to their position in their social group.[3]

Confirmation of a conclusion along these lines was soon at hand. Excavators found seashell beads in the same level. Some of the beads had been treated with red ochre.[4] The people were not merely rubbing red powder or paint on their bodies. They were also decorating themselves with necklaces and thereby saying something about themselves by means of another, but no doubt complementary, 'code'.

That discovery, with its implications of early social distinctions, was fascinating enough. But something even more significant was taking place at Stone Age Blombos. Someone, who will forever remain unknown, man or woman, young or old, took a small, flat piece of ochre that easily fitted in the palm of a hand. Then, instead of grinding its large surfaces to produce red powder, he or she concentrated on its narrow (and to us more insignificant) edge. There this person engraved a neat, symmetrical series of crosses with a line through the middle of them. Around them, he or she scratched a containing line. In this respect, the design resembles an ancient Egyptian cartouche. It seems that the whole 'composition' was a carefully balanced, bounded *unit*, not a mix of unrelated marks. The scratched lines hung together in a patterned way (Fig. 2).

Could this pattern have been simply a one-off meaningless doodle? The excavators' worry was allayed. They found a second piece of ochre, also engraved with crosses. The motif was repeatable. Like a simple utterance that melds a number of different sounds (phonemes) into a single, intelligible entity (a word), the elements of the Blombos motif (crosses and lines) joined together in a complex way to 'say' something.

When these finds were made in the late 1990s, there was already a lively debate among archaeologists, palaeontologists and philosophers about the origins of the sort of thinking and behaviour that we today recognize as 'modern', as opposed to the simpler sort of thinking and behaviour we would expect to find among pre-human hominids, beings that were almost but not quite 'there'. Do we have in these Blombos artefacts the earliest evidence for the type of thought, behaviour and (this is important) mental experiences that are the focus of this book? Have researchers turned up indications that point to a *mental*, rather than an anatomical, Garden of Eden? This biblical

1, 2 Blombos Shelter on the southern coast of South Africa looks out over the Indian Ocean. Seventy-five thousand years ago, people living here engraved enigmatic geometric patterns on pieces of ochre (left). Those patterns point to a way of thinking that is unique to Homo sapiens.

metaphor may be deceptive, so I should point out right away that becoming human was not a sudden 'revolution' (to take another metaphor) in the sense that there was a time and place where fully modern human behaviour, language and thought suddenly arrived *en bloc*. Rather, there was a long period during which certain behaviours developed and, in some instances, perhaps disappeared before returning later.[5]

No one believes that the two Blombos pieces of ochre were the very first such artefacts. The making of patterns probably started somewhat earlier, though no persuasive evidence has – so far – been found to support this presumption. The Blombos pieces therefore remain the oldest evidence that we have for a complex way of thinking. And, if we are rash enough to define 'art' rather broadly, they are the world's oldest *objets d'art*.

If the twice-repeated (or almost repeated) pattern *meant* something, *what* did it mean? Did the series of crosses merely depict, say, a net for fishing or for carrying one's possessions – in others words, a utilitarian object that everyone would easily recognize? If the people indeed made nets, the retiform pattern would have been familiar to them. This straightforward explanation seems highly unlikely, given the awkward placing of the designs on the narrow edges of the ochre pieces. In any event, why would anyone wish to make a literal depiction of (only part of) a net on a small piece of ochre – a material normally used for obtaining powder that was probably intended to say things about the people on whom it was rubbed? If the pattern did depict a net, there must have been more significance than we now realize to nets *and* to the substance on which images of them were engraved.

Something that ochre itself could represent, of course, is blood. Many researchers suggest this possibility when they find ochre in other archaeological sites throughout the world, some much older than Blombos. As we shall see, blood leaves an ambivalent, disturbing trail throughout human history. It may well have been a significant substance in the deepest prehistoric times.[6] The red colour of ochre may have signified blood, but the bodily substance itself would, in turn, have stood for other far more complex *ideas*. Life, death or a lineage spring to mind as possible significances because blood can stand for them in our thought today. But in Blombos times blood may have signified ideas and beliefs at which we cannot now even guess. There was probably a set of complex, symbolic meanings. It may have run something like this: blood → ochre → menstruation → birth → death → social relationships → a specific lineage → ancestors → gods → religious experiences → life after death. That is how human symbolic thought works: it leads us through a maze of associations, some obvious, some less so. The complex associations of blood combine to make it an emotionally powerful symbol.

This kind of representation of something – even as abstract as a mood or emotion – by a motif, object or sound that bears no resemblance to it was a huge step in the development of human thinking. Symbolic thought and language were the gateway to all that makes us human. No fully human communication would have been possible without symbolic thought. And the anthropologist Roy Rappaport goes on to argue that 'religion emerged with language'.[7]

But why was the pattern twice placed on the *narrow edges* of pieces of ochre and not on the larger flat surfaces where it could be more readily engraved and seen? Easy visibility does not seem to have been the engravers' priority. Moreover, the patterns on both pieces neatly fill, and so seem to be in some way related to, the narrow edges of the ochre. I therefore ask: do the patterns refer to something *inside* the ochre, rather as the title on a book spine refers to what is inside the volume? Perhaps this 'something' was released by the grinding of the larger surfaces and then deployed in powder form in an emotionally charged ritual. From hints such as these, slender though they may be, it is beginning to sound as if this 'something' may have been some sort of spiritual concept, power or being. Clearly, we are poised on the brink of a daunting question: do we have here at Blombos the earliest evidence not only for symbolic thought but also for the type of thinking and experiencing that would eventually grow into what we now understand by the word 'religion' – thinking that goes beyond what is visually evident in the material world and that engages unseen powers and realms? The details may remain forever unknown, but the lineaments of religion were there.

Our next two windows on the past offer glimpses of very different human communities, each of which most definitely had (and, it could be argued, suffered from) a religion.

II Blood

Silent on a peak in Darien, stout Córtez was dumbstruck by the vastness of the Pacific Ocean – at least in legend. What did this unsuspected expanse mean for his monarch's desire to find a new route to the East? Later, when Spanish Conquistadors came across the tail-end of the Central American Maya civilization, their Christian sensibilities were revolted by the religious rituals they encountered. In keeping with a current trend to 'respect' things supposedly spiritual and religious, writers today tend to emphasize the indisputably wonderful architecture, feats of engineering, mathematics, art, mythology and astronomy that the Maya produced and to downplay the bloodletting and cruelty that was so important in their life. How did the blood-centred Maya society come about?

There had been a long period of formation during which the Maya civilization gradually emerged. In its earliest form (known as Preclassic), it lasted from about 1500 BC to about AD 200. What is now known as the Classic Maya civilization lasted from about that time to AD 900, though in some regions the old way of life endured until 1541, the time of the Spanish conquest. This is a long span of time, and we should not expect Maya beliefs, myths and rituals to have remained entirely unchanged. Many aspects of Maya belief did change, though recent discoveries of early wall-paintings show that the fundamentals continued through the entire period. Some components of belief systems are remarkably resistant to the economic and political changes that take place around them, as present-day Judaism and Christianity show. A few Maya codices (folding-screen books made of fig-tree bark) were preserved and a handful of Spaniards recorded Maya beliefs, but the most graphic account of their religion and rituals is inscribed on their own buildings, stele and ceramics. These records remained unintelligible until the second half of the twentieth century, when the code of Maya glyphs was cracked in a remarkable series of breakthroughs.[8]

As with many belief systems around the world, including the Western tradition with which this book is primarily concerned, Maya cosmology was tiered. There was an Underworld, a Middleworld and an Overworld. The Underworld could be reached by means of caves or large bodies of water – flooded sinkholes in the limestone Yucatan plateau, lakes or the ocean itself. The Middleworld, on which the Maya lived their daily lives, was oriented according to the four compass directions, each having its own complex imagery and symbolism. The Overworld was the sky, especially the night sky, and learned Maya studied the movement of planets across the canopy of fixed stars to foretell the future and to construct their calendar.

In the codices and carved glyphs we meet the pantheon of gods, animals and hybrid beings that inhabited the cosmological tiers. We also see depictions of the religious techniques that the Maya used to pass from one level of their cosmos to another, from one *experiential* realm to another. As archaeologists and epigraphers continue to decipher Maya glyphs, their painstaking work reveals, to us, utterly alien religious experiences and practices and, at the same time, elements of religion that have disturbing counterparts in the modern Western way of life. Archaeology may be a glass though which we can see only darkly, but in those shadows we can sometimes discern the present prefigured.

We speak of 'the Maya', but they were in fact a number of political groups, not a single centralized state. Small polities were frequently at war with one another. Paradoxically, it was war not peace that held the Maya together. The rulers of these polities were like kings *and* like priests, though some writers prefer the word 'shaman'. They were politically powerful *and* they mediated the tiered cosmos both in their person and in the rituals they performed.

Although Maya wars were politically motivated, they also provided opportunities for the taking of captives, the more politically prominent the better. After a victory, captives were led back to the victors' settlement to be humiliated, tortured and finally despatched. Over the years, the Maya brutalized and sacrificed thousands of captives; no one knows how many. They believed that human beings were created to nourish the gods through sacrifice. Indeed, a primary role of war was to provide a supply of blood and sacrificial victims. The word for sacrifice in the Yukatek Maya language is *p'a chi'*. It means 'to open the mouth' in supplication. In these rituals people literally smeared the blood of sacrificial victims onto the mouths of wood and stone images to feed the gods within.[9]

Sacrifice was closely associated with the pyramids that have become emblematic of Maya culture (Fig. 3). The oldest date back to at least 600 BC. Rising above the baldachin of the forest, these 'mountain-pyramids' were places for accessing supernatural power and, at the same time, tombs. The door to the mountain-pyramid was a constructed cave leading to the heart of the mountain and the supernatural world. Successive rulers built mountain-pyramids on top of and around earlier ones thus making the inner sanctum more and more sacred.

As soon as people create divisions (such as tiers of the cosmos) they create a need to bridge, or mediate, those divisions. If there is no bridge, the divisions can have no practical value. Numerous anthropologists have pointed out that ritual and sacredness often seem to focus on mediators.[10] To take only one example, Christianity focuses, especially in ritual, more on Christ, the one who mediated between Heaven and Earth, than on God the Father or God the Holy Spirit. Both people and structures can act as mediators. Maya pyramids and king-priests were both links between the three cosmological tiers, and complex rituals grew up around them.

3 *At Tikal, a soaring Maya pyramid reaches for the upper level of a tiered cosmos. Beneath it lies a revered ruler, adorned with a hoard of jade, shell and pearl jewelry. Through the pyramid, his spirit ascends to the sky above.*

The stairways that are a prominent feature of Maya pyramids were mediating pathways between divisions of the cosmos. A temple stood on the summit of the mountain-pyramid: the most secret and sacred contact with the gods was, like the throne in Isaiah's vision, high and lifted up. At the foot of the stairways, the great plazas glistened like the sea. Although the plazas themselves were not thought of as entrances to the Underworld, the sea and other bodies of water were so regarded. The whole layout of a Maya city thus replicated the form of Creation, but it was the mountain-pyramid itself that mediated, joined and held everything together in a visually and overpoweringly emotional way.

Blood was another stairway. Conceptually, it joined the levels of the cosmos and, simultaneously, realms of mental experience. Sacrificial rituals 'activated' the cosmology of the mountain-pyramids and were believed to be necessary for the survival of both gods and people. Bloodletting was a climactic element in numerous rituals: the accession of a new ruler, crop-planting, birth, marriage, death, all these and other occasions required the piety of bloodletting. Bishop de Landa, the first Catholic bishop of Yucatan and a fearsomely cruel persecutor of the Maya, described such an occasion in the 1560s, and a great many stele and glyphs on buildings confirm the accuracy of his account. With such explicit evidence at hand, the archaeological glass through which we see the Maya is not nearly as dark as the one through which we see the Blombos pieces of engraved ochre with their hints of complex symbolism.

De Landa reported that huge crowds gathered on the plaza below the mountain-pyramid. On raised platforms and wearing elaborate costumes, troupes of dancers performed frenetically to drums, flutes, whistles and rattles. Groups of people selected to give blood sat in prominent positions: they wore special cloths and bloodletting paper that would show up the blood to good effect. To prepare themselves for their ordeal they had fasted for days and taken steam baths.

At the appropriate moment the king-priest and his wife showed themselves to the crowd. They appeared on the summit of the mountain-pyramid, close to the Overworld. He then cut his penis and she her tongue. To promote the flow of blood and to direct it to the strips of paper, they passed cords, sometimes spliced with thorns, through the holes they had cut in their own bodies. More than that, they became incarnations of not just one but, simultaneously, many gods, such was their power of mediation. Being both human and divine (like Christ), Maya rulers knew that the spirit realm was real and could speak about its 'many mansions' with authority because they had been there and seen it.

If, as some today would have us believe, all religions are groping for spiritual truth and should be 'respected' as such, what are we to say about the Maya? Some people may respond that, *essentially*, Maya religion was one of communion with a creator God and a unifying force within society; as such, it produced admirable art and architecture. These optimists would add that the horrifying parts of Maya religion – the torture, bloodletting and human sacrifices – were not part of *real* Maya religion. But who is to say what is 'real religion' and what are aberrations?

Similarly, we cannot say that militant and harshly fundamentalist elements of the religions with which we are today familiar are not part of *real* religion. The dreadful aspects are as much a part of the religion as the ones which we now find acceptable. All religions have awful as well as more comforting features. My reasons for this apparently hardline judgment will become clear in subsequent chapters.

III Brimstone

As settlements of the time went, Sodom was probably not much different from its smaller neighbour, Gomorrah. And both were probably not much different from any other Bronze or Iron Age Middle Eastern towns. Yet, unlike those other places, Sodom and Gomorrah remain famous thousands of years after their demise, and have entered the mythology of three major religions – Judaism, Christianity and Islam. Why?

All three religions see Abram, or, as he later became known, Abraham, as the archetypal man of faith. Yet the story of his life has some pretty unedifying episodes. Small wonder, then, that the story of Sodom and Gomorrah is usually presented in a highly bowdlerized form. It makes dramatic Sunday school fodder, but only when shorn of its central episode.

Let us take up the biblical version at the point when wealthy Abram was sheltering in Egypt from a famine that was destroying his own land. In Egypt, he passed his beautiful wife Sarai (Sarah) off as his sister. Hiding behind this false identity, she caught the eyes of Pharaoh's princes, and they commended her to the ruler as a pulchritudinous candidate for a new royal wife. Understandably enough, though to Pharaoh's irritation, Abram declined to part with his 'sister'. The Egyptians could hardly be held responsible for this impasse. Nevertheless, as a foretaste (or foreshadowing) of what would happen many years later with Moses, God sent 'great plagues' on Pharaoh's house (GENESIS 12:17). As with the gods in most religious scriptures, the God of the Old Testament does not deal in half-measures. Smitten by these God-sent plagues, Pharaoh allowed Abram to leave with his valuable belongings intact.[11]

On this first 'Exodus' from Egypt, Abram was accompanied by his nephew Lot, another rich man. Soon, as may be expected with nomadic herdsmen, there was a dispute over grazing, and they parted company (GENESIS 13:6–11). After much intertribal warfare and slaughter among the various peoples of the region, Lot settled in Sodom. One night, he invited two angels to sleep over with him, though their heavenly status was unknown to him.[12] The people of Sodom were incensed and demanded that Lot give them his two handsome guests. Lot refused but, astoundingly, offered instead to give the mob his two virgin daughters: 'Let me, I pray you, bring them out unto you, and do ye to them as is good in your eyes' (GENESIS 19:8). The crowd could, he

4 A mosaic (c. 1175-1250) in the nave of Monreale Cathedral, Sicily, shows Lot and his family making their escape from Sodom. Lot's wife unfortunately glances back to see the conflagration of the city and is punished for disobeying God's command. This dramatic incident masks Lot's offer to hand his daughters over to a mob.

said, violate them sexually if they so wished but not the two male visitors whom they passionately wished to 'know' in the biblical sense of the word. The daughters' own view of all this is not recorded. It simply did not matter. After attempts to break down the door of Lot's house (it was men, not girls, that the mob was after), the angels blinded the crowd, and Lot and his wife made their escape from Sodom.

Just in time it seems, because a devastating climax was about to unfold: 'Then the Lord rained upon Sodom and Gomorrah brimstone and fire...out of heaven; And he overthrew those cities, and all the plain, and all the inhabitants of the cities, and that which grew upon the ground' (GENESIS 19:24–5). The wages of sin were indeed death, lots of it. God destroyed Sodom and much else as a result of the people's sinfulness, but he permitted Lot to escape. During this flight from Sodom, God told Lot and his family not to look back at the town. Famously, but understandably given the pyrotechnics taking place there, his wife disobeyed, and God turned her into a pillar of salt (GENESIS 19:26) (Fig. 4).

Lot and his daughters seem to have taken this harsh transformation in their stride. If anyone deserved a shower of brimstone, it was of course Lot himself, he who was prepared to offer his daughters to the mob so that they could 'do ...to them as is good in [their] eyes'. Yet it was he whom the Lord saved. Indeed, Lot lived on to commit incest with his daughters while he was drunk. Each conceived a son; they became the principal ancestors of lineages. The Old Testament is not for the morally squeamish.

Stories of comparably bloody behaviour in the Old Testament bring to mind Maya human sacrifices, the flagellation and crucifixion of Jesus, the initiatory dismemberment of which shamans around the world speak, male and female genital mutilation, and many other grisly beliefs and practices. More specifically, the deaths of Christian martyrs illustrate Christianity's absorption with gruesome ends that parallel, though they may be reluctant to admit it, the atomizing of Islamic suicide bombers. Indeed, gross physical abuse of the human body seems often to be a part of religious experience, belief, narrative and art. We again ask: can we separate the horrifying elements of religion from the more attractive ones and at the same time remember that blood and suffering were for many centuries central, and admired, elements of Christianity? If we think that we can sort out the 'good' from the 'bad' in religions, we must still ask a more fundamental question: why are contradictory elements so frequently intertwined in religions?

These three windows on the past have allowed us to glimpse some aspects of religion that will crop up again when, in the following chapters, we consider how scientific thinking developed in a religious milieu. The first window showed that the kind of symbolic thinking that lies behind religion is of breathtaking antiquity. The second showed how ghastly the combination of religious experience, belief and symbolism can become. The third showed how the adoption of a canon of ancient scriptures can create escalating problems as centuries and millennia go by. Why is it that belief in religious revelation continues to exist alongside rational thinking?

CHAPTER 1
A New Way of Thinking

Through archaeological windows we glimpse the remains of ancient human activities. Some of these snapshots look mundane: people set about the tasks of hunting, making pots for cooking, tilling the soil or building dwellings. Others seem irrational, mysterious: people carve enigmatic images; they kill other people in specially constructed buildings; they produce texts that they revere and treat in special ways. In many instances, we can interpret these activities as manifestations of what we today call 'religion'.

Then, in more recent times, we begin to see another activity, one that emphasizes rationality and evidence. It is science. It is not my intention to provide an account of all the pivotal practitioners in the development of science. Instead, I ask two questions:

1 How did science emerge from the kind of thinking that we detect in the making of geometric patterns at Blombos, Maya human sacrifice, and violent divine intervention in the Old Testament?

2 How did the seemingly ineluctable entanglement of science with religion develop and change over the centuries?

As we grapple with these questions, we shall find that science has always been uncomfortably intertwined with religion: the present conflict is – at least in essence – nothing new. Religion was the cocoon in which science developed, but from which it has found it difficult to emerge.

Ionia and crucial philosophical distinctions

Although there are many diverse cultures in the world, there is today only one kind of science. We know it works because it makes verifiable discoveries and produces technologies that function no matter what the beliefs of the people using them may be. The moon is not a divine dwelling place; it is a large lump of dry rock. We know this statement is true because scientific theories of physics and the complex technologies derived from them have enabled people to go there and check. We also know that belief systems founded on the unpredictable interventions of supernatural beings and forces could never lead to the successes of science. How, then, did people break away from mythical thinking and create scientific reasoning?

A sixth-century BC Greek thinker, Thales of Miletus (a town on the coast of Ionia, modern-day Turkey – see Fig. 5), was one of the first people to question the traditional mythical form of explanation and to propose coherent and consistent natural explanations for natural phenomena.[1] The mythical explanations that most people of his time unquestioningly accepted did not have the qualities of coherence and consistency. Mythical explanations are necessarily ad hoc: that is, they ascribe events and natural phenomena to the capricious intervention of fickle gods. Completely different mythical explanations may be put forward to explain why mountain A exists and why mountain B, just a few kilometres away, is where it is.

Here is the crucial point: to attain consistency of explanation, Thales had to think of the natural world as a *unity* that was free from idiosyncratic, supernatural interventions – one god who was responsible for this and another who was responsible for that. For instance, he argued that, fundamentally, the whole world and everything in it was made of water. Rocks and trees may be different things, but he believed that they shared a common origin in water. Two lesser known Milesians, as the group became known, also argued for the fundamental unity of nature. They were Anaximander (*c.* 610–*c.* 545 BC), who thought that air was the fundamental substance, and Anaximenes (writing in *c.* 545 BC), who argued for an unidentifiable entity that he called *apeiron*.

At first we may think that Thales' proposition about water is little different from the mythical explanations that most people of that time accepted as truth. But, in principle rather than content, Thales' idea was not as bizarre as it may at first appear. He had reasons, not just beliefs, for thinking it. Objects made of stone and wood respond in identical ways when they are subjected to mathematical measurements and calculations in the construction of buildings and monuments; water too can be weighed and measured. In Thales' view, there was thus a commonality in the material world that existed beneath the diversity of the objects and substances that we see. The ubiquity of water suggested to him that it was the fundamental substance. For someone living in a world generally believed to be governed by capricious gods, this was a remarkable and crucial insight. In it we can see the beginnings of one of the chief characteristics of scientific thinking: overarching, unifying principles make better sense than idiosyncratic, ad hoc mythical explanations. Indeed, if we can discern laws of this kind operating in the natural world,

5 *Ancient Greece and the eastern Mediterranean, the birthplace of Western science and religion.*

there will be no need to invoke repeated interventions of gods. The retreat of the gods began in Miletus.

That Thales was later shown to be wrong about water being the fundamental substance of the whole world is, paradoxically, another of the strengths of his new type of explanation. Mythical narratives and explanations of the way in which the world works do not come accompanied by a set of instructions explaining how each may be evaluated and, if found wanting, discarded. All mythical explanations have to be accepted at face value: their underlying validation resides with the gods and their priestly representatives on earth and therefore cannot be challenged by mortals. With mythical explanations, there can be no critical thinking beyond the bounds of which god did what, and, because the gods are, like people, unpredictable, it is pointless to look for consistency. At best, you may substitute one mythical explanation for another. In stark contrast, Thales' way of thinking meant that, as knowledge about the world increased, explanations could be superseded and discounted without endangering the whole philosophical structure – the internal logic – of the new way of thought. Indeed, nothing was sacred, nothing was to be accepted simply because those in authority proclaimed it. Priests were retreating in step with the gods.

This is a crucial distinction: continuous criticism is one of the principal features that distinguish Western science from other systems of knowledge. Scepticism, not faith, is the hallmark of science. If we are asked, as sometimes happens these days, to consider 'alternative' forms of 'science' that embrace 'mystical' elements, we must at once enquire if these other 'sciences' have critical methods built into their very nature. Do they show signs of continuous evaluation and rejection of superseded explanations? Or are all the 'modernizations', theological innovations and 'broadness of thought' that they sometimes proclaim merely rewordings of old, unassailable belief systems?

Another difference between older ways of thinking and Thales' system also hinges on the unity of knowledge. The ancient Egyptians understood much about geometry: that, after all, was how they managed to build enormous pyramids and temples. But they saw their calculations of angles and triangles merely as tools for measurement in the practical construction of monuments rather than as statements of universal truths that are inherent in the very substance and functioning of the universe.

We can take one example. As every (well, perhaps not every) schoolchild today knows, the square on the hypotenuse of a right-angled triangle is always equal to the sum of the squares on the other two sides. Pythagoras, the sixth-century BC Greek philosopher and mathematician, showed this relationship to be so. Not even the gods can change it. It is a universal truth – at least in any dimension we are likely to experience. In addition to this sort of practical knowledge, useful if you are building a pyramid, the ancient Egyptians knew much about the movements of heavenly bodies, and they constructed many of their edifices to tie in with those movements. But they considered this sort of information to be religious secrets to which only priests had access, rather than statements that could be challenged and, if necessary, rejected by ordinary people. They thus recognized that there were two distinct kinds of knowledge: the one kind practical and debatable, the other sacred and the possession of an elite. Ironically, they used consistency to construct monuments to the capriciousness of gods.

Thales was unhappy with this distinction. He sought a kind of knowledge that was independent of – or that existed beyond – the inhibiting distinction between practical, discoverable information and that which was supernaturally vouchsafed by the gods. In seeking to attain this verifiable sort of knowledge, he went further than the Egyptians by devising abstract statements – to take another geometric example, that the opposite angles of two intersecting straight lines are *always* equal. He was thus beginning to detect universal 'laws' that would be true at all times and in all places, a universality that is not implied by ad hoc mythical explanations.

Perhaps Thales' greatest triumph was his prediction of a solar eclipse in 585 BC. How could he possibly have achieved this feat using only mythology?

He could not have. Rather, he had to work with laws and calculations that were inviolable, even for the gods. Here we have the first instance of the predictive value of scientific thinking, one of its chief features. Today many historians note that Aristotle considered Thales' successful prediction of a solar eclipse to be the origin of Greek science: at a single stroke it became clear that the cosmos was governed by laws that are immutable and not subject to the caprice of gods. Rational thinking could also be profitable: Thales used it to predict good weather and a bountiful olive harvest. To demonstrate his confidence in rational thought he bought up all the olive presses he could. He was proved correct.

I have emphasized the dangerous dichotomy between two kinds of knowledge because it threads its insidious way through the history of science right up to the twenty-first century and our present-day conflicts. To take an example: did God give the Holy Land to the Jews, or did Allah give it to his followers? In these kinds of disputes, there is no point in appealing to history, demographics, environmental issues, rigorous logic or verifiable evidence. Supposedly revealed knowledge takes precedence over practical, scientific knowledge. Precisely what information has to be merely (one could say, blindly) *accepted* and what can be *questioned* has been a continuing theme in the history of science and its entanglement with religion.

If we are talking about the planets and the stars, scientific astronomy, astrology, the ancient Egyptians' celestial route of the dead, and Maya beliefs about a three-tiered cosmos are not all equally valid 'ways of knowing'. Astrology and that sort of thing is, to put it bluntly, nonsense. It is not 'another way of knowing' about planets and stars. On the other hand, if we are talking about the 'meaning(s)' of *Hamlet*, we are in another realm altogether. But for the material world in which we live there is only science.

In all human history, justifications for the existence of two types of knowledge about the real world have been sustained deceits, sometimes consciously contrived, often unthinkingly accepted by those for whom they were intended. The legitimizing of different 'ways of knowing' about the real world has inhibited the advance of human understanding, retarded the development of human rights, and supported grotesque social inequalities. When today the progress of science is challenged, we shall do well if we recall the disastrous effects of sacred beliefs that are said to stand beyond human critique and that some say should control the practice of science.

Why did this momentous shift in understanding take place in Greece and not elsewhere – in ancient Egypt, for example? It seems that Greek Ionia, strategically situated as it is on the western shores of Asia Minor, differed from Babylon and Egypt in that it was not part of a powerful state that could impose its every whim on its subjects. Rather, the Ionians were a strongly independent trading people and, through commerce, they had contact with ideas in distant parts of the known world. They knew that there were people elsewhere who did not hold the same religious beliefs as they did but who nevertheless managed to get by without major trouble. Apparently it did not matter what myths people believed – life simply went on. Nor was Ionia priest-ridden. There was no privileged, priestly class with vested interests in the status quo.[2] Such classes of people insist that some knowledge is not to be questioned – the knowledge that they themselves possess and guard.

'A continuous scale of ascent'

The critical thought that developed in Ionia spread to the rest of ancient Greece, from one city-state to the next. The names of those who led the way are familiar: Pythagoras, Archimedes, Democritus, Euclid (who coined what became the Latin phrase *Quod erat demonstrandum*, QED), and many others. Today we rightly see them as the founders of Western thought. Together, they constitute a formidable phalanx. Now we need note only two of the most famous of these towering figures: Aristotle and Plato. The relationship between these two men's different ways of thinking encapsulates a tension that manifested itself through subsequent centuries, through into the Christian Middle Ages and, in different garb, to present-day disputes about science and religion.

Aristotle (384–322 BC), sometime tutor to the young Alexander the Great, came from a well-to-do family (Fig. 6). His father was a physician at the Macedonian court. He studied at Plato's Academy for as long as twenty years, and also in Asia Minor. A polymath, he wrote about ethics, physics, zoology, politics, rhetoric and, still influentially, poetics and drama. It is through Aristotle's discussion of tragedy that we today have the notions of *hubris* and catharsis.

Aristotle did not concern himself much with religion. He seems to have subscribed, outwardly at any rate, to the formal religion of his day, but he saw God more as a principle than a person. The great philosopher's open-endedness in such matters has led to ambivalent attitudes to him among world religions. Perhaps he intended it this way. His best comment on religion comes in his *Politics*, in which he explains that subjects are less apprehensive of illegal treatment from a ruler whom they consider god-fearing and pious. On the other hand, he observes, if the subjects believe that the ruler has the gods on his side, they are far less likely to rebel. Cynically, Artistotle advises tyrants to present a religious face to their subjects, an injunction that seems to be widely observed in today's world, where there is no shortage of pious rulers, and the Pope, the head of the Roman Catholic Church, is invited to address the United Nations.

Like the Pope, Aristotle also spoke of a 'soul', but he did not mean the eternal ghost believed to survive a person's death. He meant a kind of essence, the thing which a particular object or creature does by its very nature. The soul of a human being is his or her ability to think and act rationally. In Aristotelian ethics, excellence in human beings is to act in accordance with our rational 'soul'. It is manifested especially in friendships.

Here we are concerned with only one component of Aristotle's vast and varied scientific output. His greatest contribution was the attention he gave to scientific method.[3] Indeed, it could be said that he invented the discipline of science. He tried to formulate rules of logic to guide the sort of work that Thales had undertaken. In doing so, he distinguished between inductive and deductive reasoning. This was a huge step forward: it laid a foundation for rationality. What are induction and deduction? Nowadays the two words are often used loosely and interchangeably in common parlance, but the distinction between them is worth making. Let us start with induction.

Induction is the kind of logic that governs much of our daily lives. For instance, we notice that, if we approach unknown dogs, we are liable to be bitten. If this happens *a number of times* consecutively, we conclude that dealings with *all* unknown dogs will be dangerous. The more cases of being bitten, the more we feel that our generalized conclusion is justified. This is *numerical induction.*

6 *A bust of Aristotle from the ancient world. The philosopher was fond of saying that a 'likely impossibility' was always preferable to an 'unconvincing possibility'.*

For Aristotle, the next step in scientific method was deduction. The generalization or hypothesis (all strange dogs bite), having been formulated, was to be used as a statement from which testable *deductions* could be made (that unknown dog over there will bite me if I get too close to it). The argument may be formally expressed as a syllogism: if such-and-such is true for all things of a particular class, then, if X is a member of that class, it must be true for X. For instance, I may *deduce* that Nero, a particular strange dog, will bite me if I get too close to him, simply because I have concluded (as a result of numerical induction) that *all* strange dogs are bad tempered. Then, to my surprise, I find that Nero does not growl as I approach. Instead, he wags his tail. So, if I meet an unknown dog that unexpectedly welcomes my affectionate gestures, I know that my induced generalization that *all* unknown dogs are vicious is faulty. I must therefore reject my induced generalization (my 'law' or 'rule') and try to formulate a new one. Verifiable evidence is a cornerstone of scientific thought.

This, in very rudimentary form, is the mechanism of science, and it was Aristotle who tumbled to it and set it out logically.

Much more could be said about him, but one final point well illustrates his prescience and ties in with the general theme of this book. Working with Thales' notion of the unity of things and his own formulations of logic and evidence, he remarkably foreshadowed Darwin's theory of evolution (though not its driving mechanism):

Nature proceeds little by little from things lifeless to animal life in such a way that it is impossible to determine the exact line of demarcation, nor on which side thereof an intermediate form should lie. Thus, next after lifeless things comes the plant, and of plants one will differ from another as to its apparent amount of vitality.... Indeed...there is observed in plants a continuous scale of ascent towards the animal.[4]

What more can one say? Today we can only marvel at Aristotle's insights.

'Leave the starry heavens alone'

All this about induction, deduction and evidence seems fairly reasonable. The logic behind it is easily grasped. We enter a different world when we consider Plato (429–347 BC), Aristotle's former teacher.[5] Unlike Aristotle, Plato placed greater emphasis on thought than on observation. It may be a caricature to say that Plato believed that thought was more 'real' than the material world, but the assertion does give some flavour of how he worked.

Placing too much emphasis on intuition, Plato defended some, to us, bizarre notions. They include the following: there is such a thing as a human soul; it is immortal at death; it passes from one body to another.[6] Then, when considering the origins of life, he recounted the myth of Atlantis and the belief that God was the initial cause of the world and, significantly, of morality (the *Timaeus*). Associating morality with God is one of the great deceptions of religion, one to which I shall refer again in subsequent chapters. Today scientists are often considered incapable of working out what is ethical and what is not without recourse to the Church or religion. Most influentially, Plato propagated the idea that there is another realm, separate from the material world, in which pure Forms exist. Forms on earth, such as Beauty or Justice, are but poor reflections of the absolute Forms in the mystical realm. These abstract Forms can be grasped only by the soul.

Astonishingly, and to Aristotle's dismay, Plato argued that intuitive knowledge of the Forms was superior to empirical observation of the world. He declared:

Astronomy, then, like geometry, we shall pursue by the help of [logical and philosophical] problems, and leave the starry heavens alone, if we hope truly to apprehend it, and turn the natural intelligence of the soul from uselessness to use.[7]

Today, this proposition is incomprehensible. How could anyone possibly study astronomy without meticulously observing the heavens? Yet, as we shall see, the Catholic Church deployed this very notion when Galileo's telescopic observations challenged the centrality of the earth in the solar system.

Plato went on to argue that a philosopher should lead people to discover truths within themselves by means of what is now known as the Socratic method. It is the questioning technique that Socrates used to draw ideas out of a pupil. Those who attained intimate knowledge of the Forms by this method or by intuition should, Plato claimed, become the guardians of society, a spiritual and intellectual elite.[8] These guardians would have the power to exile or execute those who transgressed. In Plato's thought, knowledge of the Forms, residing (as he believed them to be) in a supernatural realm, became the foundation of draconian political and religious power.

Plato thus incorporated into his philosophy a kind of supernaturalism with which Aristotle would have no truck. Now, with hindsight down the whole sweep of history, we can see that supernaturalism inevitably leads to oppressive government and the destructive notion of benign dictatorship. Because there is no way of establishing empirically the nature of the Forms, a politically and religiously powerful elite can simply claim that it has direct access to them and therefore power to rule as it sees fit. The common run of people, who are excluded by education or limited mental ability from knowledge of the Forms, should, in the very nature of things, be ruled by an educated elite. The foundation for morality was thus necessarily situated in a spiritual realm of mystical Forms *accessible to only a few*. We can see that the interlinked notions of supernatural certitude, secrecy, power and fear are embedded in Plato's thought. For us today this is terrifying. We have seen this sort of thing happen all too often in history. Theocracy, government by God or a sacerdotal class, is probably the most oppressive and cruel kind of government that human beings have managed to devise, notwithstanding overwhelmingly cruel regimes like those of Stalin and Hitler.

Though he was immensely influential in ancient times, Plato was also resolutely challenged. In ancient Greece no one was above criticism. But Plato's way of thinking did not end with ancient Greece. He had a second chance in later centuries. Notions of a mystical realm, a soul and the pre-existence of knowledge that was lost by some sort of Fall would resonate in another religious system and eventually lead to our present-day Western confrontation between science and religion.

Philosophers like Thales, Aristotle and Plato sometimes puzzle modern readers because they seem to have been ambivalent about religion. They and others like them found that they simply did not believe the religious dogmas and pantheons of their day. But they went along with religion because it provided a social framework within which they could operate and spread their deeper message. Newton did much the same thing two millennia later. Today, atheistic scientists work as members of Oxbridge colleges with Christian foundations and chapels. They value sung evensong with Tallis and Byrd and come to terms with the recitation of the creed.

In sum, we can say that science, as developed by Thales, Aristotle and numerous other Greek thinkers (but not Plato), has a number of distinctive features:

1 Science makes statements that are *universally true* but that may not be obvious (that the sun revolves around the earth is 'obvious' but incorrect).

2 Science seeks *abstract laws* that govern the way in which the world works. It seeks explanations at a general, rather than an ad hoc, level.

3 Criticism of scientific propositions must be based on (among other considerations) *empirical observations* that can be *repeated* and *checked* by anyone with the opportunity to do so; evidence is fundamental.

4 *All* the pronouncements of science are *open to criticism*. Science has a built-in, self-regulating system of assessment. It may not be too fanciful to say that its 'holy scriptures' lie in the future, not in the past.

'The wisdom of this world'

The notion of inerrant, ancient scriptures brings us to the next stage in our overview of the development of science within its religious milieu. Here we find the beginnings of disputes that today plague the West and, moreover, create the much wider tensions that threaten global stability. Why did the progress of science, so ably initiated in ancient Greece, stall during the Dark and Middle Ages?

Soon after the beginning of our era, the young Christian Church, a scattering of independent communities, had to come to terms with Greek philosophy and other competing religions – as well as with the different interpretations of the life of Jesus that were then in circulation. From the point of view of Christian belief, it was a divisive and stressful time.

The monumental task of sifting the wheat from the chaff fell largely to a man of the first century AD named Saul. He hailed from Tarsus, an important city on the southern coast of Asia Minor: as he himself put it, he was 'a citizen of no mean city' (ACTS 21:39). He was also an educated Jewish Pharisee who proudly, so it seems, claimed to be a Roman citizen. Although the Pharisees were a sect that emphasized the observance of rules of behaviour and ritual, Saul was more cosmopolitan than many of his fellow members. As a Roman citizen, he gazed across the whole Empire, not just a parochial Holy Land.

Saul became, but did not remain, an influential member of the Pharisees. His famous experience on the road to Damascus turned him from reactionary Jewish persecutor of the new-fangled, heretical Christians into a man who tried to unite the Church under a theology that he himself formulated. In Jerusalem, he had 'persecuted this way [Christianity] unto the death, binding and delivering into prison both men and women' (ACTS 22:4; brackets added). Having thus achieved some fame because of his zeal, he set out for Damascus carrying letters from the High Priest and the Council of Elders addressed to the Jewish authorities there. These documents stipulated that he should be allowed to escort imprisoned followers of 'this way' back to Jerusalem for trial and punishment. How ironical it is to see in Saul's activities the murderous zeal of later Christian inquisitions.

But his mission was thwarted. At midday, not far from Damascus, 'suddenly there shone from heaven a great light round about [him]'. Blinded, he fell to the ground and heard a voice saying, 'Saul, Saul, why persecutest thou me?' (Fig. 7). When he asked, 'Who art thou, Lord?', the voice replied: 'I am Jesus of Nazareth, whom thou persecutest' (ACTS 22:7–8). His companions are said to have seen the light but not to have heard the voice. When Saul asked what he should do, the voice replied: 'Arise, and go into Damascus; and there it shall be told thee of all things which are appointed for thee to do' (ACTS 22:10). Saul obeyed his vision.

In Damascus, Ananias, 'a devout man according to the law', restored Saul's sight (the miracle confirmed Ananias's authority and the healed man's importance), and gave him his divine commission: 'For thou shalt be his witness unto all men of what thou hast seen and heard' (ACTS 22:15). Saul, like Oedipus and Tiresius before him, and, later, Shakespeare's Gloucester, lost his physical sight (though only temporarily) but gained moral insight. To see the 'other' world we must be blind to this one. As a result of his personality-altering experience, Saul changed his name to Paul.

Paul's radical realignment after his vision has become emblematic of a dramatic form of Christian conversion. What actually happened? His experience may have been a result of an ailment of which he writes: '[T]here was given to me a thorn in the flesh, the messenger of Satan to buffet me, lest I should be exalted above measure' (2 CORINTHIANS 12:7). It has been suggested that this 'thorn in the flesh' was epilepsy. If that was so, it would explain his seizure on the road and his fall to the ground at about noon, the hottest time of the day. The blinding light, his fall and his hearing a voice are all symptoms of temporal lobe epilepsy. The important point, however, is that it was through a *mental experience* rather than through philosophical or other teaching that Paul made his about-turn. Reasoning, based on scripture or not, had nothing to do with it. Aristotle would have expressed reservations about any knowledge claims that came out of such an experience, but Paul did not.

Immediately after his conversion, Paul started to preach to Jews in Damascus. He seems to have had scant success: in his later letters he does not refer to this part of his life. Subsequently, having returned to Jerusalem, he experienced another vision: '[W]hile I prayed in the temple, I was in a trance; And saw him saying unto me, Make haste, and get thee quickly out of

7 *Ludovico Carracci's* The Conversion of St Paul *(1587-1589) shows Paul on his way to persecute Christians in Damascus. Then terror reigns as Jesus speaks to him in a vision.*

Jerusalem' (ACTS 22:17, 18). He himself recognized what today we call an altered state of consciousness. Thereafter he travelled to his home town, Tarsus, apparently for a period of reflection and consolidation.

Revelations of the divine will through unusual mental states was thus a formative part of Paul's life and teaching. Unlike the disciples, whom he came to know, he had not enjoyed direct contact with Jesus. Some other justification of his authority was therefore needed.

Some time later, when he was returning to Jerusalem for the celebration of Pentecost, he visited Miletus on the Ionian coast, the city where Thales had lived 600 years earlier. One cannot help wondering if Paul knew about that father of scientific thought and his successful prediction of an eclipse. We do not know. As it was, he called 'the elders of the church' in Ephesus, a nearby city, to come to him in Miletus. There, he warned them not to heed 'grievous wolves'. They should accept nothing but his own teaching (ACTS 20:29–30). Paul's understanding of inviolable knowledge differed markedly from the view of his Miletus predecessor.

What were the circumstances that Paul now faced? Who were these 'grievous wolves' who threatened to challenge his teachings? After the crucifixion, Jesus' followers tried to make sense of what appeared to be an awful tragedy, the end of their hopes. Jesus himself left no coherent, developed theology. Some theological framework was therefore needed to make sense of the tumultuous events of his life and, especially, his death. Only his sayings and fragments of his discourses remained in oral traditions and in a multitude of gospels that adopted a range of conflicting views of his life and ministry (there were many more gospels than the four that the Church now recognizes). Various groups, or sects, therefore tried to formulate theologies. The Gnostics were one of them. Today they are better known as a result of the twentieth-century discovery and early twenty-first-century publication of the Judas Gospel.[9]

In the end, it was Paul's forcefully presented interpretation that became the orthodox view of the Christian Church. His theology was partly, but only partly, a response to traditionally minded Jewish Christians who were reluctant to abandon the Old Testament Law with all its minutiae. For the Jews, truth was embedded, or encoded, in ancient divinely inspired texts. It was the task of specially trained rabbis to crack God's code and to make the truth known. Why, having gone to the trouble of inspiring sacred texts, God would choose to make his truth less than clear to everyone, theologically trained or not, remains unexplained.

Through his theology, Paul turned tragedy into triumph. He made the crucifixion a central, not a final, event. There is a thrill and frisson in tragedy, as the Christians' forebears, the Greeks, realized. The Aristotelian elements of tragedy – pity, fear and catharsis – animate events like the crucifixion. The Jews later found this in the Roman siege of Masada and the suicide of the embattled Jewish defenders. Calvary and Masada are two hills that have much in common.

When Paul started his active ministry, he emphasized a break with Jewish orthodoxy. He said that the Law had been superseded: animal sacrifice, food taboos and circumcision (arguably the most cruel and bizarre of religious rituals) need no longer be observed. Many Jews were unpersuaded. He therefore concentrated his efforts on Gentiles – quarrelling Christian sects and educated, philosophical Greeks. Indeed, his attitude to Greek philosophy is profoundly and directly relevant to the science *versus* religion debate.

Paul's hometown, Tarsus, was a cosmopolitan place, but he seems to have gained only a superficial understanding of Greek philosophy. Rather, he was steeped in a strict form of legalistic Judaism, as his membership of the Pharisees suggests. Nevertheless, the great turning point in his life came about not, as I have pointed out, as a result of rabbinical discourse but as the aftermath of an overwhelming mental experience. This experience inundated and swept away all rational thought, and he began to proclaim the doctrine of salvation by faith (a mental state) rather than by rational observance of the Law (an empirical practice). It was inner certitude – faith – that mattered most to him, and that certitude came as a result of an inner, religious experience. Consequently, Paul's theological position is indisputable to those who have faith, but less persuasive to those who do not. People who have never had a great mystical experience are bemused by the accounts of such experiences that they hear others describe with such fervour.

In practice, belief in divinely revealed truth is a rather awkward position to maintain. Paul probably realized that visions and revelations as a source of authority are often ambiguous (though, for him, his own most certainly was not) and may even come to the 'wrong' people. To the young Christian communities in Galatia (central Asia Minor), he therefore wrote, 'But though we, or an angel from heaven, preach any other gospel unto you than that which we have preached unto you, let him be accursed' (GALATIANS 1:8). This is a stern warning to any 'grievous wolves' who may have felt that they had been spoken to in the same sort of miraculous circumstances as Paul. His personal authority mattered a great deal to Paul, and, as far as he was concerned, religious revelations were not for everybody, even if they seemed to come from 'an angel from heaven' – that is, in a vision. In taking this position, he was appropriating religious knowledge, claiming it to be of divine origin and asserting his own custodianship of that unquestionable knowledge. The notion of political theocracy was lying in wait just around the corner. One wonders what he would have made of the theological disputes that continued through the centuries after him right up to the present day. His own written words became sacred scripture and the foundation for endless dispute. His letters came to be dissected and interpreted even as he had, in his days as a Pharisee, studied the Hebrew scriptures.

Paul adopted a less confrontational approach when he addressed people who were outside the Jewish tradition and were familiar with logical thinking. He attempted this sort of subtle philosophical discourse in Athens, prodigal in its post-Periclean glitter and opulence. It was here that Plato had conducted his Academy and Aristotle established his Peripatetic School in the Lyceum. As the writer of The Acts of the Apostles remarks, Athens was a place that was interested in new ideas: 'For all the Athenians and strangers which were there spent their time in nothing else, but either to tell, or to hear some new thing' (ACTS 17:21). The city had all the open-mindedness of the Greek tradition, though the writer of Acts seems not to have thought too highly of it. Surely, Paul concluded, this Athenian openness was a wonderful opportunity for him to preach the gospel and to have it accepted as 'some new thing'.

He began his public address in a conciliatory manner: 'Men of Athens, I see that in everything that concerns religion you are uncommonly scrupulous' (ACTS 17:22; New English Bible).[10] Deftly, he went on to declare that the Unknown God honoured by an altar he had spotted on a sightseeing stroll around Athens was in reality the Christian God. He added that the Athenians' own Greek poets had declared human beings to be God's offspring (ACTS 17:23, 28). All this seems to have been uncontroversial, a comfortable softening up of his audience. But, when he proceeded to speak of the resurrection of the dead (his theological take on the final events of the canonical gospels), some of his hearers scoffed; others said they would return to hear him again. Even though 'certain men clave unto him and believed' (ACTS 17:34), no one could call Paul's Athenian address a resounding success.

For Athenians, Paul's ideas were pretty feeble stuff compared with the sophistication of their own philosophers (Acts names Epicureans and Stoics). It may have been experiences such as his only partial success in Athens that led Paul into a dilemma. Should he mug up Greek philosophy so that he could compete on equal terms with his hearers, or should he simply denounce *all* philosophy and cling to his Damascus road vision? The Greeks, he knew, could outwit him in debate, so perhaps the best scheme would be to reject philosophy and emphasize faith. But there he had a problem. The Greeks debated religious issues publicly and their, and the Romans', religious rituals were public affairs. Open reasoning was the sort of religious discourse with which they were familiar.

Caught in this dilemma, Paul lashed out at philosophy. Writing to the fledgling church in Rome, he said of philosophers, 'Professing themselves to be wise, they became fools' (ROMANS 1:22). It was easier to abuse than to refute the philosophers. To the Christians in Greek Corinth, he wrote, 'My speech and my preaching was not with enticing words of man's wisdom, but in demonstration of the Spirit and of power: That your faith should not stand in the wisdom of men, but in the power of God' (1 CORINTHIANS 2:4–5). The 'demonstration' of supernatural Spirit and power – miraculous deeds and words – wiped out any thought of reasoned support for ideas and teachings. For Paul, it was irrelevant if Christian teachings appeared to be rendered doubtful by philosophical argument. More than that, worldly wisdom was dangerous and should be avoided by Christians: 'For the wisdom of this world is foolishness with God. For it is written, He taketh the wise in their own craftiness' (1 CORINTHIANS 3:19). In the end, at the final judgment, faith not philosophy will win out: 'God hath chosen the foolish things of the world to confound the wise' (1 CORINTHIANS 1:27). Paul exhorted his followers to dismiss as 'craftiness' any reasoning that began to appear persuasive – an injunction that religious leaders still today urge their followers to heed.

Paul was not the first to adopt an anti-rational position. It has roots in *all* forms of faith, not just Christianity. Long before Paul, the Psalmist wrote: 'The fear of the lord is the beginning of wisdom' (PSALM 111:10; also PROVERBS 9:10). In adopting this long-standing position that privileges faith over reason, Paul was laying the foundations for the perpetual contradiction in Christianity with which we have to deal today. Although theologians from Paul to present-day professors and clergymen have tried to be logical and to formulate reasonable arguments for the truth of Christianity, *ultimately* it is always faith that matters, not logic. Unquestioning faith, they argue, is in fact a 'reasonable' path to follow. They try to use reason to show the reasonable-ness of unreason. Religious epistemology (a 'way of knowing') is very different from the sort of thinking that Thales and Aristotle developed. Absolute certitude comes from belief, seldom, if ever, from reason.

Let us remember here that we are not dealing with interpretations of *Hamlet*, but with statements that, though said to be religious, actually refer to the material, not the abstractly ethical, world: the Immaculate Conception of the Virgin Mary, the Virgin Birth of Jesus, the Resurrection, miracles of healing, and so forth. Religion makes scientific statements about virgin births, healings and rising from the dead, but pretends that they are exempt from scientific scrutiny.

Undoubtedly, there is a certain comfort for some people in the distinction between faith and reason. Those who carry Paul's epistemological banner of faith cannot be bested in an argument because they do not, and need not, argue: they merely express certitude, sometimes abrasively, but also sometimes winsomely. As the soprano so movingly proclaims in Handel's *Messiah*, 'I *know* that my Redeemer liveth.' Often, this blissful certitude sounds admirable, and many critics of pure belief feel themselves silenced, even humbled, by it. But, in the long run, it inevitably has unfortunate consequences. Faith, rather than reasonable discussion, has been the source of so much of the strife, murder and mayhem that stains Church history. It is the foundation for the present contest between religion and science. Was Paul the father of the Christian version of the conflict between science and religion? The succinct words of Ernst Haeckel (1834–1919), the German zoologist and evolutionist, are relevant: 'Where faith commences, science ends.'[11]

After Paul, the Church continued to grow within the sprawling Roman Empire. But persecutions also increased in intensity. The emperor Diocletian, who assumed that high office in AD 284, did much to consolidate the empire. He wished to centralize control of his far-flung lands, but found that his administrative policies tended to be rendered ineffective partly by the diversity of religions.[12] At first, he tried to go easy on Christians who refused to bow to the Roman gods. Realizing that many Christians relished the idea of martyrdom and instant translation to heaven, he tried confiscating their goods and religious objects, hoping that this comparatively mild measure would be sufficient to win them over or at least to deter them, but, as it turned out, they were not intimidated.[13] So, in the end, he felt himself driven to out-and-out persecution. In 304 he ordered that Christians should sacrifice to the Roman gods like everyone else. Those who refused to do so were to be put to death.

Like so many harsh policies throughout history, Diocletian's order carried within it the reasons for its, at least eventual, failure. The orgy of death that followed the bloody proclamation revolted even traditional Romans, and, in the eastern empire especially, Diocletian's edicts of persecution were only fitfully observed, if at all. Many Christians found that pagans too could be humane and willing to shelter them from the wrath of Rome. The young Church did not have a monopoly on goodness.

Diocletian expressed his conservative beliefs in a proclamation that sounds uncannily modern – and, indeed, has continuously seemed that way for a great many centuries. Its import is today well worth pondering.

The immortal gods in their providence have so designed things that good and true principles have been established by the wisdom and deliberation of eminent, wise and upright men. It is wrong to oppose these principles, or desert the ancient religion for some new one, for it is the height of criminality to try and revise doctrines that were settled once and for all by the ancients, and whose position is fixed and acknowledged.[14]

Throughout history, the Christian Church itself has repeatedly expressed an identical viewpoint. It is easy to imagine such a statement coming today from the Vatican, from some fundamentalist church or, for that matter, from a mullah – even to the word 'criminality' and its horrifying implications. By and large, the Romans had been tolerant of religions other than their own, as long as everyone acknowledged, or respected, the Roman gods. Early Christians, following the Old Testament commandment to have 'no other gods before me' (EXODUS 20:3; DEUTERONOMY 5:7), refused to be tolerant of Roman beliefs. Today (most?) Christians are willing to remove their shoes when they enter a mosque or don a yarmulke when dining with Jewish friends, but intolerance has deep roots in the Judeo-Christian tradition.

'By this sign, conquer'

In the fourth century a wily Roman emperor, Constantine (c. 288–337), realized that it would be better to have the Christians on his side than to estrange them as his persecuting predecessors had done. The Roman empire had outgrown its communications system and had begun to fall apart into western and eastern entities. Christianity was one of the few communities that crossed all boundaries: it offered some semblance of unity within the fragmented empire. By legitimizing Christianity, Constantine found, in the words of the now-famous soundbite, that the blood of the martyrs that his imperial predecessors had so lavishly spilt soon became the seed of the Church[15] – and of political systems. The great issues of the relationship between Church and State began. But Constantine mitigated such conflicts by declaring religious freedom. He did not forbid religions other than Christianity: '[N]o one whatsoever should be denied freedom to devote himself either to the cult of the Christians or to such religion as he deems best suited for himself.'[16] His edict was to give *all* religions the right to function freely. Nevertheless, Christian churches began to receive tax concessions; it soon became profitable to be a Christian, especially a priest. Later, it was the emperor Theodosius who, in 381, forbad pagan worship and sanctioned the punishment of those whom the Church Fathers defined as heretics. By this time the cracks in the Roman Empire were widening, and rigidity was the administration's response.

Constantine's conversion is today dated to his AD 312 victory at the Battle of the Milvian Bridge, a river-crossing near Rome. After the emperor Diocletian retired in 306, three generals fought for power. Constantine was one of them; another was Maxentius, who held Rome itself. Constantine decided to attack the much larger forces of Maxentius. The whole affair at the Milvian Bridge was, however, less clear-cut than the version that is today frequently accepted. It is worth examining closely because some of the elements it contains seem to have more than just a whiff of mythology about them. A theme that started long before and that was exemplified by Paul comes to the fore.

According to Eusebius (265–340), his rather sycophantic biographer and Bishop of Caesarea, Constantine's vision before that battle was similar to Paul's on the road to Damascus: a dazzling light, this time in the shape of a cross, appeared to him above the sun (Fig. 8). Eusebius wrote of this pivotal point in Western history:

Being convinced however that he needed some more powerful aid than his military forces could afford him, on account of the wicked and magical enchantments which were so diligently practised by the tyrant [Maxentius], he [Constantine] began to seek for divine assistance, as more important than even weapons, and a huge army. He considered how diverse emperors had invoked the heathen gods yet had come to destruction. On the other hand he recollected that his father, who had pursued an entirely opposite course, who had condemned their error and honoured one supreme God during his whole life, had found Him to be the saviour and Protector of his Empire, and the Giver of every good thing.

Accordingly he called on Him with earnest prayer and supplication.... And while Constantine was thus praying with fervent entreaty, a marvellous sight appeared to him in heaven, the account of which might have been difficult to receive with credit had it been related by any other person.... He said that about midday, when the sun was beginning to decline, he saw with his own eyes the trophy of a cross of light in the heavens, above the sun, and bearing the inscription 'By this, Conquer'. At this sight he himself was struck with amazement, and his whole army also, which happened to be following him on some expedition and witnessed the miracle...

8 *A fresco (c. 1523-1525) in the Hall of Constantine in the Vatican shows Constantine receiving his second vision. This time it is a divinely vouchsafed sign by which he will achieve victory at the Milvian Bridge and slaughter his enemies.*

He said, also, that he doubted within himself what this apparition could mean. Presently he fell asleep and in his sleep the Christ of God appeared to him with the same sign which he had seen in the heavens, and commanded him to procure a standard made in the likeness of that sign, and to use it as a safeguard in all engagements with his enemies.[17]

Perhaps Constantine had read or at least been told about Paul's miraculous conversion as it is described in Acts. Eusebius was certainly familiar with it. In his version of what happened, visions and dreams go hand in hand, as they do in the narrative of how Saul became Paul. Indeed, Eusebius' narrative became the model for many later accounts of leaders whom writers wished to present in a favourable light. This model is so persuasive that it masks glaring contradictions. Eusebius accuses Maxentius of practising 'magical enchantments', yet the bishop accepts Christ's injunction to 'procure a standard made in the likeness of that sign'. This new standard will replace the old pagan Roman ones and will act as 'a safeguard in all engagements with his enemies'. A 'magical enchantment'? Magic and religion are not easily separated, and the former was probably not simply an ignorant precursor of the latter. Today 'magic' is a pejorative word: they do magic; we have rituals.

Constantine's experience before the Battle of the Milvian Bridge was not his first vision. Only two years earlier, in 310, he claimed that the god Apollo had appeared to him in a vision and had given him a laurel wreath to signify the thirty years he was to reign. In keeping with this pagan vision, Constantine associated himself with a sect known as *Sol Invictus*, the unconquered sun, and had himself depicted with rays emanating from his head in a way that today recalls the halo of a saint.

His motivation in turning to Christianity – upheld as perfectly valid by Eusebius – is by today's estimate of religious conversions suspect, to say the least: he desired a bloody victory over Maxentius and the political power that it would bring him. Thereafter the Christian God was hailed as a God of Battle. The idea of 'Onward, Christian soldiers' has its roots deep in the past and is less metaphorical than some lusty singers of the hymn think. Indeed, the banners of the Crusades and persecutions were ready to unfurl. 'Conquer' was the operative word.

Even at this early time, Christianity was inextricably bound up with politics. And Constantine later exploited this nexus: he donned elaborate robes, had himself seated on a raised dais, and developed Machiavellian policing and taxation schemes, all in the manner of oriental potentates. He sought to restore the prestige that his predecessors had lost through civil wars. Despite convening the Christian Council of Nicea, Constantine was not baptized until shortly before his death.

A triumphal arch was erected rather hastily in Rome to mark Constantine's victory at the Milvian Bridge. It reused spolia from older monuments and was embellished with reliefs depicting his reign, victories and virtues. But, significantly, no Christian symbolism appears on it. Not even in the depiction of the Battle of the Milvian Bridge is Constantine's vision alluded to in any way. No one wanted to privilege the Christian religion above any other of the many cults that the Romans practised. Rather, by associating Christianity with veneration of the pagan worship of the sun, Constantine was able to please both pagans and Christians. (By and large, the Roman nobility remained pagan.) It was, of course, on *Sun*day that Jesus was said to have risen from the dead, and 25 December (approximately the winter solstice), the principal festival of the sun, became the day on which Christians celebrated the birth of Jesus.

Moreover, we should not forget that, although Constantine may have seen to the building of churches, he did not forbid lethal and cruel entertainments in circuses. The Christianity of those times, with its slavery, torture and atrocities, differed greatly from the religion we know today. Or did it?

If there is some doubt about Constantine's personal devotion to Christianity, there is even more about that of a man who is sometimes seen as a second Constantine. Clovis I (466–511), leader and unifier of the late fifth- and early sixth-century Franks, married a Christian woman. He himself was of doubtful religious leanings: some writers believe he was an Arian Christian (one who does not accept the Trinity), while others think of him as a pagan. In any event, Bishop Gregory of Tours, writing a history of Clovis about a century later and from a distinctly Roman Catholic perspective, described the moment of his conversion. The bishop apparently consciously modelled his account on Eusebius' history of Constantine.[18] In a battle against another barbarian tribe, the Alemanni, things were going badly for the Franks. Clovis there and then rejected his pagan gods and decided to try his wife's Christian God. At once the Alemanni turned and fled. Clovis, apparently transformed and speaking with the power of God (in the usual pious phrases), then easily persuaded the Alemanni to accept Christianity. Clovis was baptized amid great pomp and ceremony in a basilica on the site of the present-day Rheims Cathedral and thereafter proceeded to subdue other barbarian tribes with unabated ferocity.

As with Constantine, there were clear political implications in Clovis's conversion. As the only Catholic leader in Gaul he could call on the Gallo-Romans for support and he could align himself with the Catholic episcopate, the only political and economic power that existed throughout Gaul. Paris was designated the capital of Gaul, and Clovis encouraged the legend that St Denis, often said to have been a disciple of Paul, had been martyred there. Paris became a Christian holy city as well as a political centre. There was much in the Christianity of that time that believers today would reject as integral to their tradition.

Paul's and Constantine's visions exemplify rather than initiate a traditional way of receiving special, unquestionable, supernatural knowledge. The Bible is, of course, full of such revelations. Because of the highly personal nature of visions, the ideas of secrecy and special interpretation are at hand: visions are usually seen by one person only and are not always crystal clear. For instance, visions, secrecy and interpretation come together explicitly in the Book of Daniel. King Nebuchadnezzar of Babylon had a disturbing dream which (he claimed) he later could not recall and about whose meaning he worried. Eventually Daniel, who 'had understanding in all visions and dreams' (DANIEL 1:17) was able to remind the king what he had dreamed ('a great image' with feet of clay) and to tell him what the dream signified (the downfall of his kingdom). These insights came to Daniel by divine revelation: 'Then was the secret revealed unto Daniel in a night vision' (DANIEL 2:19). His 'night vision' was probably a dream, if it happened at all. In any event, a distinction between dreams and visions was at that time not clearly drawn. What mattered was that knowledge came to him through means other than the usual senses or rational reflection.

Not only in Christianity but in other religions as well, the *concept* of a vision as a path to otherwise unattainable insights became a mode of expressing and claiming special status, whether the revelation involved what we would today call an altered state of consciousness or not (Chapter 5). Isaiah, for instance, opens his book with a splendid claim: 'The vision of Isaiah the son of Amoz which he saw concerning Judah and Jerusalem' (ISAIAH 1:1). Similarly, Joel assures his readers that a wonderful time will come when 'your old men shall dream dreams, your young men shall see visions' (JOEL 2:28). Visions and the revelation of secrets were accepted as a 'way of knowing'. The *idea* of a vision became a manner of speaking, a trope that demanded unquestioning credence. That sort of connotation still lurks around the word. Today, someone may say, 'My *vision* for the future of this organization is...'. If this is said with enough fervour, others around the committee table may feel uncomfortable and even silenced. The metaphor of a vision is still powerful.

Despite his claims to supernatural revelations, Constantine seems to have had little interest in Christianity beyond its evident popularity and the stability it could bring to the empire, though historians still debate this rather negative view. At that time bishops decided doctrine within their own episcopates, and this led to endless disputes. To bring unanimity, Constantine convened the pivotal Council of Nicea (over which he presided to ensure that its deliberations were in keeping with his own political ambitions). The gathering was essentially an Eastern affair; the Western Church was hardly represented. It was the philosophical Greek milieu of the East that led to disputes over the nature of the Trinity, one of the principal points that the council tried to settle. The outcome was the Nicene Creed, still recited in Christian churches. Like all creeds, it was designed more to keep people out than to provide succour and teaching for the faithful. Today, faced with dwindling support, theology is more interested in ecumenical endeavours.

The reality of what happened in Nicea was different from the usual Church history account. It was not a cosy conclave of elderly clerics lovingly seeking the truth. Indeed, the Council of Nicea and its successors did not heal the split between the Eastern and Western branches of the Church, an inheritance from the Eastern and Western divisions of the Roman Empire. The schism endures to this day in the Roman Catholic and Eastern Orthodox churches.

As Paul's New Testament letters and The Acts of the Apostles show, there never was a primeval time of love and concord within the Christian Church. Even Constantine was shocked by the continual squabbling and dissent he found within Christian communities. All Christians believed that revealed knowledge could not be questioned without risking charges of heresy and, if found guilty, excommunication. After all, this was Paul's explicit teaching. The notion of revealed knowledge led inexorably to discord rather than the love and acceptance that appeared then (and today still appears) to be the Church's message. Religious unity is a mirage – virtually an oxymoron.

Christian eclecticism

Why did so many Romans and others take to Christianity? Today it is customary for preachers to emphasize the uniqueness of Christianity. There is, however, little in it that was new. Christianity absorbed the 'best' in the religions of the time: it was an amalgam. Descendants of gods were part of many ancient pagan mythologies. That Jesus should be hailed as the Son of God was, for people of that time, nothing new. Likewise, virgin births are not uncommon in ancient mythologies. Rising from the dead, too, was not an unusual mythical occurrence in ancient times. From Stoicism, the Church took over the idea of the 'brotherhood of man', while Platonism supplied mystical elements: perfection was to be found only in Heaven. Then, too, the idea of holy scriptures was taken over from Israelite religion. Scriptures implied literacy, a powerful weapon, as Christianity found especially during the Middle Ages: educated people could feel at home in the Christian fold.

The central Christian ritual, the Eucharist, also fitted in well with religious ideas of the time, especially with the Greek and Roman mystery cults.[19] These cults emphasized initiation (baptism in the Christian version), communion with God (the Eucharist in Christianity) and blessings not only after death but also in this life (a central Christian expectation). At the climax of a three-day initiation into the Eleusinian cult, neophytes were shown a sacred object in a blaze of torch light. As is usually the case in such religions and cults, the object was in fact quite ordinary – probably a sheaf of wheat. Like the elevated bread (wheat) and wine in the Christian Eucharist, something ordinary became something dazzling in its symbolic content. And symbolic associations became immanent in the object.

Growing power

Throughout this chapter, and especially into the next, we see religion becoming an industry in a capitalist sense. An elite group owns the resources (revealed knowledge) and the means of production (religious buildings, schools and so forth), while the public at large buys the product (salvation, peace of mind) and thus enriches the elite (witness the wealth of the Vatican and other major religious denominations).

I now move on to two historically important confrontations between science and revealed knowledge that followed the establishment of Christian orthodoxy within the Roman Empire. At base, they are not so much disputes about whether such-and-such a Christian doctrine is correct or false, though that is an impression easily accepted. Rather, they were epistemological crises – conflicts between fundamental 'ways of knowing'. If seen in this light, they clearly echo the disputes between Aristotle and Plato, between Greek philosophy and Paul, and between science and revealed religion. Repeated patterns of human belief and behaviour are becoming evident.

CHAPTER 2

From Skies to Species

By the seventeenth century, the ancient Egyptians' view that there was a distinction between practical geometry and priestly accounts of the universe had been taken over by the Church. Christian philosophers were free to discuss the structure of the solar system – provided they recognized that their hypotheses had no relevance to the *actual* positions of the sun and the planets. That matter was to be decided by scripture and the clergy. This stipulation came about because the Church tended to follow Plato rather than Aristotle. Simply put, Plato's philosophy was more spiritual, more accepting of supernatural realms, than other philosophical systems, and this feature was obviously attractive to an institution that valued visions and unquestioning faith above meticulous observation of nature. The Church was built on inscrutable supernaturalism.

The Middle Ages, which led up to the confrontation between the Church and scientists who took observation seriously, was, in present-day popular imagination, a period in which monasteries of saintly monks were islands of knowledge and Christianity amidst a sea of rather brutal secular regimes. On the contrary, it was a time of intense political, economic and theological conflict.[1] The Church successfully and influentially appropriated education and literacy and used them for its own advantage. Secular authorities, which were at least nominally Christian, depended on the Church for literary and numerical expertise. Furthermore, with Latin as its universal language, the Church usefully supplied political entities with a pan-European diplomatic network. Rich established families were dependent on the Church in another way: pervasive beliefs about Heaven and Hell led people to pay for priests to pray for the departed. There was a symbiosis between Church and State, with a good deal of compromise on the part of the ecclesiastical authorities. Sometimes secular and religious power coincided, as in the case of Bishop Gaudry at Laon, in France. In this instance, unity evaporated, and the peasants rebelled and burned down the cathedral.

De Civitate Dei

Plato's teachings, transformed into Neoplatonism, were incorporated into Christian theology by the tormented, tempestuous St Augustine of Hippo (354–430), one of the most influential of the early Church Fathers (Fig. 9). Hippo was a North African city near present-day Tunis. Like other prominent early divines he was noted for his erudition and subtlety of thought. In his

9 An intense, intimidating St Augustine clutches symbols of his learning, piety and episcopal power. In this painting by Piero della Francesca, now in the Museu Nacional de Arte Antiga, Lisbon, he seems to say that no one should dare to question the authority of the Church.

own day, he was a revered leader and has remained so up to the present. He is often presented as a fallible man, understandable if not immediately likable.

Yet even for a man so cerebrally disposed, there was a non-rational element in Augustine's conversion to orthodox Christianity. His *volte-face* came after years of theological experiment with the heresy known as Manichaeism. Mani (d. 276), after whom the movement was named, was a Persian. His followers, like those of many other sects, accepted that Jesus was an important prophet, but they were more concerned with the pervasive nature of evil. All matter, including the human body, was, they believed, evil. An ascetic life could help believers to escape this context of corporeal evil. As it always does, asceticism led to elitism and not to genuine humility. Insistence on extreme piety seems so often to go hand in hand with hypocrisy. Indeed, the Manichaeans' failure to live up to their own standards was something that Augustine began to note and deplore.

As in Paul's case, theological justification is nearly always post hoc, something that follows emotional conversion – and is therefore suspect. Augustine described his turning point in his *Confessions* (Book VIII, Chapter XII). One day, repenting of his early dissolute life, he cried out in anguish:

'And thou, O Lord, how long? How long, O Lord? Wilt thou be angry forever?'
I was saying these things and weeping in the most bitter contrition of my heart, when suddenly I heard the voice of a boy or a girl – I know not which – coming from the neighbouring house, chanting over and over again, 'Pick it up, read it; pick it up, read it' [tolle lege, tolle lege]. Immediately I ceased weeping and began most earnestly to think whether it was usual for children in some kind of game to sing such a song, but I did not remember ever having heard the like. So, damming the torrent of my tears, I got to my feet, for I could not but think that this was a divine command to open the Bible and read the first passage I should light on.[2]

After a long period of divine nudging, Augustine had come to a critical moment of revelation and salvation. But, like Paul, another convert well versed in theology, learning was not the lever. Rather, it was the voice of a child that he heard. Why was this so? 'Out of the mouth of babes and sucklings hast thou ordained strength because of thine enemies' was an early aphorism that Jesus himself endorsed (PSALM 8:2; MATTHEW 21:16). To that seeming paradox, Jesus added, 'Suffer the little children to come unto me, and forbid them not: for of such is the kingdom of God' (MARK 10:14; also MATTHEW 19:14 and LUKE 18:16). This notion of wisdom in simple things and people takes up the idea of the uselessness of erudite philosophy that we noted in Chapter 1. The innocence of children (notwithstanding the pernicious Christian doctrine of Original Sin) is something that attracts and awes sophisticated adults, as William Blake showed in his *Songs of Innocence and Experience*. Yet for every lamb there is a tiger.

There are three other interesting points about Augustine's account of his conversion. First, he says he wondered if the voice was real or a divine aural visitation. This expression of doubt when faced with something apparently supernatural still occurs today in reports of apparently remarkable conversions and miracles: it seems to pre-empt any misgivings a reader or hearer may entertain. The convert cannot be accused of gullibility. Psychics say that they did not seek their supernatural abilities: they had them thrust upon them.

Secondly, Augustine emphasizes his pre-conversion depravity. Like so many converts today, he feels it necessary to dwell (longingly?) on just how evil he was before being 'saved'. It is difficult to read the *Confessions* without concluding that confession can itself become self-indulgent. It is as if converts feel that exaggeration of their prior evil enhances the respect in which they will now be held by their religious community. They are often right.

Thirdly, and most strangely, Augustine turns to a rather magical way of approaching scripture that some still advocate: he opened the Bible at random and trusted God to guide his fingers and eyes to some immediately and personally significant passage. Not just the scriptures themselves but also the manner in which they are approached are believed to be divinely inspired.

Today, Augustine's personal life and struggles seem more interesting than his ceaselessly disputatious theology, and certainly less dry. In his *Confessions*, he wryly wrote, 'Grant me chastity and continence – but not yet.'[3] He was perhaps the most human of the Church Fathers, a down-to-earth man who knew the moral struggles of life as it is lived in the rough and tumble of the secular world. He was not isolated in an ascetic cloister, as was the case for many later theologians. Nonetheless, his moral wrestling did not prevent him from holding some intolerant opinions. One of his views has echoed disastrously down the centuries: 'There is no salvation outside the church.'[4] Unbaptized babies could, because of Original Sin, be consigned to Hell. This kind of exclusivity is, of course, a logical outcome of belief in a special supernatural revelation. Like most of the early Church Fathers, Augustine was interested in formulating theology not to help believers but rather to have clear grounds for the exclusion of 'heretics'. Exclusion, he and others found, is a road to power.

This visionary who came to adopt such hardline teachings faced enormous changes in his world, changes that were emphasized when the Goths sacked

Rome in 410. By the time of Augustine's death, even remote Hippo was threatened by the advance of the Vandals, the Germanic tribe that was obliterating all that remained of the Roman Empire in North Africa. The clash was not merely political: it was also religious. The Vandals, though usually known only as 'barbarians', were in fact Arian Christians who rejected the doctrine of the Trinity propogated by the Roman Catholic Church. The Vandals were on what can be seen as a forerunner of the Crusades: they wanted to put down the 'heretical' Roman Church, as well as to seize economically important land, just as the crusaders wanted to exterminate heretical Islam and, at the same time, to secure valuable trade routes.

It is sometimes overlooked that Arian Christians living as a minority in Hippo actually welcomed the Vandals. But Trinitarian Christians began to flee as the Vandals approached. Bravely, Augustine refused to join the refugees, but it is thought that he died shortly before the Vandals occupied the city after a long siege. He was thus spared the indignities that they would have inflicted on him.

The imminent fall of the Roman Empire, that far-flung institution that had not long before officially embraced Christianity, was a major challenge to Augustine's Christian faith, as is the present-day secularization of the West to many Christians. Both the invading barbarians and traditional Roman pagans made no bones about it: Rome, they believed, fell because it had forsaken its traditional gods for Christianity. How could Christians answer this explanation? Was God abandoning Constantine's legacy? Another of the early Church Fathers, St Jerome, writing in 410, gives us a vivid insight into the state of mind that pervaded those ominous times. Thinking of Rome's Christian status and legacy, he wrote,

I shudder when I think of the calamities of our time…. Indeed, the Roman world is falling; yet we still hold up our heads instead of bowing them…. Well may we be unhappy, for it is our sins that have made the barbarians strong; as in the days of Hezekiah, so today is God using the fury of the barbarian to execute his fierce anger. Rome's army, once the lord of the world, trembles today at sight of the foe.[5]

Many Christians in the crumbling world of the fifth century believed that God was enlisting evil forces to punish his recalcitrant people, a view of natural and man-made disasters sometimes still heard today. Some Christians claimed that God permitted the 2005 flooding and devastation of New Orleans to punish Americans (even those who were fortunate enough to live elsewhere) for promoting the 'wicked' notion of gay marriage.

Facing up to the theological problems of his time, Augustine proceeded within his adopted Greek philosophical tradition by drawing finer and finer distinctions, as theologians often do. One example will suffice. He found the concept of a 'City of God' in the Book of Psalms: '[T]he city of our God: God will establish it for ever' (PSALM 48:8). Later, the writer of the Epistle to the Hebrews (traditionally thought to be Paul) moved away from the notion of a special, eternal place on earth: 'But ye are come unto mount Sion, and unto the city of the living God, the *heavenly* Jerusalem, and to an innumerable company of angels' (HEBREWS 12:22; emphasis added). The apocalyptic Revelation of St John the Divine, however, seems to foretell a time when a 'new Jerusalem' will 'come down out of heaven from my God' (REVELATION 3:12). In his vision, John saw that this City of God was 'garnished with all manner of precious stones', including jasper, chalcedony, emerald, sardonyx, topaz and amethyst, all symbolic in one way or another (REVELATION 21:18–21). Now here was the question: was this City of God to be taken as a real place, perhaps a new Christian Rome, or was it a metaphor? Augustine concluded in his *De Civitate Dei* that the City of God of his title had nothing to do with temporal kingdoms and empires, Rome or any other. They were all ephemeral. The notion of a 'City of God' was not to be taken literally. Only the mystical City of God, itself not a human institution but rather the sum total of all believers both living and dead, was of any consequence. We shall encounter the concept again in Chapter 9, where I discuss a neurologically generated visual percept that mystics have interpreted as the divine city. Significantly, both Augustine and the medieval mystic in Chapter 9 see the City of God as a fortress. An embattled position standing up bravely to the rest of the world is for some people attractive.

Less educated people have to take teachings like this on faith and accept that all the ins and outs of tortuous theological thought are understood by the Church Fathers and clergy. Ordinary people, comparative outsiders, may try to understand it all and, an easier task, to know something of the intricate symbolism of Christian ritual, art and thought. Augustine, for instance, taught that the sun and the moon stood for the Old and the New Testaments: as the moon receives light from the sun, so the Old Testament is illuminated by the New. Fairly simple people who entertain little expectation of compre-hending the subtleties of, say, the Trinity can, in day-to-day life, repeatedly see the two great heavenly bodies and be reminded of their biblical significance. In this way, they can be party to at least some of the esoteric ratiocinations of deeply respected divines. But, for those outside the fold, all this is a secret.

Augustine needed an intellectually acceptable theology to go along with his notion that the City of God did not mean a 'real' city but rather something that existed in a mystical realm. He found it in Neoplatonism. The Neoplatonism of his time embraced some oriental mysticism and emphasized the existence of supernatural beings and perfections in a spiritual realm – in other words, the Forms that we encountered in the previous chapter. The Church on earth was an imperfect image of the mystical perfection of the heavenly Church – it was an earthly version of a perfect Form. If we accept this proposition, we can embrace a philosophically respectable tradition and believe in the genuineness of a Church that comprises the faithful, both living and dead, and, at the same time, we can account for its glaring imperfections. Problem solved.

Augustine's acceptance of this brand of Platonism was in some ways a counter-balance, though a poor one, to Paul's insistence on the supremacy of faith. If intellectuals were to be attracted to the Christian Church, theology would, as Augustine (and all subsequent theologians and apologists) realized, have to be seen to be on a par with philosophy. University divinity departments today adopt this position, though many agnostic and atheistic philosophers doubt its validity. It is untrue to say that even today religious belief altogether renounces rational thought and logical argument. Intricate reasoning is unquestionably a part of theology[6] but, ultimately, such apparently rigorous reasoning rests on the irrational notion that there is a supernatural realm that takes an interest and sometimes interferes in worldly affairs. Still, if a theologian loses a logical argument, it does not really matter: faith and personal inner experience can take over. We can accept Paul's injunction to be prepared to give an answer for our faith, but we have a fail-safe back-up if we lose the argument.

Because faith, not rational thought, always has the final word, Augustine rather shockingly accepted the efficacy of forced conversion to Christianity and called upon the state to act against what he considered to be heresies. Had not Christ said, 'Go out into the highways and hedges, and *compel* them to come in, that my house may be filled' (LUKE 14:23; emphasis added)? Once heretics had been forced into the Church, Augustine foolishly believed, true faith could be lovingly nurtured in them. Beat them and then love them. If you conquer your enemies and then help them to rebuild their shattered cities, they come to love you. The idea is still around. Not surprisingly, theologians and historians today debate whether Augustine can be held responsible for so many subsequent persecutions of Jews and supposed heretics who were 'compelled' to accept orthodox Christianity. He certainly made a handsome contribution to the implementation of that fearful policy, and his dangerous idea can be seen, only partially transmuted, in the West's military, rather than overtly religious, invasions of Afghanistan and Iraq in 2001 and 2003 – and, of course, in the more distant past.

Slavery was another feature of society about which Augustine was ambivalent. Certainly, Christians should be kind to their slaves, but they need not free them. Augustine saw slavery, like so much else, as a consequence of the Fall, but he did not endorse its abolition: he saw such a step as economically unviable. To what extent, we may ask, should Christians allow social circumstances to dictate morality? Can slavery ever be a necessary evil?

Inevitably, as Augustine developed his theological stance, religious behaviour and doctrine became intertwined. Piety became inextricably confused with doctrine. Augustine saw to that; it did not happen by accident. Piety outside approved doctrine was deemed worthless. 'By faith are ye saved', wrote Paul – but faith in what? Not only in the person of a benevolent God and his Son, but also in the convoluted reasoning behind the 'mystery' of the Trinity, the Virgin Birth and the bodily Resurrection of Jesus. It became worth burning and killing people for these theological intricacies. In the end, day-to-day piety counted for nothing. It was which version of divine revelation that one accepted on blind faith that mattered.

Perhaps more clearly than in any other statement, Augustine enunciated the malign dictum that unquestioning faith in (ambiguous) revealed knowledge is the road to salvation in his work entitled, ironically enough, *On Free Choice of the Will.* He wrote: 'Unless you believe, you will not understand.'[7] Thus there was no room for science.

[T]here exists in the soul...a cupidity which does not take delight in carnal pleasure but in perceptions acquired through the flesh. It is a vain inquisitiveness dignified with the title of knowledge and science.... To satisfy this diseased craving...people study the operations of nature which lie beyond our grasp, when there is no advantage in knowing and the investigators simply desire knowledge for its own sake.[8]

This pronouncement by one of the greatest (perhaps *the* greatest) of the ancient Church Fathers pulls us up short with a nasty jolt. Is science nothing but a 'cupidity', a 'diseased craving' that ,is comparable to those 'carnal pleasures' that Augustine reluctantly abandoned? An absorption with sin, so common in the Church, infects and discredits rational thinking. Augustine's hyperbole here, and indeed throughout his writing, makes us re-evaluate our initially fairly positive view of his character. We have been thinking of him as a down-to-earth man who knew the moral struggles of life. Perhaps. But he was also a bully, a bigot and an obscurantist. It could be argued that his sins after his conversion were more heinous – certainly more far-reaching – than the peccadilloes and debaucheries of his youth.

Aristotle resurgent

The cataclysmic fall of Rome, Augustine's vision and conversion, and his adoption of a form of Neoplatonism together constituted a major historical turn that, like Constantine's miraculous vision, did not bode well for science. Yet, despite these developments, the torch of Greek science was not extinguished. It was kept alight by Muslim scholars into medieval times. Constantinople and the Byzantine empire fell to the Islamic armies in 1453. The eastern Christian empire that Constantine founded in 330 thus came to an ignominious end.

That end was visible long before the fall of Constantinople: by 1071, the Byzantines were fighting a losing battle, and the western Christian empire, centred in Rome, was of little help. Startlingly, the leaders of the 1203 Crusading army, supposedly bent on recovering the Holy Land from the Muslims, changed course and sacked Christian Constantinople instead. It was not until 2007 that the Pope apologized for this 'error'.[9]

Eventually, translations of Aristotle found their way via Islam and Spain into the libraries of Christian divines. At first there was resistance. The ancient, fundamental differences between the long-accepted Plato and the newly introduced Aristotle resurfaced. It was not until the thirteenth century that influential writers such as Thomas Aquinas tried to bring Aristotelian and Platonic ideas into harmony. Aquinas in particular tried to counteract components of Augustine's legacy and to reinstate reason and rational enquiry. In doing so, he exemplified another recurring pattern in human thought: rather than choose between contradictory positions, people try to reconcile them. They do not recognize that, if the positions are indeed contradictory, no meaningful reconciliation will be possible. Theology, like politics, is a matter of compromise.

Summa Theologica

Thomas Aquinas (*c.* 1225–1272) was a Dominican friar and, like Augustine, a formidable intellect (Fig. 10). Noted as much for his obesity as his serenity (he was known as the 'angelic doctor'), he came from an aristocratic Neapolitan family, but he moved to Paris, where he wrote his most influential work. Though his monastic order's discipline demanded asceticism and poverty, it seems that he wanted for nothing throughout his life. His followers became known as Thomists. His influence on the Catholic Church, and thus on Western thought, has been enormous. Dante spotted a transfigured Aquinas in the Heaven of the Sun. In 1879 Pope Leo XIII declared him to be the definitive theologian for the Church.

Unlike Augustine, Aquinas realized that there would have to be some sort of rapprochement between faith and science, as it was encountered in the newly translated Aristotelian texts. Science was not a 'diseased craving'. In a manner that strikingly prefigures those theologians of today who once more seek some sort of accommodation with science, Aquinas tried to reach the theological conclusions of Augustine but by means of Aristotle's science and logic. As Norman Cantor says in his history of the medieval period, Aquinas's endeavour was founded on 'a magnificent paradox'.[10] The medieval divine tried to use logic to justify illogical beliefs.

Aquinas's take on philosophical thought is interesting. For him, reason was embedded in nature. It was therefore natural for human beings to be logical and reasonable. It follows that our morality is rooted in qualities that we share with animals: they, too, have orderly, intelligible communities. But, by taking note of Scripture, we go further than animals. We are able to develop distinctive kinds of rationality and social structures based on revelation.[11] In doing so, we are simply building on what is, as a result of God's creation, already in the hearts of all people. Aquinas was not the first to point this out. Many centuries before, Paul had written that 'the Gentiles...do by nature the things contained in the law.... [T]he law [is] written in their hearts, their conscience also bearing witness' (ROMANS 2:14–15). In making this claim, Paul was affirming 'natural law'. He was, however, also insisting that 'we' have more insight than 'the Gentiles' because we attend simultaneously to divine revelation.

Aquinas's great work, *Summa Theologica*, became the cornerstone of the Scholastic movement (*c.* 1100–1500) and indeed of the whole Catholic Church. In this influential book, Aquinas argued that there was not one truth in science and another in theology. Following Paul, he accepted that some truth could be reached by reason – what he called 'natural revelation' – and

10 *This detail from a psalter (c. 1350-1386) shows Thomas Aquinas, endowed with a nimbus of piety, teaching suitably subservient and receptive followers. The vignette encapsulates the Church's power as invested in selected individual human beings.*

that further truth could be attained only by 'supernatural revelation'. He held that natural revelation and special revelation were complementary: truth is necessarily a unity. Working from this premise, he formulated five 'proofs' for the existence of God. Because these proofs were achieved by reason, they should be understandable by people who did not believe in revelation. But, once the fundamental existence of God had been demonstrated, special revelation, achieved through faith, the Bible and the Church, had to take over. Today we still hear this line of thought as an argument for a human need for religion. Reason has its limits, we are told, though how we can find out just what these limits are is unclear. From this factitious foundation, the argument goes on to say: *therefore* we need to postulate a God. So specious an argument need not detain us.

The fact that Aquinas deemed it necessary to formulate 'scientific proofs' for the existence of God (theology regarding the deity's attributes could come later) is historically significant. The whole edifice of unquestioning faith in supernatural realms and divine revelations had begun to crumble, in part as a result of the new-found Aristotle. Ecclesiastical diktat was imperilled by debate. Within the established Christian tradition the end of supernaturalism was in sight. Once the efficacy of rational ('scientific') thought is recognized, Pandora's box flies open.

Some of Aquinas' 'proofs' are still around today, if in disguise. People still tax their minds by trying to figure them out. Aquinas did get one thing right. Non-believers do not have to *disprove* the existence of the supernatural. The onus is on believers to show why their beliefs are reasonable and true; it is *they* who have to produce reasons why we should believe in a supernatural realm and revealed knowledge. Those who make a statement have to supply evidence for it, evidence that can stand up to scrutiny.

Aquinas's five arguments, his *quinque viae*, are worth examining, if only briefly. Three of these are actually permutations of a single line of thought;[12] they are the Unmoved Mover, the Uncaused Cause and the Cosmological Argument. All three depend on a single argument. Take the first two. All things are caused by earlier things. For instance, the rotation of the earth on its axis causes the succession of night and day. But what causes the rotation of the earth? Gravity causes the movement of the earth. Ah! But what causes gravity? Hmm.... Well, there must be a God who causes gravity. QED. In other words, when we contemplate causality in the cosmos, we must conclude that there was a Prime Mover, a First Cause or 'Some Thing' that brought it all into existence by initiating a chain of causes. Aquinas put it like this: 'Therefore it is necessary to arrive at a prime mover, put in motion by no other; and this everyone understands to be God.'[13] His invocation of 'everyone' seems to be special pleading: 'everyone' believes it, so it can't be wrong.

Because the First Cause argument is today making a comeback in the contentions of Creationists and those who argue for Intelligent Design, it will be worth pausing for a moment to consider it more carefully. For a moment or two, an appeal to the existence of a Prime Mover sounds persuasive. But think again. The argument suggests that we must draw a line somewhere across our infinite regression of A was caused by B; B was caused by C; C was caused by D, and so on. But where? And why should that spot be termed God? There is no reason why it should. Where we draw the line is arbitrary. But perhaps not entirely arbitrary because the spot we tend to choose is determined by our prior, inherited notion of God as the creator of the physical cosmos. Our already-formulated definition of God determines that 'he' must have started a chain of causes, and that 'he' must be immune to change because he is necessarily outside that chain. Moreover, having identified a point where we think God was the initiator, why do we not ask: What caused God? What existed before God and, in some way, caused God to come into being? Again, it is our inherited *definition* of God, rather than any logic, that excludes questions of this kind. Those who today respond to the atheistic attack on religion should remember that definition is not argument.

If the Prime Mover, or First Cause, argument sounds familiar, it is probably not because of Aquinas but because it reminds us of the Big Bang theory that is today accepted by most physicists and cosmologists. This modern theory has taken the place of the older theory of a universe without a beginning or an end. It postulates that there was a moment, long, long ago, when all matter came into being in a single place as a result of a Big Bang. All matter was squeezed into an infinitesimal and exceedingly hot primal 'atom'. This atom exploded and thus created all the matter and forces (e.g., gravity) that we know today. Some creationists declare that the 'Big Bang' is no more than a reformulation of 'God the Creator'. For four reasons, the Big Bang is *not* simply a revision of Aquinas's argument.[14]

First, there is empirical, scientific evidence to show that the cosmos must have begun somewhere and at some time. What is that evidence? The constitutive parts of the cosmos are flying apart at observable speeds that scientists can now actually measure. If we 'rewind' the clock, we can see that all these flying parts (stars, planets, galaxies – the whole lot) must have come from one event – what we call the Big Bang. Improved technology has also made it possible for cosmologists to detect left-over heat from the Big Bang. Just as Galileo's innovative telescope provided empirical evidence for a heliocentric solar system, so more recent technology has provided hard evidence for what would otherwise be pure reason and theory. This is something that never happens in that other empire of the mind – religion.

Secondly, we can respond to the Prime Cause argument for the existence of God by saying that, prior to the Big Bang, there was (like everything that followed it) some *physical*, or *material*, entity and event that are, as yet, unknown. There is no need to postulate a *supernatural* entity at that point.

That thought brings us to the third and what must be the most devastating problem. How do you get from a First Cause (or a Big Bang) to the omniscient, omnipotent, loving God who takes an interest in the behaviour and welfare of each and every human being, a concept of which many religions speak? Even Aquinas's contemporary critics claimed that his argument led to a mechanistic God rather than to the omnipotent, omniscient and caring deity of Augustine. Today we can go a bit further. Once you propose a *supernatural* First Cause, you open up a vast field of implications that lands you in deep water indeed. For instance: did God withdraw after detonating his Big Bang or does he continue to tinker with his creation? If he withdrew, we cannot argue that religions are able to contact him. If he is still active, we must ask for unequivocal evidence of his activity in the material world. If God does intervene in the material world, then he exposes himself to scientific investigation.

Fourthly, the First Cause argument reminds us of another 'infinite regression' that, for religionists, demands a moment of divine intervention. The whole sweep of evolution from single-cell organisms to mammals and, eventually, ourselves is a material, or anatomical, progression. Those who believe in God need to postulate a moment at which he injected a supernatural, eternal soul (made in his own image) into the human evolutionary line. We cannot imagine a soul evolving. Could there be a 'semi-supernatural' soul? The inherited 'spirit of each human being' is either immortal or it is not immortal. Like pregnancy, there is no in-between state. There cannot be a semi-supernatural entity that gradually evolved into a fully supernatural, eternal soul.

Let us return to the rest of Aquinas's 'proofs'. His fourth argument is the argument from degree. Today, few, if any, thinkers consider it a profitable line of thought. Simply put, the argument says that we recognize degrees of goodness and evil and beauty and much else. Therefore there must be some sort of absolute by which we judge the degree. This Bernini statue, we may say, is more beautiful than that one by Epstein. Therefore there must be some Absolute Beauty (or arbiter of beauty) by which we make our discrimination. If this sounds familiar, this time it is not because the idea is still around. It is Plato revamped. Aquinas followed Plato by arguing for absolutes, the perfect Forms that must exist in some supernatural realm.

Aquinas's fifth and final argument is also familiar to us. It is the argument from design. Augustine argued that something that appears to be designed (say, the human eye) must imply the existence of a person-like Designer. As we shall see, Darwin tumbled to the elegant explanation of natural selection, by which complex, functioning things, like the human eye, can evolve without any implication of a Grand Designer. A pattern in nature is not necessarily a design.

Aquinas was not alone in grappling with such logical niceties. The Franciscans, a competing monastic order, emphasized poverty and humility, as a result becoming known as the 'Friars Minor'. They tried to achieve the desired harmony between Aristotle and Plato, not by following the route taken by Aquinas, but by downplaying Aristotle. They were not free from internal discord and divided into three branches, but theologically they all emphasized Plato's mysticism and the notion of Forms. They adopted a largely anti-intellectualist stance that emphasized piety and the majesty of God. Reason is all very well and must not be abandoned, the Franciscans allowed, but it must play second fiddle to unquestioning faith.

The founder of the Franciscan order, St Francis of Assisi (1181/2–1226), was, however, wary of the kind of intellectual gymnastics that this proposition entailed. His conversion to Christianity resulted from two serious illnesses and a spiritual crisis while (like Constantine and Clovis) on a stressful military expedition. The well-known tale of the saint subsequently preaching to birds is probably apocryphal (at least one hopes so), but it does illustrate the naivety of St Francis's thought. Mythology clusters around saints. Be that as it may, his more realistic followers paid little attention to his mystical position and tried to meet the Thomists on their own philosophically rigorous terrain.

One of the problems that Aquinas, as well as his critics, faced derived from a fundamental tenet of science: science advances by jettisoning outmoded explanations and adopting new ones. The sort of harmony between science and religion that the medieval Schoolmen tried to bring about incorporated early 'scientific' ideas about the geocentric structure of the cosmos. Eventually, the time came when those older ideas were seen as an integral part of Christianity. So, when science moved on, the Church, saddled with its acceptance of the old version, objected. When the Church took science into its nest, it turned out to be a cuckoo. Science seeks change; religion seeks an eternal God. Today it is the impossibility of joining up science and revealed knowledge in some sort of harmonious, continuous, dual-faceted understanding of life, that is attractive to some. Science and religion are contradictory, not complementary. That the medieval attempt to reconcile Aristotle and his type of logic with Christian faith-based religion would fail soon became apparent. A major clash was inevitable.

The Middle Ages are fascinating because they seem (at least to me) to be vivified by a mindset utterly different from our own. Those centuries, despite glimpses of courtly love and monastic hospitals, were a time of disputation and despoliation, pious positioning and rebarbative cruelty. I can easily imagine enjoying a dinner with the Roman poet Catullus (c. 84–c. 54 BC). We could laugh together about napkin-purloining guests, but we could also face and share thoughts about the utter finality of death, as the poet himself did at the grave of his brother: *Ave atque vale* (Hail and farewell). No topic of conversation would be taboo. Augustine, Aquinas and the self-absorbed, claustrophobic Church Fathers would, I think, prove uncomfortable table companions. The niceties of doctrine and a persistent desire to define *true* Christians so that they themselves would be at the centre of the pious congregation constantly exercized their minds. Other, equally sincere (if indeed sincerity comes into the matter at all) people would be resolutely excluded. In theology, logic was perverted to become a weapon of exclusion, not clarification. If we wished to have a relaxed medieval evening, we would have to avoid many topics for fear of igniting a conversational *auto-da-fé*. Another country, another world, and a distasteful one at that. Its shadow falls across our own times of religious revival.

De Revolutionibus

The first really public and widely noted confrontation between science and religion was the one with which we associate the names of Nicolas Copernicus and Galileo Galilei. As is well known, it concerned the physical structure of the solar system. Is the sun at the centre, as astronomers claimed, or is the earth at the centre, as the Bible appears to imply? Today it seems odd that such a point could be of considerable theological, rather than scientific, import. But it was. Augustine would probably have denounced the whole issue as 'vain inquisitiveness', but by the time of the early Renaissance he would not have got away with such obscurantism. In any event, Galileo was not the first to posit a heliocentric solar system. Aristotle, that immense mind of ancient Greece, had discussed the problem.[15]

About a century before Galileo, Copernicus (1473–1543) had published his view that the sun was at the centre of the solar system, which was, in turn, surrounded by fixed stars.[16] He pointed out, perhaps in self-defence, that the idea was not absolutely brand new. Indeed, he gave credit to Pythagoras for thinking of a moving earth. But the time when a new idea is vigorously put forward is, of course, crucial. With post-Constantinian Christianity enforced, the world was very different from the philsophically open-minded one that Aristotle and Pythagoras inhabited.

In taking the position that he did, Copernicus was breaking with the Ptolemaic cosmology that the Church had taken on board by arguing that his findings were not merely an elegant set of calculations to predict the movements of heavenly bodies but rather a statement about the actual, physical nature of the solar system (Fig. 11). His claim was unfortunately somewhat obscured by a preface to his ground-breaking book *De Revolutionibus Orbis Coelestium* (1543). It was written by Osiander, a Lutheran theologian who insisted that Copernicus's model was, in accordance with the theological thought of the time, simply an abstract notion, not a statement of fact. Copernicus's calculations were useful in predicting the movements of heavenly bodies but they had nothing to do with the actual structure of the universe. In a situation that in some ways recalls the distinction that the ancient Egyptians drew between priestly astronomical knowledge and mathematical statements that could be used in practical work in the material world (Chapter 1), the Church held that the concepts produced by writers such as Copernicus were entirely abstract mathematical formulations. This is why Copernicus's work did not have the same impact that, later, Galileo's did. Copernicus died soon after the publication of his book, so he was unable to reply to Osiander, and the matter rested there. He could not have been happy with the gloss that the preface placed on his work.

One old and rather charming Ptolemaic idea that Copernicus retained was the fixture of the planets each on the surface of its own crystal sphere. All the spheres were nested. The tenth sphere was the Empyrean, a region of elemental fire and the seat of the deity. Pythagoras had calculated that each planet moved at its own rate and emitted a musical note. Not to be outdone in fancy, Plato elaborated this idea by saying that each planet bore a siren singing a delectable song. These songs harmonized to produce wonderful music. To this 'discovery' Copernicus added another: he deleted the notion of sirens and argued that each musical note is created by the rapidity of the vibrations of its planetary orbit. Because he also believed that nature is harmonious in all its parts, it was a short step to the pleasing notion of the 'music of the spheres'.

By the Middle Ages, Christian theology had added its own gloss to these ideas of cosmological harmony. The tenth sphere became known as the *Primum mobile.* It was solid and formed a barrier between the universe that human beings inhabited and the nothingness beyond and the Empyrean, the seat of God and his angels.

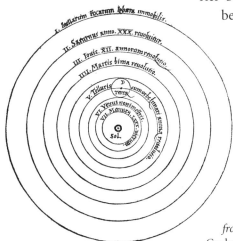

11 *Copernicus's assertion that there is no one centre of all the celestial spheres can be read literally or metaphorically. This diagram of the solar system comes from his* De Revolutionibus Orbis Coelestium *(1543).*

Through the centuries, the music of the spheres intrigued many a writer. Shakespeare wrote:

> There's not the smallest orb which thou behold'st
> But in his motion like an angel sings,
> Still quiring to the young-eyed cherubins.
> Such harmony is in immortal souls.

The Merchant of Venice, V:1:69–72

It fell to Shakespeare's contemporary, Galileo Galilei (1564–1642), to take the notion of a heliocentric solar system further than musical spheres (Fig. 12).[17] By means of an improved telescope he was able to detect the moons of Jupiter and to put forward evidence for the literal centrality of the sun. His observations of nature, rather than the cerebral ruminations of philosophers, would have delighted Aristotle and dismayed both Plato and Augustine. In 1632 Galileo published his sensational *Dialogue Concerning the Two Chief World Systems*. In this book, comparable in its impact on Western thinking to Darwin's later *Origin of Species*, he presented a full argument for a heliocentric solar system, basing his conclusions on observations made through his telescope. In doing so, he pitted one kind of knowledge against another. He challenged the Catholic Church's view that its brand of knowledge, derived as it was from revelatory scriptures and faith, was supreme and immutable. Many of its errors could actually be *seen* through a telescope.

A disconcerting, though amusingly ironic, outcome of Copernicus's and Galileo's work was that a heliocentric solar system made it easier for the clergy to calculate the correct dates for religious festivals such as Easter and Pentecost than those that were possible with notions of a geocentric solar system. After all, it was science rather than mythology that allowed Thales accurately to predict a solar eclipse. This matter of the ecclesiastical calendar and its dependence on calculations made from a heliocentric solar system was a dilemma for the Church. But it had much weightier matters on its mind that called for swift, decisive action.

The Catholic Church was already in deep theological and economic trouble. It did not need another dispute. North of the Alps, the Protestant Reformation was challenging the authority of the Pope in Rome. One of the issues that Luther, Calvin and others raised was the authority of scripture. Should everyone have access to the word of God, or was the interpretation of scripture the exclusive preserve of the Church in Rome? This issue seemed

12 *Galileo's dictum that 'two truths cannot contradict one another' is as true of religion as it is of science. This portrait by Justus Sustermans hangs in the Royal Society, London.*

fundamental to both sides of the dispute. What was more, passages in the Bible seemed to bear directly on the structure of the solar system.

We may take one example. When Joshua was leading the Israelites on their bloody conquest of the Promised Land, God caused the sun to stand still: '[T]he sun stood still and the moon stayed, until the people had avenged themselves upon their enemies.... And there was no day like that before it or after it' (JOSHUA 10:13–14). There are two issues here: the working of the solar system and the actions of the Israelites. The first appeared to condone the second. The linking of these two issues was highly significant because it could be argued that God created and controlled the solar system in the interests of one Chosen People. The consequent murderous terror that the Israelites visited upon their enemies at God's behest does not seem to have troubled early seventeenth-century Christians as much as the matter of a stationary sun:

[P]ursue after your enemies, and smite the hindmost of them; suffer them not to enter into their cities: for the Lord your God hath delivered them into your hand. And it came to pass, when Joshua and the children of Israel had made an end of slaying them with a very great slaughter, till they were consumed, that the rest which remained of them entered into fenced cities. (JOSHUA 10:19–20)

The Promised Land flowed with blood as much as it did with milk and honey.

For Catholic theologians, already with their backs to the wall as a consequence of the Reformation, the moral issues raised by the biblical narrative could easily be brushed aside. After all, they wished to inflict (and indeed did inflict) comparable atrocities upon Protestants – who were not slow to retaliate in like manner. So it seemed to them better to concentrate on the indisputable scriptural authority, as interpreted by themselves, for the movement of the sun around the earth. A geocentric solar system was something to be defended at all costs. After all, if they let this go, what dreadful apostasy would follow? Weaken the authority and secrecy of scripture and all is lost.

As is now well known, Galileo was tried and eventually brought before the Inquisition in 1633.[18] Present-day events have highlighted this seventeenth-century *cause célèbre*, and today we tend to see it, not without reason, as emblematic of our own debate between science and religion.

Before his trial, indeed as early as 1604, Galileo had been warned not to dabble in the theories of Copernicus. In 1616 he was again warned off. Sixteen years earlier, Giordano Bruno, a lapsed Dominican and an intellectual who argued for infinite worlds and, it seems, a heliocentric cosmos, was burned alive after many years in prison. He was tied to the stake, and then a spike was thrust through his tongue so that his last words would be stifled. As the flames rose, a priest held up a crucifix for his edification. It is said that he averted his gaze.[19]

Galileo paid no attention to the threats directed at him or to the unbelievably cruel events of the times. But, in the end, he was forced to recant, though he is said to have muttered as he stood down, 'Eppur si muove' ('All the same it [the earth] does move'). The longevity of this possibly apocryphal statement was assisted by a 1643 painting by Murillo. It shows Galileo pointing to these words on the wall of his prison cell. It now seems that the famous phrase was in fact invented by a Catholic supporter.[20]

Galileo lived out the rest of his life under house arrest, though he managed to write further scientific works and to enjoy the high life of Italian noble courts. Among these books was *The Two New Sciences*, a work on which Isaac Newton was later to build.

It is important to remember that Galileo was not an atheist. On the contrary he was a believer. But he wrote some telling words on the position that the Church was adopting at that time. In a letter to Christina of Lorraine, Grand Duchess of Tuscany, he said,

But I do not feel obliged to believe that the same God who has endowed us with senses, reason, and intellect has intended to forgo their use and by some other means to give us knowledge which we can attain by them.[21]

In the same letter, he quoted what has now become a famous phrase:

I would say something that was heard from an ecclesiastic of the most eminent degree: 'That the intention of the holy ghost is to teach us how one goes to heaven, not how heaven goes.'[22]

Some historians present Galileo as a rather abrasive person, a natural controversialist. They suggest that, if he had been more diplomatic, he might have persuaded the Church to accept a heliocentric solar system.[23] This take on Galileo is interesting because it reflects the attitude of many people today to writers like Richard Dawkins and Daniel Dennett. These incisive atheists, it is sometimes said, may be right in what they say, but they should be more tactful and not appear to ridicule the deeply held beliefs of their opponents. Their supposed lack of tact and 'respect' is taken as sufficient reason to ignore their rational arguments. Many people feel that religious beliefs should be sacred – again, two kinds of knowledge, challengeable and unchallengeable – even if they do not themselves hold any.

The Galileo affair has not been laid to rest. Hitherto unknown documents that cast new light on the issues involved continue to surface. For instance, the historian Pietro Redondi has argued that there was more at stake than helio- centrism. He claims that Galileo's heresy really concerned the Eucharist. Galileo's view of the physics of light cast doubt on the Catholic doctrine of transubstantiation – the actual transformation of the sacramental wine into the blood of Christ.[24] With the powerful Reformation challenging this ortho- dox view of the Eucharist as one of its main thrusts, it may well have seemed a good idea to the Catholic authorities to divert attention to astronomical matters. If, by the simple expedient of quoting scripture, they could discount the possibility of a heliocentric solar system, the even more uncomfortable implications of Galileo's work for doctrines about the Eucharist could be conveniently sidelined. The common people (and many clergy) would find it easier to grasp arguments about the movement of the sun around the earth – something they could actually see – than the physics of light and the impact of such abstruse ideas on the invisible transubstantiation of Eucharistic wine. Perhaps Redondi is right and that is what happened. If readers today are bemused by Church concerns over heliocentrism, they will be completely baffled by the intricacies of disputes about transubstantiation. Fortunately, there is no need for us to pursue them. Nevertheless, we should not lose sight of the fact that it was on the scientific rather than the theological issue of heliocentrism that Galileo was overtly convicted.

All along, the Catholic Church had proclaimed its support for science, and still does. The Church claimed that it was not simply anti-science. But it nevertheless confined its science to areas that were unlikely to lead to major theological disputes. We observe something of this today, though not especially in the Catholic Church. Chiefly Protestant Intelligent Design and Creation Science groups claim that they are not opposed to science, just to false science. They too, so they say, do scientific work, and Intelligent Design should therefore be taught in science classes, not merely in religious classes.

Principia Mathematica

The early Protestants had fewer problems than Catholics with science. They were, after all, opposed to the Catholic Church, and that church's obsession with a stationary, central earth invited Protestants to protest on scientific as well as theological grounds. There were, however, conservative Protestants who were wary of the new-fangled view of the universe. Both Luther and Calvin were uncomfortable with the heliocentric notion, as was Osiander the Lutheran writer of the detracting preface to Copernicus's book. Calvin is often quoted as asking: 'Who will venture to place the authority of Copernicus above the Holy Spirit?'

Nevertheless, the leaders of the Reformation prudently paid little attention to the matter. Given this situation, it is not surprising that scientific advances initiated south of the Alps moved north to Protestant lands. Now Tycho Brahe in Denmark, Johannes Kepler in Swabia, together with Isaac Newton, Robert Boyle and Francis Bacon, all in England, and others took centre stage.

Some of these advanced thinkers were in fact still not thorough-going Copernicans. There was much debate among them.[25] The famously wealthy[26] Tycho Brahe's most significant contribution, for instance, did not concern the overall structure of the solar system. In 1572 he detected a 'new star', what today we would call a supernova. The significance of this discovery was that Brahe was able to show that the new star lay well outside the solar system in a cosmological region that the Church believed was changeless, being the seat of the deity, the Empyrean. And Brahe managed this without a telescope. Instead, he designed and had made a series of accurate but expensive instruments for pinpointing the positions of heavenly bodies. Using such means, he constructed a model of the universe that seemed, at least to him, to embody the best features of both the geocentric and the heliocentric versions – again the desire for compromise. The earth, he claimed, was at the centre of the universe and the sun and the moon revolved around it. The other planets, he said, orbited around the sun. The road to a full heliocentric system was curiously circuitous and controversial.

Johannes Kepler (1571–1630) was caught up in those controversies. Brahe hired him to defend his version of the universe against the one proposed by Nicolai Reymers Ursus, who, Brahe believed, had plagiarized his ideas. When Brahe died in 1601, Kepler was delighted to inherit the marvellous instruments that the older man had designed and had had made. Although Kepler was a Copernican, he retained some odd ideas: he was, to begin with, at least, an advocate of the 'harmony of the spheres' and used musical concepts to explain the 'harmony' of the solar system. In his book *Mysterium Cosmographicum* (1596), Kepler claimed that he had gained insight into the mind of God via his study of the solar system. Using mathematical calculations, he believed that he had determined a nest of five regular solids, each related to one of the then-known planets. Thus Saturn was associated with a cube, Jupiter with a tetrahedron, Mars with a dodecahedron, Earth with an icosahedron, Venus with an octahedron, and Mercury with a sphere.[27] This geometric complexity was, Kepler believed, designed by God. Somewhat surprisingly, it was within the framework of such complicated theories that Kepler eventually came to accept that the planets did not move in circular orbits, but in ellipses. This was his major contribution to astronomy. Clearly, it cast doubt on the notion of nested, harmonious, crystal spheres. How could a set of *elliptical* 'spheres' move smoothly inside one another?

The momentous innovations of this time – the sixteenth and seventeenth centuries – took place in a world that was in some but not all respects very different from our own. Even highly intelligent scientists remained infected by beliefs about supernatural elements that structured and made sense of life and the world within which they lived. In the minds of many of these researchers and writers magic played a curious role. The changeover to a heliocentric solar system was not a simple, swift event, a decisive routing of the supernatural. Mystical ideas remained embedded in astronomy and other sciences for some while. Some of these ideas even appeared to be confirmed by scientific discoveries. One such was the discovery of magnetism. Magnetism, researchers found, was an invisible force, and they believed that it opened up the possible existence of supernatural influences. Of course, as it all later turned out, the inexplicability of magnetism was temporary and certainly did not warrant recourse to belief in supernatural forces. This is a lesson we should bear in mind today.

For some scientists, the occult, alchemy and astrology seemed to promise significant insights. Among these was Sir Isaac Newton (1642–1727), like Galileo, another controversialist. Alexander Pope wrote a witty but, alas, unused epitaph for him:

> Nature, and Nature's Laws lay hid in Night.
> God said, 'Let Newton be,' and all was Light.[28]

Newton was born in the same year that Galileo died, 1642.[29] Some of his most important thinking was done in 1665 in a country village where he took refuge from the great plague that was ravaging London. He also wrote the highly influential but complexly mathematical *Principia Mathematica* (1687) and opened up the study of optics and the constitution of light. In his book *Opticks* (1704) he posed a series of deeply probing questions about the nature of the universe, many of which he left for others to engage with. Devising fruitful, penetrating questions is as important as answering them, a principle especially important in an investigation such as the one we are presently following.

Newton was apparently a rather curmudgeonly and absent-minded soul, but he rose to a position of immense influence and respect.[30] He was knighted, though it has to be added that Queen Anne bestowed this honour on him more for political than scientific reasons. She desired his influential support (he was at the time Master of the Mint) in forthcoming parliamentary elections; the ruse failed, and the party of which Newton was a member was defeated.[31] Today, Newton is seen as an archetypal mathematician and scientist rather than an alchemist and politician. Wordsworth (another Cambridge man) wrote:

> The statue stood
> Of Newton, with his prism, and silent face:
> The marble index of a mind for ever
> Voyaging through strange seas of Thought alone.
> *The Prelude*, 3:1:60

Newton is, of course, most famous for his working out of the idea of gravity, a force that explained the 'harmony' of the universe without invoking crystal spheres and music, let alone sirens. Gravity – without the notion of a God who kept everything in order and functioning – explained how the universe held itself together and how it all worked. The force of gravity was, arguably, the greatest discovery ever made up to that time. It was not equalled until Darwin published *Origin of Species* in 1859. Together, these two astonishing insights continue to provide the framework for our understanding of the world and, indeed, the whole cosmos without recourse to religion and supernatural beliefs.

Yet, as I have pointed out, Newton was not free from belief in some curious things. Although his work on gravity rendered the notion of continuous divine support for the universe unnecessary, he believed in a God. Theologically a (secret) Arian, or Unitarian, rather than an orthodox Trinitarian Anglican, he was much interested in the kingdoms of the Old Testament and the prophecies of Daniel. To facilitate these studies, he went to the trouble of learning Hebrew. He believed, heretically, that the early Church had obscured the truth of the scriptures and of the old philosophers as well. At the same time, Newton pursued the quest of alchemy – how to turn base metals into gold – and conducted experiments in his Cambridge rooms.

All in all, Newton was a man of mixed beliefs. Interestingly, today people who advocate belief in both science and the supernatural do not hail the great Isaac Newton as an intellectual ancestor: he unfortunately believed in the wrong sort of supernatural things. Today we wonder how it can be that people are able to entertain what appear to us blatantly contradictory beliefs, an ability that is sometimes called 'cognitive dissonance'. Certainly, Newton and others were consciously trying to construct a new science, but they still had a foot in the old camp. To revert to a metaphor that I used earlier, they were emerging from the cocoon of a superseded form of knowledge, but some of its shreds still adhered to the new creature. During what has become known, rightly or wrongly, as the Scientific Revolution, the equal acceptability of both divinely provided knowledge and knowledge derived from observation was seriously questioned – though not completely abandoned.

A battle lost and won

How the Christian Church faced up to the incontestable conclusion of heliocentrism and the subsequent discoveries of Newton and his *confrères* of the Scientific Revolution is a key to the overall argument of this book.

Try to see the problem from the point of view of the devout. If the earth is a peripheral speck and not the centre of the universe, how can anyone believe that human beings are the real reason for God's creation? The new astronomy just did not make sense when it was viewed from a religious perspective. It clearly clashed with the very fundamentals of Christianity: God's grand scheme of Creation, the Fall of Man, and Redemption brought by Christ do not make sense if the site of supposed divine interest is relegated to an insignificant planet spinning far from the focus of the universe. Scientific knowledge had soundly defeated revealed knowledge: the earth was *not* the centre of things.

Yet the Church survived, despite the utterances of its leading bigots. Luther, it will be recalled, famously said, 'Reason is the devil's harlot.' And John Wesley believed that 'Sin is the moral cause of earthquakes.' The Church lost the geocentric battle but won the war – or so it appeared at the time. It achieved its survival by a shift in emphasis. Prelates announced that the new astronomy simply did not matter. The contradiction between a divine focus on humankind and cosmological marginality was of no significance. It had mattered passionately for centuries, but no longer. It was again business as usual. The huge contradiction was simply ignored.

What could have – what should have – brought the Church down (if we take seriously Christian teaching about the reason why God created the cosmos) did not, after all, matter. Or so it began to seem. But that sort of dismissal of the religious implications of scientific work is facile. All the passion that the clergy had put into combating the new heliocentric astronomy was not only wasted: it was, in retrospect, manifestly misguided. The Church's arguments and strategies designed to defeat the astronomers could not possibly have been inspired by a God who created the cosmos with its peripheral earth. The shameful magnitude of the Church's error and the human suffering it caused were conveniently glossed over. In today's parlance, the Church simply said 'Oops!', and moved on. Of course science, too, changes its mind and moves on to new understandings, but there is a big difference. Science openly seeks to challenge received ideas; it expects – hopes – to move on. Unlike the Church, it does not get knowledge from on high on the understanding that it is vouchsafed by an unchanging God. Though individuals may be discomforted when they have to abandon long-cherished scientific hypotheses, the whole enterprise is not discredited. On the contrary, it is vindicated. Because the Church claims and defends divine, revealed knowledge, it should be embarrassed when that knowledge proves to be wrong.

The change to a heliocentric astronomy was not the end of the Church's woes. Worse was just around the corner.

'Let there be light'

The scientific work of Isaac Newton, Francis Bacon, John Locke and other brilliant men (and indeed the Reformation with its challenge to Catholic supremacy) laid the foundation for what has become known as the Enlightenment.[32]

The Enlightenment was a hugely important seismic shift in people's attitudes to the two kinds of knowledge that we are considering, scientific and revealed. The movement flourished across Europe, different nationalities emphasizing different components of thought. Clashes of opinion within the movement were an indication of its virility and a vindication of its principles of free thought. In France, where Enlightenment thinking was fiercely directed at the Catholic Church and authoritarian government, the foundations of the French Revolution were laid, though its excesses were certainly not on the Enlightenment agenda. In America, ideals of the Enlightenment were incorporated in the Declaration of Independence. The far-flung Enlightenment was a highly complex movement, and this chapter cannot do justice to its currents and crosscurrents, its influence not only in philosophy but also in art and music.

Fundamentally, the Enlightenment urged people to use their own minds and not to depend on the instruction of others. The effect of this dictum was that people were not to restrict themselves to scientific matters; they should also challenge all political, ethical and religious authority. Today we sometimes forget that the Enlightenment was not a coldly impersonal insistence on reason, as some came to think of it: rather, it sought political and religious liberation. It was, in a modern phrase, a liberation movement. Above all, it was a humane movement, however badly it may have fared during the French Revolution. Spinoza, the Dutch philosopher, put the message of the Enlightenment thus: 'I have laboured carefully, not to mock, lament, or execrate, but to understand human actions.'[33]

One of the more intellectual missions of the Enlightenment was to reconcile objective observation and experiment with deductive reasoning. This philosophical (perhaps esoteric) task takes us back to Aristotle and the distinction between inductive and deductive thinking (Chapter 1). Yet, all in all, whatever their differences, Enlightenment thinkers did not question the overriding power of logical, rational thought – not divine revelation – to get at the way things actually are in the universe *and* in human behaviour. Enlightenment thinkers would have no truck with Newton's notion of a God tinkering with his created universe, though they gratefully accepted his work on gravity and other issues. For them, the Heavens did not proclaim the glory of God, as Joseph Haydn's *Creation* so splendidly announced in 1798. All talk of the supernatural was merely obfuscation, though today, as then, we can accept that the glory (however we now construe that word) of the universe stands independently of any supernatural underpinnings.

Inevitably, the Enlightenment clashed with religion. The clergy, many of whom were corrupt and decadent, were what we would today call 'sitting ducks' for the barbed shafts of highly intelligent men. But at that time blasphemy was illegal, and critics were careful not to commit their most scurrilous sallies to paper – and, when they did, to keep well out of the way. Their criticisms of religion were not purely philosophical. Advances in scientific understandings at the time opened up a broad avenue of attack on the Church, which was already embarrassed by the ambiguities and contradictions within what had so long been considered sacred, inerrant scripture. It became clear to Enlightenment scholars that the Bible should be approached like any other book: its inconsistencies and palpable contradictions of the straightforward facts of the world should be squarely confronted. The Church, too, took up this challenge, but, unlike the Enlightenment writers, the clergy did not question the fundamental notion of divine inspiration; they stayed with the safer (literally, not just intellectually) issues of interpretation and explication.

Out of this sort of discussion came Enlightenment deism. This recension of religious belief in God was attractive to thinking persons. It rejected miracles, prophecy and conventional Christianity, together with its corrupt priesthood. Reason, it was argued, provided the only approach to religion that did not end in superstition and a life of fear. In place of a personal, often wrathful, God, deists accepted the existence of a 'god' who created the universe and then withdrew, thus leaving human beings to rely on their own rationality, not on miraculous interventions. God was a principle rather than a person. Enlightenment thinkers went on to argue that viewing the universe through the spectacles of clear thought induced the decency and honesty that should be the foundation of ethics. But a 'god' who does not intervene is a contradiction, and it now seems inevitable that Enlightenment deism should, by the exercise of its own principles of reason, eventually give way to atheism.

The role of miracles rather than reason in conversion to Christianity was, as we have seen, a common feature of the Christian tradition. The Scottish Enlightenment philosopher David Hume (1711–1776) saw the whole affair though the eyes of the new movement:

The Christian religion was not only first attended with miracles, but even at this day cannot be believed by any reasonable person without one. Mere reason is insufficient to convince us of its veracity: and whoever is moved by faith to assent to it, is conscious of a continued miracle in his own person, which subverts all the principles of his understanding, and gives him a determination to believe what is most contrary to custom and experience.[34]

Here, in an engaging, ironic phrase, Hume sums up a theme of this book: in theology, reason is always post hoc.

In history, too, the Enlightenment challenged the view that had developed under the influence of the medieval Scholastic movement. Historical events should no longer be seen simplistically as the unfolding of God's plan or as portents of a coming apocalypse. Enlightenment writers claimed that, contrary to what early Christian historians thought, Rome did not fall because God became disillusioned after his initial success with Constantine. Rather, there were political and geographical factors that led systematically to the fall of the empire. In England, Edward Gibbon (1737–1794) caused a sensation when he adopted this line in his tome *The Decline and Fall of the Roman Empire*. Christians resented Gibbon's rejection of sin as the reason for Rome's decline. Today historiography is much debated, but no serious historian advocates a return to pre-Enlightenment approaches.

The diverse teachings of the Enlightenment and the scientific advances of the time were influentially proclaimed in *L'Encyclopédie*, the many volumes of which were for twenty years edited by the famous French savant Denis Diderot. Ambitiously, this monumental work set out to explain rationally all aspects of existence, thus placing itself from the beginning on a collision course with the Church. In Scotland, the Scottish Enlightenment led to the comparable publication of the *Encyclopaedia Britannica*, but, because anti-clericism had been somewhat defused in Britain by the establishment of the Anglican Church, the English-language encyclopaedia was not as iconoclastic as the *L'Encyclopédie*.

The Romantic movement of the early nineteenth century was in part a reaction against what it saw as the sterile rationalism of the Enlightenment. Wordsworth, for instance, condemned Voltaire's *Candide* as, 'This dull product of a scoffer's pen'.[35] Today, after the disillusionment of two world wars and other calamities, voices are still raised against the Enlightenment. In the twentieth century Michel Foucault challenged what may now seem to be the unwarranted optimism of Enlightenment philosophers and also the universality of the principles of the Enlightenment.[36] It is true that the aims of the Enlightenment have not yet been fully realized, but that is no reason for discounting them. A counter-revolution that sees 'enlightenment' in mystical terms is now gathering force. Why, we may well ask, do people nowadays wish to fudge rational knowledge and slide back into the feel-good mysticism of the New Age and ecstatic Christianity? But we must defer consideration of this alarming tendency to later chapters.

'By Means of Natural Selection'

After Galileo's sensational trial and the great surge of the Enlightenment, the next open and very public confrontation between science and religion took wing in the second half of the nineteenth century. Its focus, rather than its initiation, was the publication in 1859 of Charles Darwin's *On the Origin of Species by Means of Natural Selection, or the Preservation of Favoured Races in the Struggle for Life.* The debate between religion and science moved away from the early astronomical concerns, these having been indisputably settled.

Superficially, the new controversy was whether humankind, along with all other creatures, evolved from simpler forms of extremely ancient life or whether, as prominent members of the Church claimed, the earth and all that is in it was created by God in six days not so very long ago. I say 'superficially' because the fundamental issue was more subtle and, ultimately, more disastrous for religious belief. Instant, miraculous creation rather than slow evolution was an outpost to be defended lest the battle come too close to the inner citadel wherein dwell all things supernatural. Hold the barbican and you may not have to defend the keep. The erupting debate could be construed as concerning the way in which human beings came to be as they are, not about the very existence of *anything* supernatural. There was, therefore, a partial replay of the Catholic Church's seventeenth-century emphasis on a geocentric solar system rather than on the more difficult issues of the physics of light and transubstantiation. Evolution could be confronted rather than the absolutely fundamental issue of whether or not there are any supernatural realms, beings, influences and occurrences.

By 1859, the year in which Charles Darwin published *Origin*, the idea of evolution had been around and widely discussed for some decades. The basic notion of evolution was not new. Erasmus Darwin, Charles's progressively thinking physician grandfather, accepted the idea of gradual development (Fig. 13). In those pre-1859 days, evolution was known as 'transmutationism'. Confronted with fossil marine life in the hills of the English Midlands,

Erasmus Darwin adopted as his personal motto the words *E conchis omnia* (Everything from shells). It is said that he had this prescient phrase on the door of his carriage and on his bookplate.[37] But, rather than fossils, it was observable similarities between different living species that Erasmus Darwin thought pointed to common ancestry and hence to transmutationism. In following up this idea, he observed the effects of domestication and planned breeding on animal species. It is surprising how close he came to what was to be his grandson's great contribution to knowledge. As early as 1794, he wrote:

The final cause of this contest amongst the males seems to be, that the strongest and most active animal should propagate the species, which should thence become improved.[38]

In another publication, six years later, he wrote:

From the sexual, or amatorial, generation of plants new varieties, or improvements, are frequently obtained; as many of the young plants from seeds are dissimilar to the parent, and some of them superior to the parent in the qualities we wish to possess.... Sexual reproduction is the chef d'oeuvre, the master-piece of nature.[39]

He was on the brink of discovering genetic variation and natural selection. One could say that the idea of evolution evolved within the Darwin family. Nevertheless, Erasmus resisted the more materialistic ideas of the Enlightenment and allowed that the sort of development of which he spoke was purposeful and 'creative'. He believed in a kind of spontaneous creation that took place at the beginning of everything. For 'spontaneous' the pious

13 *Charles Darwin recalled of his grandfather, Erasmus: 'He used to say that "Unitarianism was a feather-bed to catch a falling Christian".*[40]

could read 'God-engineered', if they so wished, even as today they read 'creation' for 'the Big Bang'. In one of his widely read poems, Erasmus Darwin wrote:

> Hence without parent by spontaneous birth
> Rise the first specks of animated earth.[41]

He did not believe in a six-day creation; nor did he accept the notion of the Fall of Man and Redemption by Christ. In these post-Enlightenment years, such challenges to Church authority were common. But he did believe in a Prime Mover, a 'God' who set creation in motion and was now watching the success story of his design. For Darwin, transmutationism meant progress. Living as he was during the Industrial Revolution, he saw 'progress' in technology as entirely beneficial. Indeed, 'progress' could be said to be his philosophy of life. If God could be shown to be behind all 'progress', that was all to the good. The British empire with its colonial exploitation and, at home, Blake's 'dark Satanic mills' could flourish with God's blessing. This concept of a divine Initiator of Progress is, of course, rather different from the notion of God that traditional Christians believed, but it nonetheless afforded some sort of compromise, and, at this time, numerous people were looking for compromises. It could be argued that Thomas Huxley's later coining of the word 'agnostic' created some middle ground between absolute belief and complete disbelief.

All these ideas were passionately discussed at the time. Despite compromise on the one hand and resistance on the other, the Enlightenment with its rejection of mythology and superstition and its insistence on materialism was achieving practical effects, though more slowly than its great eighteenth-century protagonists had hoped.

Then, in 1844, 42 years after Erasmus Darwin's death, came *Vestiges of the Natural History of Creation*, a sensational book. In it the anonymous author tackled the problem of how one species could evolve into another. An intrigued public suggested that the book had been written by such unlikely figures as Prince Albert, Queen Victoria's Consort, and, perhaps more reasonably, the geologist Charles Lyell.[42] In fact, the author was Robert Chambers (1802–1871) of the publishing firm that still bears the family name. Like Erasmus Darwin, Chambers did not wish to jettison the idea of God unequivocally, so he argued that the law of 'like produces like' in the procreation of species could be interrupted by a 'higher law'. An embryo could be 'bumped up' to the next level of development and a new species could thus come into being. This notion helped to resolve a dilemma created by trans-mutationism: did God control evolution by a long series of miracles rather than by one six-day spectacular? Did his will manifest itself in the working out of laws that he himself designed? The notion of a hierarchy of divine laws seemed to accommodate the misgivings of many believers. It also chimed well with the Victorians' desire for a highly regulated society in which virtue was equivalent to abiding by the law. Today, the idea of sporadically intervening 'higher laws' unknown to science continues to lurk in Intelligent Design and Creationism.

Apart from the way it papered over cracks between science and belief by postulating 'levels' of laws, *Vestiges*, widely read as it was, can be seen as the starting point of the really public debate about religion and evolution: it opened up the grand theme of creation for no-holds-barred discussion.[43] Indeed, some of the negative reactions to *Vestiges* were intemperate. The famous British geologist and clergyman Adam Sedgwick (1785–1873), the man with whom a youthful Darwin was later to walk the hills of Wales, wrote apoplectically to Lyell in 1845 that *Vestiges* begins

from principles which are at variance with all sober inductive truth. The sober facts of geology shuffled, so as to play a rogue's game [i.e., to cheat]; phrenology (that sinkhole of human folly and prating coxcombry); spontaneous generation; transmutation of species; and I know not what; all to be swallowed, without tasting and trying, like so much horse-physic!! Gross credulity and rank infidelity joined in unlawful marriage, and breeding a deformed progeny of unnatural conclusions! ... If the book be true, the labours of sober induction are in vain; religion is a lie; human law is a mass of folly, and a base injustice; morality is moonshine; our labours for the black people of Africa were works of madmen; and man and woman are only better beasts![44]

There is a bitter pill here: much of what Sedgwick deplores in the second half of this quotation eventually turned out to be true – but this did not destroy civilization, as Sedgwick feared it would. Unlike Sedgwick and others of his time, the ancient Greek philosophers would have been pleased by the challenge that, even in its inchoate form, the theory of evolution presented to the Church. Sacred revealed knowledge, so long held to be beyond any debate that did not imply notions of heresy, excommunication and, sometimes, torture and death, was coming under increasing fire. People were openly questioning and arguing about fundamental religious teachings more freely than ever before. Like Paul speaking in the Agora, the Church found itself a voice crying in an encroaching wilderness.

Thanks to *Vestiges*, the general public was, by the 1850s, already familiar with the notion of evolution and its grim implications. Nearly a decade before the publication of *Origin*, Alfred Lord Tennyson (1809–1892), a master of melancholy, famously wrote of 'nature red in tooth and claw'. No doubt having read Lyell, in 1851 he less famously contemplated the bleak evidence of geology:

> Are God and Nature then at strife,
> That Nature lends such evil dreams?
> So careful of the type she seems,
> So careless of the single life;
>
> …
>
> I stretch lame hands of faith, and grope,
> And gather dust and chaff, and call
> To what I feel is Lord of all,
> And faintly trust the larger hope.
>
> 'So careful of the type?' but no.
> From scarped cliff and quarried stone
> She cries 'a thousand types are gone:
> I care for nothing, all shall go…'
>
> *In Memoriam*, LIV, LV

Tennyson knew the slough of religious despond before Darwin published his theory. Geological despair pre-dated Darwin. Indeed, it could be argued that Darwin echoed Tennyson's verses in his peroration to *Origin*: 'Thus, *from the war of nature, from famine and death*, the most exalted object which we are capable of conceiving, namely, the production of the higher animals, directly follows' – not, be it noted, from benign design.[45]

Generally, the Church did not like Darwin's ideas. Some Christians tried to come to terms with the notion of natural selection; others hyperbolically rejected it out of hand.[46] Among the more accepting was the Reverend Charles Kingsley, author of *Westward Ho!* and *The Water-Babies*. He wrote to Darwin, who had sent him a complimentary copy of *Origin*, that he had 'gradually learned to see that it is just as noble a conception of Deity, to believe that He created primal forms capable of self development' as it was 'to believe that He required a fresh act of intervention to supply the lacunas which he himself had made'.[47] Perhaps Emma, Darwin's devoutly religious wife, adopted this perspective. Had she not reached some sort of accommodation with her husband's ideas of evolution, he would not have asked her to see to the publication of his early short version of his acceptance of natural selection, should he die prematurely.[48]

What, then, did Darwin achieve? Even if *Vestiges* was correct in its general transmutational theme, there remained an unanswered question that offered some solace to believers. They saw what they thought was a God-shaped gap. No one seemed able to explain fully and comprehensively *how* one species could evolve into another. It remained for Darwin to provide a *mechanism* for evolutionary change and to present the whole idea of evolution in so fully argued a form that it could be ignored by neither scientists nor religionists. Religionists are unwise to put their faith in scientists' inability to explain this or that.

Very briefly, Darwin initiated the idea of random (not divinely guided) mutations that produced slightly altered offspring – even though he had no idea of genetics.[49] Many of these mutations were eliminated because they were ill suited to their environment. But some mutations were advantageous. They were 'selected' by breeding. They then led on to the evolution of new species. This is the now well-known theory of natural selection.

Darwin's cogent argument in *Origin* started with the common practice of breeding among domestic animals – a comfortable and familiar starting point for his readers. For example, pigeon breeders watched for the spontaneous appearance of a new trait. Then, by careful selection of mates, they could capitalize on this new feature and produce a new type of pigeon. This process was effected in nature, not by divine intervention but by the elimination of non-adaptive traits and the promotion of advantageous ones – hence *natural* selection. Prodigal nature spawned many variations but few survived: species lived, had their apogee, and faded from the face of the earth. One could say that natural selection is wasteful and bumbling, not economic and purposeful.

Matters came to a head with the Church in the celebrated 1860 debate in Oxford between Darwin's supporter Thomas Henry Huxley and Bishop Samuel Wilberforce. The bishop facetiously asked Huxley whether it was through his grandmother or his grandfather that he was descended from a monkey. Famously, Huxley replied that he would rather be descended from a monkey than from a bishop who prostituted the gifts of culture and eloquence in the service of falsehood – or words to that effect.

It is astonishing that, 150 years after *Origin* and despite an overwhelming accumulation of evidence, there are still people who reject evolution. Ignorantly, some say, 'Teach the debate!' There is no debate. Practising scientists who produce valid, high-level research accept evolution. Without the theory of evolution they would be able to make only small-scale, particularist observations. Still, believers who reject evolution are in at least one way more perspicacious than their *confrères* who accept it and then try to paper over the cracks it has caused in religious belief. Conservative believers realize, if dimly, that evolution, with its dead-ends and random muddles, undermines their foundational belief in a benign supernatural realm.

Today, evolution is part of every educated person's thinking – like it or not. There is as much evidence for it as for the heliocentric solar system. We may wonder how far along the early twentieth-century psychologist William James's scale we have come:

First…a new theory is attacked as absurd; then it is admitted to be true, but obvious and insignificant; finally it is seen to be so important that its adversaries claim that they themselves discovered it.[50]

Another battle lost and (almost) won

It is illuminating to compare the two major confrontations (heliocentrism and evolution) between science and religion that I have identified. They neatly complement one another. Both shunted humankind away from anything that could be called a centre of interest. Copernicus removed the earth from the centre of the solar system, and Darwin showed that human beings are no more than a sport of nature, the result of a blind process. The ineluctable workings of that process (natural selection) *necessarily* produced something that, for some people, resembled a design but was not. Overall, purposeful design – 'creation' – became a redundant notion. Quite simply, it is not only unnecessary: it also masks the processes of nature that are there for all to see.

Yet, by and large, the Christian Church managed to reach accommodations with both scientific positions, though only after much debate and, certainly in the case of the first confrontation, much cruelty. Today the relationship between the sun and the earth is doubted by no one, and, as far as I know, is not questioned by any significant religious leader. It is a non-issue, long forgotten.

The second confrontation (evolution), on the other hand, has not been as completely resolved.[51] Again, try to see it from the point of view of a devout believer. The question of the centrality of the earth in the solar system was decisively laid to rest. But religious belief in the unique status of human beings was (apparently) salvaged. Despite cosmological marginality, many people continued to believe that humankind is the cynosure of God's love and salvation. But now the devout hear something worse: they are told that human beings are in no way unique. Rather, they are simply one of the more recent evolutionary forms of animals. No clear hiatus between them and animals, no moment of 'ensoulment', can be discerned. The implications of evolution are thus truly devastating.

Nevertheless, many Christians have managed to believe that both evolution and religious teaching about humankind being the only reason behind God's decision to create the entire universe may be simultaneously true. For instance, the nineteenth-century evangelist and theologian Henry Drummond neatly turned evolution to his own advantage: 'No man can run up the natural line of Evolution without coming to Christianity at the top.'[52] By and large, most Christian denominations are quietly accepting the evolution of humankind. The strident opposition to evolution that we now often hear emanating from the USA is, for many Christians, an embarrassment. They say: if we evolved, so be it. As some put it: religion tells us who made the universe and why; evolution tells us how he did it.

This compromise is another papering over of cracks. Believers have not taken on board the full implications of science's victory. It is a case of religious business as usual – but not quite as usual. Although believers may discreetly avert their gaze, scientific knowledge has indisputably again defeated what was at one time taken to be revealed religious knowledge. The two fields of science and religion are therefore less distinct and autonomous than some well-meaning folk would today have us believe. Science *does* impact upon how believers understand revealed knowledge. And religion continually makes insupportable scientific statements.

Throughout history, both science and religion have changed. Today scientists no longer believe in nested crystal spheres, and few theologians believe God literally threw fire and brimstone at Sodom and Gomorrah. But there is a difference between these two trajectories. Science advanced when its practitioners were brave enough (I use the phrase deliberately) to ignore supposedly revealed knowledge. Theology, on the other hand, has had to change its stance again and again in the light of scientific discoveries. Science has *never* adjusted its findings in the light of theological positions. If believers no longer think of a white bearded God or of a God who desires the burning of witches, we must ask how and why these changes came about. Were they really a result of further God-revealed knowledge in a long progressive revelation or were the changes a result of the nudging of science and the rational society that science has created? Rational knowledge has consistently defeated revealed knowledge.

That being so, another question at once arises. Where, in today's still partly religious world, does science end and religion take over? In trying to maintain a separate-but-equal view of science and religion, diplomatic peacemakers generally avoid awkward questions, such as how a supernatural soul could have gradually evolved. Indeed, like it or not, we are edging closer to what will be the most fundamental clash between scientific and religious knowledge. The backlash of the American religious right should not surprise us. They sense that defeat is just around the corner.

If God does intervene in the affairs of the world, we should expect to find scientists compiling lists of phenomena that can be explained only by supernatural intervention. Over the centuries they have found no such phenomena. They find things that they cannot immediately explain but no indication that something supernatural is happening.

The historical review in this and the previous chapter gives some under-standing of the long and chequered history of how we came to be where we now are in the matter of science and religion. The history of conflict between Western science and religion is a starting point for working towards the single most fundamental question of all about religion and its role in human life. The question is *not*, 'Do you believe in God?' The ebb and flow of debate that we have followed over the centuries point to something deeper, something more comprehensive. The most important question is:

Can we today, after all that has happened since Thales managed to predict an eclipse, believe that there is a supernatural realm peopled by beings and forces who are interested in the lives of those who live on planet earth and who, at least from time to time, intervene in both natural and human affairs?

Both science and religion are constituted not in books and theories but in individual human beings. People react in different, often poignant, ways to this question. We consider the lives of two such people in the next chapter.

CHAPTER 3

A Tale of Two Scientists
Charles Darwin and Alfred Russel Wallace

CHARLES DARWIN:
I gradually came to disbelieve in Christianity as a divine revelation.[1]

ALFRED RUSSEL WALLACE:
While admitting to the full extent the agency of the same great laws of organic development in the origin of the human race as in the origin of all organized beings, there yet seems to be evidence of a Power which has guided the action of these laws in definite directions and for special needs.[2]

Today, any discussion of science and religion will, sooner or later, bring in Darwin. Just over 150 years after the publication of his sensational book, his ideas retain their hold on our minds and remain central to the debate. Yet, for all his achievements, he was an ordinary human being. When we consider science and religion, we need to go beyond logic to the human condition, its emotions, loyalties and complex (often fraught) personal relationships. Scientists are not isolated, emotionally sterile robots who employ some magic formula called 'scientific method' to achieve inevitably true results. Fundamentally, they are just like the rest of us. Darwin interacted with other people, some very close and dear to him; others were his scientific *confrères*. One of these was Alfred Wallace.

Charles Darwin (1809–1882) and Alfred Russel Wallace (1823–1913) knew each other well. Together, they illustrate the tensions and anxieties that many people experience today when they contemplate the challenges that science poses for continuing religious belief. The psychological traumas of people like Paul and Augustine were not for Darwin and Wallace; they were too 'ordinary' for high spiritual drama. But their divergent, intellectually irreconcilable lives help us to identify and understand with some compassion this perpetual root of dissension. Despite his acceptance of evolution, and to the horror of his fellow scientists, Wallace came to believe in spiritualism and participated in mediumistic séances. Indeed, Darwin may well have contributed, albeit unintentionally, to Wallace's retreat into irrationality.

'A damnable doctrine'

The general outline of Charles Darwin's life is well known and I need not rehearse it in detail here.[3] There are, however, aspects of it that are often overshadowed by his image as a 'scientist'. It is these that I highlight.

He was born into a remarkable and distinguished family (Fig. 14). In many ways the Darwins were the epitome of respectable Victorian values, but the family also had a strain of free-thinking. On the one hand, Charles's stern, teetotaller father, Robert, had his son marked out to follow in his own foot-steps and become a respectable and respected physician. As we saw in the previous chapter, his father's father, Erasmus Darwin (another physician), was not only an observer of nature but also a free-thinker. Shockingly, he believed that species were not immutable. Charles's mother was a daughter of Josiah Wedgwood, founder of the celebrated pottery. She died when Charles was eight years old. The Wedgwoods were a family with advanced ideas; their guests included Coleridge, Byron and Wordsworth. In 1839, Charles married his first cousin, Emma Wedgwood. Charles's background and marriage meant that money was never a problem. He could devote himself to research independently of the worries of academic appointments, tenure and jealousies.

Even before he went to school in the English West Country town of Shrewsbury, where his father practised, Charles had a 'taste for natural history, and more especially for collecting': as he put it in his private autobi-ography, he 'tried to make out the names of plants and collected all sorts of things, shells, seals, franks, coins, and minerals'.[4] Shrewsbury, the great public school to which Charles was sent as a boarder, was only about a kilometre and a half from the Darwin home; his elder brother Erasmus was already a pupil there. Charles often managed to sneak home and then sprint back to be in time for roll-calls. Fear of being late sometimes led the boy to pray that God would assist him to run faster. It seems that he always managed to make it back in time because he later wrote ironically, 'I attributed my success to the prayers and not to my quick running, and marvelled how generally I was aided.'[5] Occasionally, one senses that Darwin had a sense of humour.

His youthful piety was, however, less than perfect. A common daily requirement of schoolboys at that time was to memorize '40 or 50 lines of Virgil or Homer'. Charles, who must have been a quick learner, managed to do this surreptitiously 'whilst I was in morning chapel'.[6] In church schools today, morning chapel still affords dilatory pupils a last chance to prepare for the day's academic challenges.

14 *Charles Darwin: 'I have no great quickness of apprehension or wit which is so remarkable in some clever men…for instance Huxley … My power to follow a long and purely abstract train of thought is very limited.'*[7]

Although Charles did not lack for friendships while at school, he came to love long walks and solitary angling. Prolonged musing was one of his most positive traits. Yet, in his autobiography, he is at pains to emphasize that he was 'a very ordinary boy, rather below the common standard in intellect'.[8] Perhaps his probing mental life led him far beyond the confines of Virgil, Homer and stultifying classrooms. So his schoolwork began to suffer. He recalled his father as saying, 'You care for nothing but shooting, dogs, and rat-catching, and you will be a disgrace to yourself and all your family.'[9]

To avoid such unspeakable humiliation and because Charles was, in any event, not doing well at Shrewsbury, Dr Darwin packed him off to medical school at Edinburgh University. This was in 1825 when Charles was only 16 years old. It was at this time that he realized that his father would bequeath him a sizeable inheritance and that he would never have to work to make a living. He admitted that, as a result of this prospect, he began to neglect his university studies.[10] Some components of the courses bored him, others disgusted him: 'I also attended on two occasions the operating theatre in the hospital in Edinburgh, and saw two very bad operations, one on a child, but I rushed away before they were completed...this being long before the blessed days of chloroform.'[11] It is hard to imagine the effect of such horrors on a 16-year-old boy.

Soon, the young student's lack of interest in medicine came to the attention of Dr Darwin, who then suggested that his son become a clergyman. This idea was fostered not by any consideration of Charles's personal piety (as we have seen, he was not a religious enthusiast) but because his father realized that, as a country parson, his son would be able to pursue his secular interests from an eminently respectable base – as Dr Darwin no doubt noticed numerous clergymen were doing. By the time of Charles's youth, Anglican Christianity had become a comfortable, eminently respectable enclave in early nineteenth-century society.

Charles took the suggestion very seriously. This was his first facing up to the implications of religion. True to his thoughtful nature, he would not be rushed into such a decision. Indeed, the way he tackled the issue is illuminating and the conclusion to which he came, if seen from our present perspective on his life and work, surprising:

I asked for some time to consider, as from what little I had heard and thought on the subject I had scruples about declaring my belief in all the dogmas of the Church of England; though otherwise I liked the thought of being a country clergyman. Accordingly I read with care Pearson on the Creeds and a few other books on divinity; and as I did not then in the least doubt the strict and literal truth of every word in the Bible, I soon persuaded myself that our Creed must be fully accepted. It never struck me how illogical it was to say that I believed in what I could not understand and what is in fact unintelligible. I might have said with entire truth that I had no wish to dispute any dogma; but I never was such a fool as to feel and say 'credo quia incredible' [more correctly, 'credo quia incredibilis': 'I believe because it is unbelievable', from Tertullian, one of the early Church Fathers].[12]

At this time, Darwin's attitude to Christian dogma was like that of many people today. When, in Chapter 6, we consider the notion of belief, we shall find that Darwin *acknowledged* dogmas rather than *believed* them in any committed sense. He had grown up with them; his whole environment embodied them. All he had to do was to associate himself with those dogmas. It simply never struck him that it is nonsense to say that one believes in something that is 'in fact unintelligible'. Perhaps most importantly, he had not had a religious experience such as those that came to Paul, Constantine, Augustine and others. He had no experience of revealed knowledge in his own life.

So it was that young Charles Darwin, who believed in 'the strict and literal truth of every word in the Bible' but was in no way pietistic, set off to Christ's College, Cambridge, to train as a clergyman. But all along he knew that if he could later be appointed to a quiet rural parish, he would be able to continue his private collecting interests and supplement a parson's meagre benefice with his family inheritance. The prospect was enticing.

At Cambridge, he enjoyed reading 'the clear language' of Reverend William Paley (1743–1805). This clergyman's books *The Principles of Moral and Political Philosophy* and *A View to the Evidences of Christianity* set up an argument for the existence of God that was based on the idea of a watch being found abandoned on a heath.[13] Following in Aquinas's footsteps, Paley argued that creation was as complex as a ticking watch. Surely, then, we must assume the existence of a watchmaker-creator. This line of thought chimed well with Darwin's interest in natural history. Already he was wondering how the world in which he saw and collected so many interesting things came to be as it is. He found Paley's notion of the Creator as a watchmaker attractive, as indeed it is – provided one does not think too deeply about it.

An interesting point about Paley that is generally overlooked today is that he was a prominent clergyman who was committed to the abolition of slavery – a goal with which Darwin also associated himself. There was a moral bond between the two men. Perhaps more importantly, Paley was thoroughly up-to-date with his knowledge of biology and human anatomy: he did not argue simply from the revealed knowledge of scripture. Indeed, he dwelt upon the complexity of the human eye and how, in his view, it could not have evolved from something simpler. The eye is a seemingly persuasive example that is still used today, even though evolutionists have adequately dealt with the matter and shown that the allegedly missing intermediate forms can still be found in creatures living today.[14]

Over 200 years after the publication of his books, Paley's powerful analogy and his notion of 'irreducible complexity' still have their attractions for many people. Their persistence led the present-day evolutionary biologist and combative advocate of atheism Richard Dawkins to entitle one of his books *The Blind Watchmaker*.[15] How harmoniously interacting patterns could come about in nature was and still is of great interest to many people. Is a pattern necessarily a 'design'? The watch on the heath is one of those misconceptions that seem not to go away, no matter how often they are refuted.[16]

As with medicine in Edinburgh, Darwin soon began to find much of the theological work in Cambridge desperately boring. Shrewsbury and Edinburgh were repeating themselves on the banks of the Cam. Paley may have been mildly exciting, but the Old Testament narratives and endless laws seemed to have little relevance to the world in which Charles was living. The biblical lists of 'begats' begat tedium not divine illumination. He probably also scratched his head over such stories as the one about Lot's generosity with his daughters' virginity.

What he read during that time was important, but the people he met at Cambridge were far more significant in shaping his life. Charles's family background, together with his enquiring mind, put him in contact with some of the most influential and intellectually diverse people at Cambridge. He was not an anonymous, ineffectual student on the fringe of things. Early on, Charles was introduced to the Reverend John Henslow, 'a man who knew every branch of science', as Darwin later put it.[17] He was a well-known Cambridge botanist, and Charles accompanied him on field excursions. Encyclopaedic knowledge, such as Henslow's, was always attractive to Darwin. At about the same time he was introduced to entomology by his cousin W. Darwin Fox, who was also at Christ's College. In a university vacation, Charles followed up yet another interest when he went with Adam Sedgwick, Professor of Geology at Cambridge, on a collecting trip through North Wales. The multi-component foundations for his future contributions to science were being well and truly laid: practical skills in botany, entomology, palaeontology and geology.

On his return home from his and Sedgwick's tramp through Wales, he found two letters awaiting him. One told him of the possibility of accepting the position of naturalist on a sea voyage round the world. The other, from Henslow, urged him to accept the invitation. Charles was eager, his father reluctant. Perhaps guessing what would happen, Dr Darwin declared, 'If you can find any man of common sense, who wishes you to go, I will give my consent.'[18] Charles managed to secure the support of his uncle, Josiah Wedgwood, and together they spoke to Dr Darwin. 'As my father always maintained that [Josiah] was one of the most sensible men in the world, he at once consented in the kindest manner.'[19] Like his son, father Darwin was a sensitive man. Charles then suspended – rather than abandoned – his theological studies.

The captain of HMS *Beagle*, the Royal Navy ship on which he was to sail, was at first not well disposed towards Darwin (Fig. 15). He seriously doubted that he should accept Charles as a 'gentleman companion' on what was to be a long voyage around the world (it lasted from 1831 to 1836). The trouble was Darwin's nose. Captain Robert FitzRoy (1805–1865) was a believer in the hugely unscientific study of character from physiognomy, a field that a French clergyman, J. C. Lavater, had popularized. For FitzRoy, the shape of Darwin's nose suggested that he did not have the moral fibre necessary to complete so gruelling a journey.[20] As it turned out, FitzRoy relented and took Darwin on board. In doing so, he set the scene for a protracted and difficult relationship, one that was played out over the next five years in a single, cramped cabin on the *Beagle* and indeed for many subsequent years. Perhaps the shape of the young man's nose was, after all, significant.

Today we tend to think of FitzRoy as much older than Darwin, but he was only twenty-six at the start of the voyage. Darwin was twenty-two. These were in fact two young men setting out on an adventure. The difference between

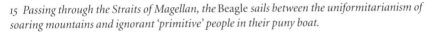

15 Passing through the Straits of Magellan, the Beagle *sails between the uniformitarianism of soaring mountains and ignorant 'primitive' people in their puny boat.*

them was that Darwin was a well-to-do country doctor's son, while FitzRoy was an aristocrat who came from a famous family – and it seems that he never forgot his superior social status.

Throughout the voyage, Darwin was a nominal Christian believer. FitzRoy, on the other hand, was not only a follower of Lavater's bewildering ideas. He was also a stern, evangelical Christian who believed in flogging sailors. Much to Darwin's disgust he was also a supporter of slavery. As captain of the *Beagle*, FitzRoy conducted divine service every Sunday morning. They must have been oppressive affairs. His religious views led him to believe that it was sinful to ask too many questions about creation. God created the world and all the various species as they now exist, and that was that.

The effect of cramped living conditions aside, FitzRoy seems to have been an unstable personality from the beginning. He flew into frequent storms of temper, sometimes directed at Darwin. These rages were followed by periods of morose sulking. But he always calmed down and renewed his friendship with the young naturalist. And, for his part, Darwin remained magnanimous (at least in his autobiography). Still, it is surprising that he and FitzRoy were still on speaking terms by the end of their circumnavigation of the world.

Years later, FitzRoy was greatly dismayed by Darwin's sensational book *Origin of Species*. As he grew older, he became more and more religious and prone to fits of depression. This unhappy combination was perhaps exacerbated by feelings of guilt, for it was he who gave Darwin an opportunity to gather evidence for what eventually became his devilish theory of evolution by natural selection. To make matters worse, Darwin became immensely famous, and FitzRoy, whom Darwin (understandably) seldom acknowledged, felt himself marginalized from the great debates going on around him – though he was clearly intellectually and emotionally ill-equipped to participate in academic and philosophical disputes. He did, however, succeed in other ways. He was promoted to the rank of Vice-Admiral and was appointed governor of New Zealand from 1843 to 1854. He was also appointed chief meteorologist in 1854 and through his interest in meteorology developed a type of barometer.

After the voyage, FitzRoy's response to Darwin's unsettling ideas was to sink deeper and deeper into obstinate fundamentalism. Like many people today, he was unable to handle the implications of evolution and his response was stubbornly to espouse wildly improbable notions of divine creation. In the end, despite the consolation of his considerable achievements, FitzRoy died by his own hand, cutting his throat with a razor. Such extreme measures seem to have run in his family, for his uncle, the famous and equally tempestuous politician Robert Stewart, Viscount Castlereagh, also committed suicide, though he resorted to a less dramatic and effective implement – a penknife. Castlereagh, sometime chief secretary for Ireland and influential participant in the Congress of Vienna after the Battle of Waterloo, was held responsible by many for the 1819 army attack at St Peter's Field, Manchester, on supporters of parliamentary reform (the popular name 'Peterloo Masacre' was in imitation of Waterloo, the Duke of Wellington's bloody 1815 defeat of Napoleon). Between six and ten protestors lost their lives – a negligible total by today's standards. Nevertheless, this event led Shelley (never one to miss an opportunity for hyperbole) to pen a neatly rhyming couplet in 'The Mask of Anarchy':

> I met Murder on the way –
> He had a mask like Castlereagh.[21]

Poor FitzRoy had an uncle's formidable reputation to live up to.

Despite nagging troubles with this moody captain of the *Beagle*, Darwin described his voyage as 'by far the most important event in my life'.[22] His interests in what his father had called 'shooting, dogs, and rat-catching' waned, and he came to believe that 'the pleasure of observing and reasoning was a much higher one than that of skill and sport'.[23] The Cape Verde Islands, St Helena, Patagonia, Chile, numerous tropical islands, and, especially, the now-famous Galapagos Islands all had a profound impact on the young man. Assiduously, he collected specimens and had them shipped back to England, where his growing collection was already exciting interest.

As the voyage progressed, he was able to see with his own eyes the truth of the famous geologist Sir Charles Lyell's ideas on 'uniformitarianism' – the principle that the forces we can now observe at work in the world fashioned the surface of the globe over extended periods of time. This view was in direct conflict with 'catastrophism', the belief, favoured by the Church, that the world was created by major, divinely occasioned upheavals, one of which was Noah's Flood. Lyell's notion caused a great stir, but Aristotle had in fact hinted at it centuries before. The ancient Greek thinker wrote:

But the whole vital process of the earth takes place so gradually and in periods of time which are so immense compared with the length of our life, that these changes are not observed, and before their course can be recorded from beginning to end whole nations perish and are destroyed.[24]

Lyell's great book *Principles of Geology* was published in 1830, and Darwin took a copy of the first volume to read on his *Beagle* voyage. The second volume reached him when the *Beagle* berthed in Montevideo. Strangely, Lyell was unhappy when Darwin later applied his views on gradual geological change to biological evolution. He found it hard to abandon his belief in humankind's special place in creation. Decentred humanity is a bitter pill.

At first, Darwin himself seems to have experienced a comparable, if less entrenched, reluctance to abandon Christian beliefs. Five months into the voyage, among exotic palms and perhaps more than a little homesick, he still imagined taking up an appointment in a rural Church of England parish: 'I find I steadily have a distant prospect of a very quiet parsonage, & I can see it even through a grove of Palms.'[25] On their way from Australia to the Cape of Good Hope, FitzRoy and Darwin co-authored an article that requested the British Government to provide more support for missionaries in the Pacific. It was published in 1836 in the *South African Christian Recorder*. Understandably enough, it seldom features today in a Darwin bibliography, though at the time its publication it may have seemed to the young man a useful career move.

The voyage of the *Beagle* brought home to Darwin more than an understanding of species, their distinctions, interrelationships and mutations. Inevitably, he also encountered people. And many of those people practised religions that were very different from the decorous Anglican Christianity with which he was familiar. The diversity of religion that the voyage allowed him to observe presented him with a slow-burning challenge.

This part of his *Beagle* experiences does not feature prominently in his diary. Religion was not on his research agenda, but what he saw and heard was nevertheless being filed away in his memory. He collected ideas as well as objects. His glimpses of strange religious beliefs and practices would forever colour the way in which he would eventually come to see *all* religions. It is unfortunate that writers usually overlook or minimize this aspect of his circumnavigation of the globe. It seems that, at first, he saw peoples of other, for him incomprehensible, religions as simply in need of Christianity, education and civilization – as his and FitzRoy's article shows. Once civilized, surely these pagans would almost automatically become Christians.

This belief seemed, though only initially, to be supported by an experiment that FitzRoy had undertaken. On a previous voyage three years earlier to South America, he had taken on board four young people (three men and a girl) from Tierra del Fuego, with the idea of bringing them back to England.

There he educated them, converted them to Christianity and showed them off to various dignitaries, much to his own benefit and fame. His plan was to return them to Tierra del Fuego where they would, so he thought, convert their home communities to Christianity and civilization. One of these, a man FitzRoy named Boat Memory, died in England of smallpox. FitzRoy named the two surviving men Jemmy Button and York Minster, and the young girl Fuegia Basket (Fig. 16). Jemmy Button was so named because FitzRoy – as we have seen, a man not averse to slavery – had bought him for a few buttons. To allow them to retain their own original names was apparently unacceptable to FitzRoy. No one thought it worthwhile to record their real names. On the voyage back to Jemmy's homeland the three Patagonians were accompanied by a young missionary named Richard Matthews. His intention was to remain in Tierra del Fuego and assist Jemmy, York and Fuegia in converting and civilizing Patagonia.

When the *Beagle* reached Tierra del Fuego, the three unfortunate aboriginal people were returned to their communities, complete with fine Victorian clothes and useless artefacts that included tea-trays, white linen and chamber-pots. Already Matthews's heart had begun to sink. Needless to say the bizarre scheme was a failure. The returned Fuegians' communities were suspicious of their former fellows; they themselves found it hard to fit in again. The vegetable garden that the sailors prepared (much to the bafflement of the local people, who were not agriculturalists) was soon destroyed. Matthews sensibly abandoned his plan to remain in Tierra del Fuego.

After a year-long back-tracking visit to the Falkland Islands, the *Beagle* returned to Tierra del Fuego to see how the three recently converted young people were getting on. FitzRoy was devastated. Jemmy Button and his two companions had abandoned their European clothes and, though cordial, did not wish to have much to do with him. Darwin, who observed all this lunacy, wrote: 'We could hardly recognize poor Jemmy. Instead of the clean, well-dressed stout lad we left him, we found him a naked thin squalid savage.'[26] FitzRoy begged him to return to England with the ship, but he refused. He had had enough of civilization.

Westerners could not accept what they saw as 'primitive' religions as in any way alternatives to Christianity. Following Augustine, they held that there was no salvation outside the Church. It was, they believed, because such religions were 'primitive' that they had been superseded in 'civilized' parts of the world. Darwin may not have consciously arranged these diverse religions in an evolutionary sequence – or, more probably, in a devolutionary order the farther

FUEGIA BASKET. 1833. JEMMY'S WIFE 1834.

JEMMY IN 1834. JEMMY BUTTON IN 1833.

YORK MINSTER IN 1832 FUEGIANS. YORK IN 1833.

16 FitzRoy's ill-begotten experiment in the transformation of human beings is expressed in starkly contrasting portraits. Top, Fuegia Basket in 1833 and Jemmy Button's wife in 1834; centre, Jemmy Button in 1834 and in 1833; below, York Minster in 1832 and 1833.

away they were from the Holy Land and its special revelations of the divine. Nevertheless, he seems to have thought of Western Christianity as the apogee of civilized religion. Unregenerate Patagonians had a long way to go.

The religion of these people was indeed incomprehensible. Part of Darwin's diary entry for 11 August, 1833 reads:

Shortly after passing the first spring we came in sight of the famous tree, which the Indians reverence as a God itself, or as the altar of Walleechu. It is situated on a high part of the plain & hence is a landmark visible at a great distance. As soon as a tribe of Indians come in sight they offer their adorations by loud shouts.... Being winter the tree had no leaves, but in their place were countless threads, by which various offerings had been suspended. Cigars, bread, meat, pieces of cloth &c. &c. Poor people only pulled a thread out of their ponchos. The Indians both pour spirits & mattee into a hole & likewise smoke upwards, thinking thus to afford all possible gratification to Walleechu. To complete the scene the tree was surrounded by the bleached bones of horses slaughtered as sacrifices. All Indians of every age & sex make their offerings; they think that their horses will not tire & that they shall be prosperous.[27]

Darwin interpreted what he saw in terms of his own experience of Christianity. So he wrote of reverence, adoration, sacrifice and the practical benefits to be derived from such states of mind and activities. As a result of their 'offerings', the Indians' horses would not tire – perhaps even as he did not tire when, as a young boy, he prayed as he ran from his home to be back at Shrewsbury in time for roll-call. As the Bible itself proclaimed, if God could be gratified, all life would be better. There was therefore some point of contact between Darwin's notions of religion and what the South American folk believed and did. For both groups, religion had material, practical rewards. We wonder what Darwin's and FitzRoy's reaction would have been had they come across a Maya bloodletting ritual or one of the numerous other societies through history that have practised human sacrifice and cannibalism rather than used it merely as a metaphor, as Christianity does in the Eucharist.

As the voyage of the *Beagle* proceeded, strange religions continued to impact on Darwin's thinking.[28] Later, looking back on those experiences, he wrote, 'The sight of a naked savage in his native land is an event which can never be forgotten.'[29] Part of his private autobiography gives us some idea of how he saw what was for him a growing problem:

But I had gradually come, by this time, to see that the Old Testament from its manifestly false history of the world, with the Tower of Babel, the rainbow as a sign, etc., etc., and from its attributing to God the feelings of a tyrant, was no more to be trusted than the sacred books of the Hindoos, or the beliefs of any barbarian. The question then continually rose before my mind and would not be banished – is it credible that if God were now to make a revelation to the Hindoos, would he permit it to be connected with the belief in Vishnu, Siva, &c, as Christianity is connected with the Old Testament. This appeared to me utterly incredible.[30]

Darwin was beginning to realize that many Old Testament narratives were as incredible as the beliefs he was encountering in remote parts of the globe. If the beliefs of the Israelites, Patagonian Indians and many other peoples were to be accepted, then God must be a tyrant who had to be appeased by sacrifice and strict obedience. This sort of thing was, for Darwin, 'manifestly false'. There was no possible way of squaring such beliefs with what he saw in the fossils and with the immense age of the earth – or, for that matter, with the generally more peaceful teachings of the New Testament. Scientific thought pointed to universal laws (like Lyell's uniformitarianism), not to the idiosyncratic whims of vengeful supernatural beings.

Other tendencies were also at work in Darwin's mind. Even before the *Beagle* returned to England, he was becoming famous in scientific circles. Sedgwick, with whom Charles had walked the mountains of North Wales, told a well-pleased Dr Darwin that his son would 'take a place among the leading scientific men'. Moreover, Charles's old friend Henslow had read some of the young man's letters before the Philosophical Society of Cambridge, and his collection of fossils had been shown to palaeontologists. A delighted Darwin received this good news while he was on Ascension Island: 'After reading this letter I clambered over the mountains of Ascension with bounding step and made the volcanic rocks resound under my geological hammer!'[31] The 'very ordinary' Shrewsbury boy, 'rather below the common standard in intellect', had found his *métier*.

In the three years following his return to England, Darwin meticulously catalogued his vast collection of specimens. He no longer thought of that 'very quiet parsonage', and his religious belief had faded. Soon, he wrote out a private 250-page document that encapsulated his ideas on evolution by natural selection. His friend the botanist Joseph Hooker was the only person to read the manuscript. Darwin did not publish his ideas. That he showed them to only one close friend suggests that he recognized their importance, but for private as well as scientific reasons he did not feel ready to expose them to public scrutiny. Instead he worked on various projects and a biography of his grandfather, Erasmus Darwin, who had died seven years before Charles was born. Charles presents his revered ancestor, who suffered from a speech impediment, as a likeable man. He recounts an amusing anecdote:

A young man once asked him in, as he thought, an offensive manner, whether he did not find stammering very inconvenient. He answered, 'No, Sir, it gives me time for reflection, and saves me from asking impertinent questions.'[32]

If you bear in mind that Sedgwick wrote his splenetic diatribe against *Vestiges* in 1845, you will form some idea of what Charles Darwin would have expected if he went ahead and published his first summary of *Origin*. Was he troubled by the prospect? Recent research and newly published volumes of his correspondence[33] have called into question the long-held notion that Darwin feared both public ridicule and rejection by the scientific community and therefore delayed publication of *Origin*.[34] It seems more likely that he was simply engaged on other projects that he wished to complete before he set about the really big one – evolution by natural selection. His study of barnacles, for instance, gave him a good empirical understanding of natural variation and, just as importantly, ensured his standing in the eyes of the scientific community before he moved on to more controversial matters.

Darwin may not have feared public reaction to evolution by natural selection itself, but he does seem to have had misgivings about the implications that his theory had for the origin of human beings. He makes this clear at the beginning of his subsequent book *The Descent of Man*:

During many years I collected notes on the origin or descent of man, without any intention of publishing on the subject, but rather with the determination not to publish, as I thought that I should thus only add to the prejudices against my views.... Now the case wears a wholly different aspect.... In consequence of the views now adopted by most naturalists...I have been led to put together my notes.[35]

Darwin's delay in publishing directly on the origins of human beings was, at least to some extent, strategic: in 1859 he felt people were not ready for his views. At home, too, he seems to have been reaching some sort of understanding with his devout wife Emma. The views on religion that he expresses in the private autobiography intended for his family are in no way muted, though Emma did ask for certain passages to be deleted before publication.

It was not until 1856, and on Lyell's advice, that Darwin started to write *Origin of Species*. Then came what he later (and now famously) called 'a bolt from the blue':

But my plans were overthrown, for early in the summer of 1858 Mr Wallace, who was then in the Malay archipelago, sent me an essay on the tendency of varieties to depart indefinitely from the original Type; and this essay contained exactly the same theory as mine.[36]

Darwin had shown interest in Wallace's work and to send his article to Darwin seemed the natural thing to do.

Darwin was deeply shocked, but, being a generous man, he did not want to appear to pre-empt Wallace. Nor did he wish to lose the primacy of being the first to think of the idea of natural selection. It was a real dilemma.

Some writers think that our usual estimate of Darwin as entirely benign and solicitous of Wallace's interests is not the whole story.[37] Perhaps we should remember that Darwin sought Lyell's advice on the same day as he received the letter from Wallace. Clearly, he thought this was a matter of great urgency. Should he publish at once or should he wait a couple of months so that Wallace's reaction and possible plans could be sought? Part of the problem was that neither Darwin nor Lyell knew if Wallace had sent his work to any other people. If they rushed out Darwin's work without mentioning Wallace, they could be acutely embarrassed if it was then revealed that Wallace had generously sent his work not only to Darwin but to others as well. Another cause for concern was that, if Wallace had thought of natural selection, someone else unknown to any of them might also have thought of it and be on the verge of publication. Another, and tragic, complication was to come. Ten days after Darwin received the letter from Wallace, his and Emma's infant son Charles Waring died of scarlet fever. Grieving, Darwin was relieved to be able to leave the Wallace matter in his friends' hands. They decided to go ahead with their plan without seeking Wallace's agreement: Darwin's and Wallace's work should be made public simultaneously and as soon as possible.

So it was that, after negotiation with Lyell and Hooker, who read both papers before the Linnaean Society, Alfred Wallace's essay was published simultaneously with a hastily put together account by Darwin in the *Journal of the Proceedings of the Linnaean Society* (5 September, 1858). The simultaneous publications received only one review. In it, Professor Haughton of Dublin declared 'that all that was new in them was false, and what was true was old'.[38] Much later, Wallace wisely reflected on this sort of rejection:

Truth is born into this world only with pangs and tribulations, and every fresh truth is received unwillingly. To expect the world to receive a new truth, or even an old truth, without challenging it, is to look for one of those miracles which do not occur.[39]

Undismayed, Darwin set about writing what was to become the most sensational and widely read scientific book of all time: *On the Origin of Species by Means of Natural Selection, or the Preservation of Favoured Races in the Struggle for Life*. It was published amid great acclaim and debate in 1859. The first, rather limited, printing of 1,250 copies was sold out on the first day.

Darwin wrote *Origin of Species* at Down House in Kent. He and his family had settled in this country retreat in 1842. Charles had married Emma Wedgwood in 1839, and they remained at Down House for the rest of his life. Protected and cared for by his devoted wife, he became something of a recluse. During this time he suffered for extended periods from a mysterious illness that made it impossible for him to work.[40] Unlike Lyell and other famous scientists, he was never knighted. His ideas were simply too radical for the time.

Inevitably, Darwin's work on evolution led him to reconsider his own religious beliefs – and no doubt to marvel at how close he himself had come to being ordained a clergyman. But he found it hard to abandon the tradition into which he was born, as indeed do many people today. It was not only science that led him to question Christianity; the deaths of his non-believing father and his innocent, beloved daughter were strong influences, as such events are for many people. Darwin described his gradual move away from religious belief honestly and simply. His words are worth pondering:

Thus disbelief crept over me at a very slow rate, but was at last complete. The rate was so slow that I felt no distress, and have never since doubted even for a single second that my conclusion was correct. I can indeed hardly see how anyone ought to wish Christianity to be true; for if so the plain language of the text seems to show that the men who do not believe, and this would include my Father, Brother and almost all my best friends, will be everlastingly punished.

And this is a damnable doctrine.[41]

Darwin's single-sentence paragraph is startling and unequivocal – as he intended it to be.

Charles Darwin died in 1882, content to be called an agnostic, the word that his tireless supporter Thomas Henry Huxley had coined. Huxley, the naturalist, physiologist, anatomist and educational reformer who became known as 'Darwin's bulldog', was a powerful advocate of evolution. Thanks largely to his energy and influence, Darwin was buried in Westminster Abbey, next to Sir Isaac Newton. The irony of his interment in the Abbey has not escaped notice. It highlights one of the paradoxes discussed in later chapters of this book: although we may no longer believe in all the dogmas of our traditional religion, we are reluctant to abandon much of it – its music, poetry, art and those comfortable rituals that mark stages in our lives and that we wish to celebrate within the tradition that we share with our families and friends. We live within our heritage.

Darwin's funeral was notable. Among the distinguished pallbearers was William Spottiswoode, the President of the Royal Society. Other pallbearers included Sir John Lubbock, President of the Linnaean Society. Although he was a banker, Lubbock had written *Prehistoric Times*, an influential book about archaeology that challenged biblical history. Then there were Darwin's great friends, the botanist and explorer Sir Joseph Hooker and, of course, Huxley himself. Bringing up the rear was Alfred Russel Wallace,[42] the man who had independently conceived of evolution by natural selection but whose life and thought had, by that time, taken a very different – some would say alarming – direction.

'An Overruling Intelligence'

Unlike Darwin, Wallace was born into a family of modest means (Fig. 17).[43] At the time of his birth in 1823, they were living near Usk in South Wales.[44] Not for him the assured wealth of the Darwins and the Wedgwoods. The Wallace children had to earn their own livings. When Alfred was five years old, the family moved to Dulwich and then to Hertford, where he attended Hertford Grammar School, an adequate establishment but by no means as famous as Shrewsbury. From there Wallace moved to Leicester and was subsequently forever on the move.

His father, Thomas, a conventional churchman, laid a foundation for Alfred's interest in books. He read to his children from Daniel Defoe's *History of the Great Plague* and, more significantly, from Mungo Park's *Travels in the Interior Districts of Africa, in the Years 1795, 1796 and 1797*, a book that described the explorer's exciting journey up the Niger River. Park was an Edinburgh-educated surgeon. While the Wallace family was in Hertford, Alfred became librarian of a small library. This gave him an opportunity to devour Smollett, Fielding, Milton, Homer and Dante. But it was Lyell's *Principles of Geology* that made the greatest impact on the young man's mind. This great work was as important to him as it was to Darwin. It was uniformitarianism, rather than biblically based catastrophism, that made sense of the natural world.

17 Alfred Russel Wallace, evolutionist and spiritualist, believed that '[t]he moral and higher intellectual nature of man is as unique a phenomenon as was conscious life on its first appearance in the world.'[45]

Later, in London, Wallace attended lectures by Richard Owen (1804–1892). Owen was an anatomist, overseer of the Natural History section of the British Museum and agnostic. As a result of his general reading and of hearing these lectures, Wallace struggled with the concept of a benevolent and omnipotent God who permitted evil to plague the world. He concluded that 'the only beneficial religion was that which inculcated the service of humanity, and whose only dogma was the brotherhood of man'.[46] He thus rejected the received – and respectable – beliefs of the Anglican Church in favour of a form of humanism.

From 1837 to 1843 (when his father died) Wallace was a pupil surveyor to his brother William. This suited him well for he had abundant time in the evenings to read, and during the day he explored the geology of the area in which the brothers were working. He collected fossils in the chalk and in the gravels along the Ouse River, all the while enjoying the outdoor life that practical surveying demanded. His interests were wide and included palaeontology, botany, zoology, entomology and instrument making. Like Darwin, he was an autodidact with wide interests.

In 1844 Wallace became a schoolmaster at the Collegiate School in Leicester. He taught reading, writing, arithmetic and surveying. At this time he read Alexander von Humboldt's *Personal Narrative of Travels to the Equinoctial Regions of America* and Darwin's *Journal of Researches into the Geology and Natural History of the Various Countries Visited by H. M. S. 'Beagle'*. Of the latter he wrote: 'His style of writing I very much admire, so free from all labour, affectation, or egotism, and yet so full of interest and original thought.'[47]

Wallace's stay in Leicester, however, had a darker side. He became interested in phrenology (the study of human character by 'bumps' on the cranium), spiritualism and mesmerism (a mixture of spiritualism, spurious 'magnetism' and hypnosis propagated by Franz Anton Mesmer). Having abandoned traditional Anglican religion, he eventually allowed spiritualism to take its place.

In 1844, once more back at surveying, Wallace was intrigued by the anonymous and much-discussed publication entitled *Vestiges of the Natural History of Creation*. As we have seen, this tract broached the notion of evolution but did not explain how it worked. The writer outlined the evolution of the world from a gaseous mass through various stages to the transmutation of apes into human beings. At the head, the apex of 'development', stood the Caucasian 'race'. Although Robert Chambers, the anonymous writer, was privately a sceptic, he deemed it prudent to allow for the presence of God in his published account. He realized that the notion of evolution without a divine guiding hand would be anathema to his Victorian readers – a risk he was not prepared to take.

Wallace considered the idea of evolution

an ingenious hypothesis strongly supported by some striking facts…. It furnishes a subject for every observer of nature to attend to; every fact he observes will make either for or against it, and it thus serves both as an incitement to the collection of facts, and an object to which they can be applied when collected.[48]

He realized that the 'blind', or random, collection of observed facts would lead nowhere; some sort of guiding *principle* was needed. The collected facts were not – indeed, could not be – entirely independent of one another. Wallace had, even at this early stage, a well-developed concept of the relationship between facts and theory. Whatever the danger of finding only what you wish to find, it is a theory that makes the collection of facts meaningful.

After a visit to the entomology galleries of the British Museum, Wallace wrote to his friend Henry Walter Bates, 'I should like to take some one family [of insects] to study thoroughly, principally with a view to the theory of the origin of species.'[49] He began to hanker for a trip to the tropics where he could find 'facts' to test the 'theory' of evolution. He must have seen the impact that the *Beagle* voyage had on Darwin: distant shores beckoned.

Wallace recognized serious problems with the theory of evolution as it was set out in *Vestiges of Creation*. The 'how-did-it-happen?' question was the key to the whole matter. Nonetheless he was so taken with the idea of substantiating evolution that he sold his surveying business. Accompanied by his friend Bates, he set off in 1847 for South America and the Amazon – not quite a circumnavigatory *Beagle* voyage but, as far as *his* life was concerned, as significant as Darwin's great adventure. Nor was Wallace's and Bates's journey paid for, as was Darwin's by the British Admiralty and his own father. The two young men hoped to cover their expenses by shipping their finds home to England and selling them to museums. This they did from time to time.

Finally, on the way back to England, the ship on which Wallace was sailing caught fire and he lost most of his notes and all of the collected items, drawings and animals he had on board – except for one parrot that clambered out on the bowsprit and then dropped into the water from which it was rescued.[50] After ten days of drifting in an open boat, Wallace, his companions and the parrot were picked up 320 km (200 miles) from Bermuda. He arrived back in England in 1852. He had not solved the problem of speciation (how new species come into being) as he had hoped to do, but, like Darwin aboard the *Beagle*, he had garnered masses of information that he believed to be relevant to the question he was addressing.

Wallace then spent two years in London studying his own as well as museum collections. In was during this time, in 1854, that he first met Charles Darwin in the British Museum. Neither made any significant impression on the other. The two men were like Longfellow's ships that pass in the night.

Wallace's next trip was to South-East Asia. After six months in the Malay Peninsula, he commenced a series of taxing journeys around the islands of the Malay Archipelago. He had plenty of time in the lonely evenings to contemplate the problem that had long bothered him – speciation. A combination of his collecting and thinking led to a series of publications. In an article entitled *On the Law which has Regulated the Introduction of New Species* (1855) he reflected on Lyell's principle of uniformitarianism:

It would be most unphilosophical to conclude without the strongest evidence that the organic world so intimately connected with [the inorganic world], had been subject to other laws which have now ceased to act, and that the extinction and production of species and genera had at some later period suddenly ceased.[51]

Here, he put his finger on a key point: the world of life must surely be governed by the same natural laws as the world of stones, mountains and oceans. Although his writing style was not as elegant as Darwin's (as he himself realized), Wallace's thoughts were clearly running parallel to Darwin's: both men saw the complete illogicality of supposing the gradual formation of the earth and, at the same time, the sudden, supernatural appearance of separate species. Both had to reconcile the stratification of rocks, the fossils that the strata contained, and the geographical distribution of species.

In his 1855 article, Wallace stated his natural law: 'Every species has come into existence coincident both in space and time with a pre-existing closely-allied species.'[52] But *how* this happened he did not know. The article was greeted by general academic silence, though one commentator remarked that Wallace was 'theorizing' too soon: 'what was wanted was to collect more facts.'[53] The question of when a researcher has 'enough' facts to induce some 'theory' is still a problem in science. Observation follows observation, but how many observations are needed before a researcher can infer some generality or law? As we have seen, the logic involved in this issue was first broached by Aristotle.

Darwin was pleased to read Wallace's article and wrote to him to say so. No doubt thinking of his early, still-unpublished 900-page draft, Darwin added, 'I believe I go much further than you; but it is too long a subject to enter on my speculative notions.'[54] He was unwilling to give away his secret. It must have been a tantalizing time for both men.

Then, in February 1858, Wallace was laid low by a tropical fever in Ternate in the Maluccas Islands. Sweating it out on his rudimentary bed, he had time to recollect and ponder what he had read years before in *An Essay on the Principle of Population* (1798 and 1803) by Thomas Malthus, a writer who had also influenced Darwin. (Malthus was an economist and English country clergyman, such as Darwin nearly became.) Wallace wrote: 'Then it suddenly flashed upon me...'.[55] He realized that Malthus held the solution to the problem. Animals all produce more offspring than ever live through to maturity. Why do some survive and others perish? It must be that those best suited to the environment will be the ones to survive: 'in every generation the inferior would inevitably be killed off and the superior would remain – that is, *the fittest would survive.*'[56] Wallace had tumbled to the theory of evolution by natural selection. It was a matter of seeing connections between facts and theories to which others were blind. Moreover, he came close to coining the famous apophthegm 'survival of the fittest', usually traced to Herbert Spencer's use of it in 1865.

At once he wrote his article *On the Tendency of Varieties to Depart Indefinitely from the Original Type* and sent it off to Darwin. We know how Darwin received it and the events that followed – the joint publication of summary articles, the presentation before the Linnaean Society, and Darwin's swift publication of *Origin of Species*. The theory of evolution by natural selection was launched. It would prove unstoppable.

Both Darwin and Wallace were magnanimous, each acknowledging the other. There was no personal animosity. Wallace later wrote: 'The one great result which I claim for my paper of 1858 is that it compelled Darwin to write and publish his *Origin of Species* without further delay.'[57] Such generosity of spirit – especially in matters as far-reaching as how evolution works – is rare today.

Then Wallace's thought took its inexplicable turn. While Darwin continued on his agnostic road, Wallace, though still rejecting conventional Christianity, turned more and more to spiritualism. His initial acquaintance with the supernatural in Leicester blossomed. He travelled widely, especially in America, attending séances and lecturing. Among those whom he met was Senator Leland Stanford, a convinced spiritualist who founded Stanford University in memory of his dead son, with whom he believed he was in constant touch through several mediums.[58] In San Francisco, Wallace delivered a highly successful lecture on spiritualism to an audience of more than a thousand.

After a séance in which the medium was supposed to have caused tables to levitate, Wallace wrote:

However strange and unreal these few phenomena may seem to readers who have seen nothing of the kind, I positively affirm that they are facts which really happened just as I have narrated them, and that there was no room for any possible trick or deception.[59]

He was completely taken in. The simple fact that many conjurors' entertaining stage presentations are inexplicable to people in the audience, even though they know that trickery is in some way involved, shows that the absence of an explanation does not mean that supernatural forces are at work. Bearing conjuring performances in mind, we may wonder how well equipped Wallace was to detect the trickery of legerdemain. 'Brilliant' is not incompatible with 'gullible'.

After a séance in England, Wallace reported further 'strange and unreal... phenomena':

From a pile of small slates on a side-table four were taken at a time, cleaned with a damp sponge, and handed to us to examine, then laid in pairs on the table. All our hands were then placed over them till the signal was given, and on ourselves opening them writing was found on both slates.[60]

'Collecting facts' as unproblematically as if they were no different from beetles was leading Wallace down murky avenues; his inductions from such 'facts' were worthless. Beetle taxonomy may be complex, but the creatures do not try to deceive entomologists.

How could Wallace reconcile such bizarre ideas with his brilliant thoughts on natural selection? He was able to devise a way, albeit a devious cop-out. Natural laws, he asserted, could account for material evolution but not for the appearance of consciousness in humankind. That is where a 'higher Power' intervened. In a manner reminiscent of Paley's watch on the heath and today's Intelligent Design supporters, Wallace argued that 'the moral and higher intellectual nature of man is as unique a phenomenon as was conscious life on its first appearance in the world, and the one is almost as difficult to conceive as originating by any law of evolution as the other'. He waxed eloquent on the subject:

Let us fearlessly admit that the mind of man (itself the living proof of a supreme mind) is able to trace, and to a considerable extent has traced, the laws by means of which the organic no less than the inorganic world has been developed. But let us not shut our eyes to the evidence that an Overruling Intelligence has watched over the action of those laws, so directing their variations and so determining their accumulation, as finally to produce an organisation sufficiently perfect to admit of, and even to aid in, the indefinite advancement of our mental and moral nature.[61]

As soon as a writer, then or now, uses capitalized phrases like 'Overruling Intelligence', vague words like 'higher', emotive modifiers like 'fearlessly' and the jussive subjunctive ('let us...'), our alarm bells should start to ring. This sort of inflated writing indicates that the author is moving into a realm of discourse that is characterized by special pleading and weak evidence. Fundamentally, Wallace was no longer talking about his observations of nature but rather of his own (supposed) experience of a spirit realm with its outward manifestations of table-floating and other trivia. His knowledge was now coming from a special, personal revelation, one in which others could not participate and which they could not corroborate – unless they too 'believed'.

Many within the scientific community were appalled, but there were some, like Sir Oliver Lodge, who were attracted by supposed psychic phenomena. On the other hand, as Wallace himself had once realized but seemed to have forgotten, intelligent people saw the contradiction: we cannot envisage one set of laws operating on the material world and another on the 'spiritual world'. The two realms (if indeed both exist) are far too closely intertwined to allow such a dichotomy.

Denunciations were published, and even Darwin went along to see the spectacle. After attending a séance at which tables appeared to move, he wrote to Huxley: 'The Lord have mercy on us all, if we have to believe in such rubbish.'[62] When writing to Wallace himself, Darwin was a bit milder: 'I differ grievously from you, and I am very sorry for it.'[63] Having had enough of 'such rubbish', Darwin remained immured in Down House, so Wallace pleaded with Huxley to attend a séance with him so that Darwin's bulldog could see the wonders for himself. He did so at one held in Queen Anne Street, London. He then reassured Darwin that the medium was an impostor.[64] Darwin replied to Huxley: 'I am pleased to think that I declaimed to all my family, the day before yesterday, that the more I thought of all that I had heard happened at Queen Anne St, the more convinced I was it was all imposture.'[65]

Robert Chambers, author of *Vestiges of Creation*, was also taken in by the spiritualists. He hailed Wallace as a man of science who admitted 'the verity of the phenomena of spiritualism' and congratulated him on having 'leapt the ditch'.[66] Chambers took a position that is still sometimes aired today: 'My idea is that the term "supernatural" is a gross mistake. We have to enlarge our conceptions of the natural, and all will be right.'[67] Well, yes, I suppose it would be. The fallacy of this semantic trick will, however, become evident in later chapters. But it provided a prop for Wallace's beliefs.

He became as well-known as a spiritualist as he had been as a scientist and testified in a number of law cases on behalf of spiritualists.[68] One of these was the American medium Henry Slade, for whom spirits were supposed to write messages on slates. He had been dramatically exposed by E. Ray Lankester, along with Huxley. While Wallace was giving evidence in court, Darwin was quietly contributing to funds to support and cover Lankester's legal costs.[69]

At this time, public debate over spiritualism and other supposed supernatural phenomena raged. Numerous prominent (and sometimes unlikely) figures were convinced. Others, such as the popular conjurors John Nevil

Maskelyne and Harry Houdini (Ehrich Weiss), the man who made escapology famous, saw through it all. They exposed the frauds one after the other. 'Spirit writing' on slates, for instance, was and still is well-known in the conjuror's repertoire.

As early as 1865, Maskelyne and his friend George Cooke staged an exposé of the famous visiting American spiritualistic mediums and conjurors known as the Davenport Brothers, Ira and William (1839–1911, 1841–1877). In the style of the period, Maskelyne's flier read:

> MESSRS MASKELYNE AND COOKE
> The only Successful Rivals of the DAVENPORT
> BROTHERS will give a GRAND EXPOSITION
> of the ENTIRE PUBLIC SÉANCE[70]

The Davenports produced their supposed spiritualistic manifestations while handcuffed and bound. They initiated the practice of a medium being tightly trussed up and placed in a cabinet, in which there were hung musical instruments. As soon as the doors of the cabinet were closed the instruments emitted sounds; when the doors were opened, the medium was seen to be still bound. It was these skills that gave Houdini the idea of escapology.[71] The Davenports provided Houdini with useful material as he pursued his parallel career as the great exposer of spiritualists. But the Davenports themselves, perhaps cunningly, avoided claiming or denying supernatural powers: they left it to their admiring audiences to argue about the matter. They plugged into the debate, not the belief.

Among those who preferred the supernatural explanation was Sir Arthur Conan Doyle (1859–1930). So infatuated was he that he came to believe that his superb creation of Sherlock Holmes was preternaturally predestined so that he, Doyle, could become famous and well liked, and thus equipped to spread the gospel of spiritualism.[72] Discoveries in the fields of radioactivity and electricity seemed to support the idea of invisible worlds, and Doyle seized on them. But it was the death during World War I of his son, Kingsley, and his apparent return at the behest of a medium named Evan Powell, that led to Doyle's conversion to spiritualism. He thought of the Davenport brothers as 'probably the greatest mediums of their kind the world has ever seen'.[73] Though Doyle and Houdini became friends (for a few years), it was beliefs about the Davenports that eventually led to a rift. Houdini claimed (apparently rightly) that Ira Davenport had, in his old age, explained to Houdini how he and his brother had effected their escapes and spirit manifestations.[74]

Doyle would not accept Houdini's word for this claim. On the contrary, he seems to have come to believe that Houdini himself possessed supernatural powers and that the conjuror was lying when he said he achieved his effects by sleight of hand. In statements that encapsulate the opaque depths of the desire to believe the unbelievable, he wrote to Houdini: '[W]hy go around the world seeking a demonstration of the occult when you are giving one all the time? ... My reason tells me that you have this wonderful power, for there is no alternative.'[75]

It was in such absurd circumstances and among such people that Wallace continued to be a determined and contented spiritualist. He found an accepting, comforting alternative to the society of sceptical scientists; among such people, he could relax the rigour that science demanded of him and luxuriate in a world of fantasy and fame. Nothing could dissuade him and true believers like him. They remained unmoved, as indeed they are today when Randi,[76] the now-famous nemesis of spiritualists, Derren Brown,[77] the stage and popular television performer, and other critics expose the mediums' tricks.[78]

In some ways Wallace recalls Newton: both men seem to have been able to hold what appear to other people to be diametrically opposed, indeed contradictory, beliefs. Acceptance of the action of supernatural forces in the empirical world did not affect their commitment to science. In Newton's day that dichotomy was still possible, as it was for Plato and others through to medieval times. A few decades ago we might have felt that Wallace was living in the twilight years of contradictory beliefs, but today we see that the world of Newton, Plato and Aquinas is still with us.

Despite his belief in spiritualism, Wallace remained, like Conan Doyle, an amiable, likeable person. He received numerous awards and medals before he died in 1913. He was not buried in Westminster Abbey, though a wall-plaque was dedicated there in 1915. But he left the world and us today wondering how so brilliant a mind could be so deluded as to believe in floating tables. Yet, if we consider Wallace's predicament from a sympathetic point of view, we may see in his spiritualism a subconscious reaction to the success that Darwin was enjoying. The phenomenal first-day sales of *Origin* must have been the writing on the wall for Wallace. He knew that he could not compete with Darwin in the field of evolution: he lacked Darwin's dogged personality and also the highly placed and respected friends that his rival cultivated. He saw that Darwin had decisively won the race, such as it was, for recognition as the originator of the idea of natural selection. Wallace would have to find another avenue for his life. He returned to his early interest in spiritualism and found that he soon became as famous as Darwin, albeit in a very different field. He did not reject evolution by natural selection. Rather, he took a compromise line still common today: he claimed to be exploring the other side of the human coin, man's spirituality.

To misquote Shakespeare's Lysander, the course of true science never did run smooth. Beliefs about the supernatural, in various guises, have continuously harried the march of advancing science. Darwin's and Wallace's lives illustrate a pattern that recurs through history: both Wallace and FitzRoy were driven deeper and deeper into irrationality by the successes of rationality. There seems always to have been a profound human desire to distinguish between the kind of knowledge that can happily be subjected to scientific scrutiny and another kind – call it religion or faith – that people try, sometimes desperately and violently, to keep beyond critique. Now we can ask: what is religion? What is this intuitively recognizable but hard-to-pin-down component of human societies?

CHAPTER 4

Explaining Religion

In previous chapters I selected a number of historical highlights and prominent figures to show two things: first, that science developed in the cocoon of religion and, second, how people with contrasting personalities reacted to the dilemmas with which history presented them. With what were the people in these chapters wrestling? There is no simple answer to this question, but, fundamentally, they were all struggling with a rickety structure of interlocking religious experience, belief and practice. Moreover, their dilemmas were of their times, each in some way unique yet, at root, the same: as is still the case today, each had to decide whether to accept the legitimacy of two different ways of knowing about the world and life – science and supposedly revealed knowledge.

For most of the centuries covered in Chapters 1 and 2, religion was overwhelmingly, if not exclusively, taken as an indisputable part of human experience. The gods were real; their interventions in daily life were real; they really listened to human pleading. Devils, sin and torment in Hell were also real – very real. Information about all this had supposedly been revealed to humankind. Religious belief and practice therefore concerned themselves with finding out what the supernatural realm was all about, what it had to say to humanity and how humanity could contact it.

Today there is a body of received opinion about religion. People believe that religion exists to do certain things and that those functions explain why we have religion today – despite the development of science. In this chapter, I challenge what has come to be taken for granted, what is considered 'obvious'. I argue that these apparently self-evident assumptions about religion are in fact the result of 'spin' put on religion by religious institutions and that the 'explanations' were, and still are, uncritically accepted not only by believers but also by historians and other writers. These are what religion itself claims are, at least in part, its raisons d'être: they should not be taken at face value. Together, these common assumptions constitute what we may call a masking ideology. They are a set of beliefs that conceals what is really going on in religion. At best they are half-truths. The trick of the spin is to make a half-truth appear to be the whole, or a substantial part of, the truth.

Theology

Many readers will notice that I seldom refer to theology – what religion says about itself and what is therefore the product of 'spin doctors'. In earlier chapters we looked briefly at Augustine's exclusiveness, Aquinas's 'proofs' for the existence of God and Paley's enduring watch abandoned on a deserted heath – all theological justifications for belief in a supernatural dimension that revealed itself in mysterious ways to selected people. To some, theology of this kind will seem the natural starting point, so my omission needs clarification.

An example will illustrate my point. If we wish to understand mythology, something we find all over the world, we do not sift through compendia of world mythologies searching for a mythological account of why and how people came to tell myths to one another and then claim that this 'origin' myth adequately explains why people recount myths. Instead, we look at the narratives and symbols of myths (as social anthropologists and folklorists do) and then turn to sociology and psychology (as many writers do) for explanations of the functions of mythology.

So it is with religion itself. If you yourself believe in God and the teachings of a religion (as most theologians do in one way or another), you cannot study religion objectively – or as objectively as it is possible to study anything. As the word 'theology' implies, the study is based on acceptance of a god of some sort (*theos*, Greek for god). Theology therefore cannot be said to 'explain' religion; rather, it *reproduces* religion by formalizing it and making it seem reasonable in successive ages. As we saw in Chapters 1 and 2, this is what the medieval and subsequent Christian theologians did. Aquinas, for instance, struggled to incorporate the newly recovered teachings of Aristotle into Christian theology without having to diminish notions about the existence of God. Once we allow for the possibility of divine revelation of special knowledge (which must, in its very nature, be fundamentally unchallengeable) we are adrift on a logically stormy sea. All that theologians can then do is try to explain and clarify supposed divine revelations and make them acceptable to their own times – as they still do today. Unlike measles, you cannot study religion objectively if you have got it.

If we claim that religion exists because there really *is* a spirit realm inhabited by a god or gods, we are using mythology (beliefs in supernatural beings and their interventions) to explain mythology. An example is the argument that God inspired selected ancient people to write holy scriptures, and the existence of these scriptures (as understood by the Church) shows that there is a God. The circularity and futility of such an approach is patent.

A simple observation points to a way out of this logical trap. If a spirit realm exists, it must impinge in some way on the material world. How else could we know of its existence? Examples abound: Paul's experience on the Damascus road, Constantine's vision and victory at the Milvian Bridge, Augustine's childlike voice, and Wallace's floating tables. *We must look to the material world for evidence that there is a spirit realm.* We cannot simply declare science and religion to be two independent ways of knowing about the world and then leave it at that: religion repeatedly impinges on the domain of science. And for all its convoluted justifications, theology is not a science, though it claims to study matters that in large measure take place in the material world (answered prayer, virgin births, resurrections). If there is evidence for a spirit realm, that evidence must exist in the material world and therefore be subject to scientific investigation. Theology may once have been regarded as the Queen of the Sciences, but it is no longer, and it stands no chance of being reinstated in that exalted position.

Still, theology remains a flourishing field of discourse, massively fertilized, I may add, by the current science *versus* religion debate. We cannot ignore it altogether. I therefore return to theology in the final chapter where I consider some responses of present-day religion to science and secularism.

Secular religions

Secularism includes the belief that the traditional functions of religion are being taken over by institutions that do not accept the existence of a god. The second step in this line of thought is that the existence of certain secular institutions somehow demonstrates that humankind has innate yearnings that can be fully answered only by 'real' religion.

For example, sometimes it is said that baseball is a religion in the United States. It has a pantheon of 'worshipped' heroes; its practice incorporates elaborate rituals; its practitioners have symbolic accoutrements; it displays sacred relics of heroes; it is socially cohesive for a team's fans; it even provides emotional storms and catharsis. One could add that, like religion, it sometimes leads to fisticuffs. But to conclude that baseball is therefore a 'religion' is surely a distortion. Rather, it is *like* religion in some ways. It may well perform some of the functions of religion, but that does not make it a religion. If we claim that it is, we are using the word 'religion' loosely and metaphorically.

Today political parties and philosophies often go further than sports like baseball and speak in pseudo-religious language. Communism, for example, is sometimes said to be a secular religion. Political parties look forward to a time when society will be perfect as a result of their policies. Then, in this Utopia, the supposed innate goodness of human beings will lead them to accept what the politicians themselves believe. 'Utopia', Sir Thomas More's neologism, aptly means 'nowhere land'. As with baseball, there are superficial parallels here with religion. A future period of political peace and good governance is comparable to the beatific millennium that the Church taught (and in some manifestations still teaches) will follow Christ's return to earth (REVELATION 20:1–5), though that apocalyptic Utopia will probably not include the freedom of religion that More envisaged. There are other parallels: the emotional fervour displayed at party rallies, rituals of incorporation into party membership, faith in a leader's pronouncements, mythical narratives about a party's foundation, heroes and achievements, and so forth. But, by the account that I develop in the following three chapters, political parties and philosophies are not the same thing as, or even descendants of, religion.

The key point is that most (not all) modern political parties in secular states do not officially believe in a supernatural realm that approves their plans (though individual members may do so). It is tempting to play down this matter of supernatural sanction when one is branding a political philosophy, like communism, a secular religion. To expand our definition of 'religion' to include secular movements and philosophies is therefore at best unhelpful; at worst, it underestimates the truly awful effects of belief in a supernatural realm and a divine being who has graciously provided us (but not 'them') with answers to humanity's ills. For those who do not themselves believe in divine intervention and guidance, who have not experienced Constantine's, Paul's, Augustine's and many others' emotional transformations, it is easy to underestimate the power of such beliefs – a power that exceeds the acceptance of most materialist philosophies.

In sum, baseball, political parties and other secular institutions are, I argue, markedly *unlike* religion in that all religions have some orientation to unseen realms, beings and powers. Religion posits the existence of supernatural things.

Explanations for religion

We should not lose sight of an important distinction. Writers often do not distinguish between function and origin. They list useful things that religion does for people (its functions) and then imply that religion came into being to perform these benign functions. This is the logical trap known as teleology: reasoning that explains developments by saying that they are caused by the purpose they serve.

I argue that religion originated independently of the functions that it is frequently said to perform. The functions, as we see them today, developed after, or as, religion came into being *for other reasons*. This brings me back to a point I made earlier. The functions that people routinely ascribe to religion are what religion claims for itself and therefore stand in need of rigorous scrutiny. Indeed, if writers appreciated that there was an independent origin of religion, they would find the commonly accepted claims to benign functions fragile, to say the least.

The young Charles Darwin saw religion through the filter of these received understandings of the value of religion. He interpreted the rituals he saw in Patagonia in terms of his own religion (Chapter 3). He spoke of 'reverence', 'adoration', 'sacrifice', and striking a bargain with gods. Still today, there is a tendency for us to see all religions in terms of ones with which we are most familiar. Like the great evolutionist, we find it hard not to use words such as 'worship', 'holy' and 'sacred' when we discuss other religions. We assume that these words denote innocuous components of *all* religions. Then it is an easy step to claim, first, that notions like 'the sacred' or 'the numinous' are defining elements of religion,[1] second, that they are in some way built into the human psyche (brain) and, third, that they therefore inevitably led to the development of religion. There is no evidence for these claims.

A far-reaching result of seeing other religions as imperfect variations of our own is that we come to consider all religions to be fundamentally benign, well-meaning and worthy of protection. We see all religions, even the most 'primitive', as fulfilling the same functions as we believe (probably incorrectly) the present-day great 'world' religions do (Judaism, Christianity, Islam, Buddhism, Hinduism).

The functions that people commonly think religions perform include:
- fostering unity within a social group,
- providing peace of mind and
- explaining puzzling aspects of life.

These functions are social, psychological and aetiological. I consider them in turn.

Social Cohesion

Many writers see religion as a social phenomenon – which, in part, it undoubtedly is – but they then tend to leave it at that, thus assuming that a social explanation of religion is adequate. I argue that the social nature of religion sits on the shoulders of a neurological substrate and the experiences that derive from it – but we shall come to that later.

Emile Durkheim (1858–1911), the highly influential French sociologist and anthropologist *manqué*, advocated a view of religion that is still attractive because it seems (to materialistically minded Westerners) to embody much common sense. He summed up his position in a passage that is worth quoting in full because it encapsulates a number of ideas that are still prominent in people's minds, be they academics or not:

For the author, religion derives from a double source: firstly, the need to understand; and secondly, from sociability. We would say, at the outset, that these factors should be inverted, and that sociability should be made the determining cause of religious sentiment. Men did not begin by imagining gods; it is not because they conceived of them in a given fashion that they became bound to them by social feelings. They began by linking themselves to the things which they made use of, or which they suffered from, in the same way as they linked each of these to the other – without reflection, without the least kind of speculation. The theory only came later, in order to explain and make intelligible to these rudimentary minds the modes of behaviour which had thus been formed. Since these sentiments were quite similar to those which he observed in his relationships with his fellows, man conceived of these natural powers as beings comparable to himself; and since they at the same time differed amongst themselves, he attributed to these exceptional beings distinctive qualities which made them gods.[2]

Durkheim, in translation at any rate, is not the clearest of writers. He seems to support his social explanation for the origin of religion by saying that people projected human relationships onto 'the things which they made use of, or which they suffered from', and then, because relationships between those things seemed similar to, but at the same time different from, relations between human beings, people conceived of a parallel supernatural realm occupied by anthropomorphic beings and forces. From here he leads on to the 'sociability' that religion seems, and indeed claims, to foster.

The assertion that religion is a glue that holds society together is widely encountered. One of the laments of the Church today is that the violence and social fragmentation that afflicts Western society, and indeed the whole world, results from a diminution in people's commitment to religion. Calls for more regular church attendance and commitment are therefore common.

Politicians, too, try to harness religion in their promotion of social cohesion. So we need to ask: does religion indeed hold society together, as so many people, religious or not, believe?

The first two chapters of this book have already provided a short answer to this question. It is no. Religion fragments societies and makes any wide-ranging unity between societies with different religions extremely difficult, if not impossible. This statement may seem rashly iconoclastic in today's milieu of much-vaunted religious tolerance and multiculturalism. We must therefore remember that any religious entity, small or large, is created *vis-à-vis* other people. It is the existence of 'others' outside the fold that makes a religious social entity meaningful. If there were no 'others' (heathens, infidels, heretics, atheists), there could be no recognizable religious group. When people argue that religion holds society together, they mean that it holds *their section* of society together, and then only in specific circumstances. So let us bear in mind that unity can be conceived only *vis-à-vis* other unities. Allegiance to a particular religious stand automatically creates divisions not unity. Denial of this rather gloomy conclusion is one of the blindnesses of our times.

The argument that follows this conclusion is frightening. Most religious groups seek to convert others to their way. They believe that peace will come only when everyone believes as they do. But we have seen that there has never been a time of unity. Think of Paul's squabbling post-crucifixion churches, Constantine's attempt to unify the Roman Empire, the failure of the Nicene Creed to bring unity to Eastern and Western Christendom, Augustine's claim that there is no salvation outside his conception of the Church, Aquinas's 'logical' theology, the Reformation, and, of course, present-day disputes concerning AIDS, homosexuality and women priests. If, after 2,000 years of trying, the Church still cannot bring unity between its own factions, there is no point in looking to it to bring peace between Christianity and other religions – despite the Pope's publicized prayers for peace in the Middle East and elsewhere. Given the history, past and contemporary, of the Catholic Church, those prayers are insultingly ironic.

Why has there been such spectacular, sustained failure? Wide social unity is intrinsically impossible because religion is founded on supposedly revealed knowledge of supernatural realms and beings, not on empirically verifiable observations. Religious revelations are notoriously contradictory. Open debate is the vital essence of science and, over the centuries, it has demonstrably led to greater understanding. In marked contrast, unfettered debate is inimical to a Church that claims to be founded on divine revelation.

But what about the small religious group, the house church, the devout prayer meeting? Even here we inevitably find a social hierarchy: there are leaders and followers. The leaders are seen to possess more revealed knowledge than their followers. As a result, tension and fission are inevitable. Moreover, the tight-knit prayer group exists because its members see themselves as different from other people. If the devout group enjoys internal unity at all, it does so by separating itself from others. As our first two chapters showed, disputation and schism are the history of the Christian Church writ large or small.

Contrary to the common claim, the unity that religion appears to bring is illusory. This is a complex and much debated matter. It does, however, seem that religious bonding is greatest when people feel that their reputation is at stake; comparable bonding is equally achieved by secular institutions.[3] Embracing religious leaders probably know that their public (and widely publicized) avowals of amity will never filter down to their followers, the true believers, whose empathy and compassion will remain inwardly directed. And the leaders themselves, as the Pope admits, harbour the conviction that, despite talk of tolerance, they are more right than anyone else – and 'more right' is absolute. We have to see through the masking ideology that religion projects and that many people uncritically accept. This is an unavoidable conclusion to be drawn from a long view of history: if we seek peace and unity, we should downplay religion, not promote it. This does not mean that societies will necessarily be more unified without religion, only that a major devisive factor will be absent and integration consequently easier.

By contrast, science does not claim that its practice will bring about social harmony. Science exists solely to find out about the world. Therefore science cannot be accused of failing to create a social paradise on earth and of being no better than religion in this respect. On the other hand, agnostic and atheistic scientists, in their personal capacities, may work for social harmony. A great many do. When they ponder social ills, they try to hold in mind the sort of rational thinking that science advocates rather than be guided by supposedly supernatural revelations. Indeed, many scientists are as concerned about social harmony and ethics as anyone else, even though some have engaged in highly dubious projects, such as population control through eugenics. If I had to choose between a world-governing body packed with religious luminaries and one composed entirely of scientists, I know for which one I would vote.

PSYCHOLOGICAL COMFORT

When we move on to the psychological needs that religion is said to fulfil, we are on even more controversial ground. To be sure, there are individuals who sincerely and thankfully derive comfort from their religious beliefs. For instance, I do not deny that some bereaved people find solace in their expectation of an afterlife. We saw that Conan Doyle, raised a Catholic, turned to spiritualism and believed that mediums could contact his son who, wounded at the Somme, died in London in 1918. He was comforted and reassured; he felt a hand on his head and a kiss on his brow.[4] Such, too, is our observation of people whom we know personally. And we refrain from disturbing their solace. Isaiah says that God will not break the bruised reed, a statement that Jesus endorsed.[5] Even though history and our experience of life show that God does indeed break bruised reeds by the thousands, we ourselves hesitate to do so. When the bereaved look forward to being reunited with loved ones in another realm, we remain silent, almost 'hoping it might be so'.

Still, allowing for such reticence, we must sidestep the masking ideology and ask if religion actually *creates*, or at any rate exacerbates, the psychological needs it is said to satisfy. Four points need to be pondered.

First, we must realize that believers are members of some religious group or other. Their solace depends therefore on the continued existence of their group *vis-à-vis* other groups. Sadly, they are deriving psychological comfort from an intrinsically divisive belief system.

Secondly, we must wonder if acceptance of untrue beliefs in a supernatural realm is wise, even if they bring comfort. If comfort is built on a falsehood, it must be in constant danger of evaporation.

Thirdly, a more terrible facet of the psychological comfort that religion can bring is that it can and has been used to shore up oppressive regimes. On the plantation the slave dreams of crossing over Jordan and entering the Promised Land. In brutally waged Scandinavian mythological battles, mounted Valkyries swoop to gather up the bodies of slain heroes to carry them off to eternal joy and feasting in Valhalla. Muslim suicide bombers look forward to the delectable maidens that await them in Heaven. The sufferings of today are temporary; the joys of Heaven are eternal. Whenever a regime sponsors religion, we should closely examine the effects of psychological comfort in that particular context. Marx had a point: religion can indeed be 'the opium of the people', and Charles Kingsley, himself a clergyman, agreed: 'We have used the Bible as if it was a constable's handbook – an opium-dose for keeping beasts of burden patient while they are being overloaded.'[6]

But does religion always provide psychological comfort for the oppressed in order to keep them oppressed and to preserve the political *status quo*? It has been argued that Liberation Theology reveals the shallowness of Marx's verdict. This brand of theology flourished in South America when some Roman Catholic priests, appalled by the brutality of fascist regimes, worked for their overthrow. A point to keep in mind when we consider this issue is that, at various times in history, religion has adopted a Liberation Theology persona. But Liberation Theology aims at the overthrow of an oppressive regime (political or religious), only to be replaced by another servitude, that to the Church with all *its* political and economic implications – and we know enough about the history of the Church to be nervous about that.

Fourthly, we must enquire about the role of the Church and religion in general in the *generation* and *exploitation* of psychological needs. The Church is built on concepts of sin and guilt. Without an abiding sense of guilt and enduring personal inadequacy, believers' reliance on the Church diminishes. The very existence of the notion of confession (either personal in the Reformed tradition or to a priest in the Catholic Church) engenders a sense of guilt. Moreover, the Church magnifies trivial lapses, often sexual, and thus ensures devotion to the institution that offers forgiveness. Not just adultery, but even thoughts that could be considered natural and understandable are prohibited: Jesus said, '[W]hosoever looketh on a woman to lust after her hath committed adultery with her already in his heart' (MATTHEW 5:28). A continuing sense of guilt and a need for God's forgiveness is pretty well ensured by inescapable human bodily functions.

The psychological damage caused to people through the ages by the Church's definition of sin and its claim that salvation from sin is obtainable only within itself must be placed alongside the common belief that religion brings psychological comfort. The Church creates, nurtures and feeds on a sense of sin and inadequacy. Similarly, it seems clear that Maya religion created, rather than satisfied, the anxiety aroused by a supposed need to satisfy the gods' demands for gruesome sacrifice. If the ghastly Maya rituals comforted people, it was only fleetingly: soon they needed to repeat the whole cycle of capture, torture, bloodletting and death. So too did *autos-da-fé*, the public burnings of witches and heretics that the Catholic Inquisition and also Protestant authorities staged.

All in all, these four points suggest that the notion that religion fulfils psychological needs is questionable. The reality is more complex than the commonly accepted platitude suggests. Some religions – for example, those that populate the world with evil spirits (which is to say most religions) – actually generate psychological *dis*comfort. Writers who consider shamanism, as still practised in parts of the world, to preserve some features of what was humankind's *ur*-religion often emphasize its supposed healing and hence adaptive properties; adaptation has unfortunately become a *sine qua non* in discussions of religion. These arguments are tautological: without digging any deeper, we can say that they merely assert that something (in this case, a religion) does what we think it does (e.g., creates social solidarity) and then add the word 'adaptive' to the supposed function. For instance, Michael Winkelman of Arizona State University writes:

Shamanism was originally an ecological adaptation of hunter-gatherer societies to mammalian social bonding rituals [and] elaborated primate healing capacities to address humans' psychosocial bonding and therapeutic needs.[7]

Winkelman thus brings together social and psychological functions. Rather than romanticize shamanism and advocate a shamanism revival, as often happens nowadays, we must remember that shamans cannot heal broken legs. Nor can they make rain. The 'therapeutic needs' that they are said to heal are largely of their own making. Like other religious systems, shamanism fills the world with potentially harmful spirits that cause great anxiety in believers. Shamans then use a variety of techniques to banish these spirits and apparently to cure their patients. Though the placebo effect of religious rituals should not be ignored, such societies would, in general, be better off without that kind of religion.

Above all, we must remember that religion, with its shaky foundation on supernatural things, is not the only source of social cohesion and psychological comfort in distress. Today many people get by without religion.

WHAT EXPLAINS WHAT?

We saw that 'the need to understand' was one of Durkheim's cornerstones. His notion of religion as explanation for natural and human events requires discussion because it is so commonly advanced today. Long after Durkheim, it is being enthusiastically propagated by believers themselves.

The question we must address here is not so much about ultimate explanations for the cosmos that are being sought today by astrophysicists and cosmologists. We are considering a time long before science had made its devastating forays into the unknown. Did the earliest hunter-gatherers at Blombos and, subsequently, Upper Palaeolithic Magdalenians, more sophisticated Babylonians, Egyptians, Maya, Chinese, Greeks, Polynesians, Africans and many other peoples throughout the world invent beliefs about supernatural beings to explain natural phenomena? Were they *innately* curious about such things? Did the ancients ask one another why the sun rose so regularly, and then, having thought about it, conclude that it must be because there are supernatural beings who are responsible *and whose existence they did not previously suspect*? Is the idea of invisible supernatural beings inescapably implied by nature?

Why do we think that seasonal changes and other natural events demand explanations in the first place? If no one believed in a supernatural realm (there probably never was nor ever will be such a time), why would people not believe that natural events were explicable in natural (or 'material') terms? Some people, such as the southern African San, get by without mythological beliefs about the changing seasons, even though seasonal changes are marked in their environment and have major social effects. For the San, the seasons just happen. Why certain natural events acquire complex supernatural explanations in some societies and not in others is a problem worth following up, though not here.

There is another way of looking at the notion that religion is there to explain a whole range of mysteries. Durkheim's argument does not simply propose that religion supplies satisfying explanations. It goes further in a way that has overtones of psychology: human beings are said to have an innate psychological desire, or genetic need, to explain their environments. That, we are told, is why religion is universal: everyone everywhere has a need to explain deep in their mental makeup. In this view, religion is essentially aetiological *and* psychological.

The existence of a whole realm inhabited by spirits, one that is set apart from the material world in which those spirits intervene to keep it going, is a powerful, ubiquitous concept. And yet, on the face of it, it is a highly unlikely proposition. There is no evidence in the material world that there are beings living, for instance, beneath the surface of the land or up beyond the blue of the sky. If people interpret thunder as the voice of a god, they do so because they *already* have a conception of invisible beings and spirit realms. The same applies to other natural phenomena, such as the shapes of clouds, lightning, tornadoes and earth tremors, all of which are occasionally put forward as triggering religious beliefs in other worlds. They do not singly or in concert point to the existence of invisible anthropomorphic beings.

A more subtle variation on the notion of religion as explanation is Paley's watch abandoned on a heath. Paley, it will be recalled, was a theologian whose works Darwin, as a young student, found intriguing but whose conclusions he later rejected. How, Paley and his present-day followers ask, can we explain the intricate workings of creation without some notion of divine design and intervention? There are numerous cogent responses to Paley. I mention only one now.

His watch analogy is based on a misleading assumption. Creation is not a well-oiled, beautifully and intricately designed machine. The world and all that is in it is actually a higgledy-piggledy, wasteful mess. Each evolutionary stage necessarily built on one, and only one, stage that went before it and used random genetic mutations as they popped up. If no useful mutation showed up in a newly fraught environmental situation, an evolutionary line simply became extinct: evolutionary history is littered with wasteful, meaningless dead-ends. Perhaps human beings are another.

Moreover, the creatures that resulted from useful mutations carried with them unnecessary baggage from their earlier forms (male nipples are a well-known example). Each new species did not start, spick and span, from scratch, as creationists believe. Chance and ad hoc muddling-through characterize the evolutionary story. To some enlightened modern religious believers this may seem a shocking take on evolution; they see evolution (if they believe in it at all) as following an a priori, divinely concocted Grand Design.[8] The cosmos seems to be 'just right' for the emergence of intelligent life, but this, of course, need not imply a Designer-God.

Before we accept that people are innately explanation-seeking animals and that they virtually everywhere invented supernatural beings and realms to explain the world around them, let us open up another line of thought.

I argue that early people attached *independently derived* religious beliefs to recurring and powerful natural events. They did this to give their already-existing 'abstract' supernatural beliefs observable substance and immediacy. If this was so, we can ask: do naturally recurring events (e.g., the cycle of the seasons) and unusual features of the landscape (e.g., a big rock in a sandy desert), by their very existence and durability, help to explain why people *continue* to believe in whole systems of supernatural beliefs, myths and beings, rather than why they believed in them in the first place? In other words, do the things in nature that are apparently explained by myths and religious beliefs really themselves do the 'explaining' by showing everyone that the already-existing protagonists and narratives of myths must be real and true?

I argue that it all worked the opposite way round from the view we usually accept. Gods were not invented to account, for instance, for the cycle of the seasons. Rather, people harnessed the inexorable seasonal cycle to *confirm* the existence of supernatural beings and influences in which they already believed *for other reasons altogether* (reasons with which I deal in the next chapter).

A question that must immediately follow: *who* harnessed the seasonal cycle to demonstrate the power of the gods? Was it everybody in a community? Or did some people press the point and, as a result, act as intermediaries with special, revealed knowledge about the activities of gods? We may take the Irish Neolithic tomb Newgrange as an example (Fig. 18).[9] We know that the sun shines along a significant axis at the winter solstice. It shines through the entrance and into the depths of the huge mound to illuminate a normally dark apse at the end of the passage. The annual winter solstice could be said to provide what Othello called 'the ocular proof' of the power of the gods *and* the status of those who have the ear of the gods. That the solstice also has economic implications for, especially, farming communities simply increases the status of those who claim to be party to the important economic event.

Sites such as Newgrange and Stonehenge were not 'astronomical observatories' designed to predict the changing seasons. As in small-scale societies around the world, the people who frequented these Neolithic monuments detected the changing seasons in a myriad tell-tale signs in nature around them. Rather, the monuments were built to harness natural events in such a way that they underwrit belief in supernatural realms and the power of a special category of people.

Belief in supernatural beings and revealed knowledge situates explanations in a realm to which some people, in the nature of human brains (some are more labile than others) and their inevitably social settings, have more access than others. Religious explanations thus provide material for political power: select people can claim to have revealed knowledge about this world *and* another, far more puissant, world. They can then use their supposed knowledge to explain not only regular natural events but also unusual or unexpected ones, such as earthquakes. Once again we see the divisiveness of specially revealed knowledge.

As this process of appropriating natural events to demonstrate the power of an elite proceeds, a form of coordinated explanation results. Regularities

18 *The power of the seasons. At the winter solstice, the sun shines down the passage of the great Newgrange tomb to illuminate a double spiral in the dark depths, the realm of the ancestors. Priests probably drew power from the annual phenomenon.*

and irregularities alike can be explained by the whims of anthropomorphic gods. Rather like a scientific theory, belief in supernatural realms satisfyingly unifies people's relationships with the natural world by bringing everything under a single umbrella. This coordination of explanations is largely handled by myths. Gods, often capricious, cause everything to happen. Inevitably, tales about them proliferate.

In medieval times, most people believed that, if they wished to know about the relationship between the earth and the sun, they should seek answers in supernaturally revealed contexts – the books of the Bible as expounded by their trained interpreters, the priests. What happened was that the Church lighted on recorded events that, when they were written down in the Bible, were not designed to explain the material world. The hint that the sun moves around the earth that is contained in the tale of Joshua's supernatural control of the battle with the Amorites is an example (JOSHUA 10:12, 13). Today educated believers have abandoned this approach. But they find themselves in a halfway house. They have had to concede that science has explained problem after problem without recourse to beliefs in divine beings. At the same time, they want to preserve the foundation of their faith. So they are constantly on the lookout for things that they believe science cannot explain.

The Big Bang is their last resort. They claim that, when we puzzle over the origin of the cosmos in a Big Bang, we must allow that there was a God who lit the fuse. Divine intervention as explanation, having suffered so many setbacks since Copernicus, has been driven back and back to defence of a First Cause, one of Aquinas's (faulty) 'proofs' for the existence of God (Chapter 2). There is, of course, no reason at all to suppose that what preceded the Big Bang was supernatural. It is more reasonable to suppose that what preceded the Big Bang was, like what followed it, natural and material.

Before leaving the matter of explanation, I need to remark on an issue that is often discussed. People say that science can answer neither the 'why' questions nor the 'deep inner longings' that we are all supposed to have. Why are we here on earth? What is the meaning of life? This, we are told, is where religion comes into its own. Like many other writers, the Canadian archaeologist Brian Hayden sets the supposed search for meaning in an evolutionary context:

Religion satisfies an inner craving for meaning, a feeling of wholeness or union with greater forces, and the inner satisfaction that comes only from ritual life, just as music and rhythm satisfy an inner emotional craving deep within our souls and minds for the trances, the ecstasies, and the profound experiences that only they can produce. These are the fundamental adaptations of our biological heritage.[10]

Science does not attempt to answer questions about the meaning of life and the supposed 'inner emotional craving deep within our souls'. One reason is that the 'meaning-of-life' questions themselves are tendentious: they pre-suppose that there are indeed answers awaiting discovery. In fact, the questions are meaningless. There does not have to be any 'meaning' to humankind's existence on this minute speck of the cosmos. That conclusion does not make life worthless or empty, as religious people claim. Many go about their lives in a humane fashion without any notion of the meaning of life. It is better to seek fulfilment than meaning.

In sum, we can say that the common notion that religion came into being to explain otherwise baffling things is not merely simplistic: it is wrong.

'Where is thy sting?'

A major area of life in which people, all of us, need comfort and explanation demands special consideration. It brings together points I have so far made in this chapter.

Some writers argue that death was an important factor in the invention of a spirit realm. An emotionally charged desire to continue human relationships beyond the grave led people to devise a supernatural realm where the dead continued to live. Today people still exclaim: surely there must be more to life than death, something meaningful, something lasting!

The earliest *Homo sapiens* communities experienced death not only among their own number but also all around them in nature: animals and trees died, decayed and eventually vanished. Death itself did not set human beings apart from other animals or plants; on the contrary, it showed people that they were no different from them. So, while they may well have grieved, there was no evidence before them to suggest that the deceased lived on in another dimension. Beliefs in life after death could not flow simply from *awareness* of death.

When communities began to make land claims, first for hunting rights and subsequently for agricultural purposes, the idea of a lineage became impor-tant. It was not just that the present population lived on a tract of land, nor that their loved but deceased parents did so, but that countless generations of their ancestors had done so. Under such circumstances, it is advantageous to claim that one's ancestors are in fact still alive: they are living in a supernatural realm that infiltrates the material world. If one already has notions about a supernatural realm, it is easy to see that the concept of a living lineage can be advantageously incorporated. Survival after death thus takes on economic and political implications beyond personal grief and loss. Still today, people believe that God gave them rights to land and those rights justify killing other people. Religion, economics and politics are intertwined.

Again, it seems to me that we are dealing with an area of human experience where religion, created by other factors, can be brought into play. If people came to believe in a post-mortem existence, they must have had something other than death itself to differentiate themselves from animals. To get at that 'something', we go back to the nineteenth-century British anthropologist Sir Edward Tylor (1832–1917). He believed that trances, visions and, especially, dreams suggested to early people the existence of a soul or spirit, something immaterial. This is the well-known dream theory of the origin of religion. Once in possession of the notion of a soul, early people found belief in its survival of death a logical next step.

In taking that step in various parts of the world, they produced highly variable notions of what life after death was like. Mostly, those concepts of an afterlife were closely related to the sorts of lives that they lived on earth and to their economic and political ambitions. Death became a potent agent in a heady and very dangerous brew.

Darwin's legacy

If we question the social, psychological and explanatory functions of religion, what are we to say in response to the related claim that, in aeons gone by, religion served an evolutionary function?

The ubiquity of religion in space and time suggests that it is so fundamental to being human that it must have been involved in some positive way in the evolution of humanity from pre-human species. Evolutionary theory states that useful mutations are selected and those that are maladaptive are excluded from the evolutionary trajectory. What survives the millennia is, we must conclude, useful in the sense that it helps species to pass on adaptive genes from one individual to the next. Religion must therefore be adaptive; it must help human communities to survive. That is why religion has endured for so long and is so widespread. Such is the generally accepted reasoning.

For many archaeologists, Durkheim's sociological emphasis is attractive because it can be readily brought into line with their evolutionary notions of adaptation. Religion exists because it is adaptive in that it helps people to survive in sometimes hostile environments by providing a foundation for social cooperation against the elements. Scattered groups often come together for the performance of religious rites and the revelation of religious insights. An example is Göbekli Tepe, a pre-Neolithic site in south-eastern Turkey that has circular 'crypts' and huge, beautifully carved stone pillars. There is no evidence for any large local settlement. Pre-farming hunter-gatherer communities came together, probably seasonally, to Göbekli Tepe to build the impressive monuments and, presumably, to communicate with the supernatural realm that the carvings represented.[11]

Religious alliances and rites, such as those celebrated at Göbekli Tepe, would, of course, have both economic and kinship implications. At religious gatherings, people often perform economic activities and conduct marriage brokering. Deals are still struck after church or synagogue attendance. The location sets a comforting seal on the agreements. This is something that social anthropologists have found again and again but which seems to make some archaeologists queasy when they find evidence for economic activity at religious sites: they seek to downplay the religious foundation and emphasize the economic activity. They do not realize that religion can have a powerful influence on societies.[12]

So it seems reasonable to conclude that religion created useful social solidarity and satisfying psychological needs at Göbekli Tepe and, for that matter, in the Maya towns and for the Old Testament Israelites. Religion thus comes to be thought of as, at base, a means to survival, even if some of its beliefs may seem to us today to be irrational. An argument like this can be expressed in diagrammatic form:

> Religion $\rightarrow$ Social solidarity $\rightarrow$ Adaptation to environment
> Therefore:
> Religion = Adaptive mechanism

That this equation is wrong will become apparent as we proceed.

The evolution of religion

Social anthropology presents belief in supernatural things as a useful phenomenon. Social anthropologists have traditionally studied exotic religions, principally those that they came across in small-scale societies throughout the world. The development of social anthropology was thus associated with Western imperial expansion.[13] While researchers from this discipline acknowledged the social and psychological functions of religion, they went on to show that each religion they encountered was logically consistent within its own parameters: the 'primitive' religions with which they were primarily concerned were not conglomerations of crazy, illogical, super-stitious beliefs and rituals. To demonstrate this consistency, social anthropologists analysed myths and rituals in terms of the symbols that they incorporated. They were able to demonstrate the existence of coherent systems of symbols beneath what at first glance appeared to be random superstitions. These symbolic systems, some thought, were emotionally so powerful that they gave rise to beliefs in supernatural entities.[14]

There have been some highly influential social anthropologists. Tylor was one of them. Following ideas put forward by Sir John Lubbock (1834–1913) and working in the milieu created by Darwin's *Origin of Species*, Tylor proposed an evolutionary line starting from animism, the belief that 'souls' – the Latin *anima* – existed in the world and bedded down in all manner of creatures and things. He went on to claim that totemism (that animals or plants have souls that are associated with social entities) developed out of animism. Then came shamanism (in which remote deities, rather than immanent souls, are contacted by special people, the shamans) and, later, theism (belief in a god or gods).

A succinct formulation of the evolutionary account of religion is:

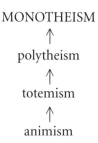

MONOTHEISM
↑
polytheism
↑
totemism
↑
animism

Something like it was made famous by Sir James Frazer (1854–1941) in his widely read book *The Golden Bough*.[15] Preferring his Cambridge study and voluminous correspondence with missionaries and district commissioners to fieldwork, he did not himself venture into remote parts of the globe. It is said that, when William James, the American psychologist and philosopher, asked Frazer if he had met any of the savages about whom he wrote, he replied, 'God forbid!'[16] In a comparable evolutionary formula, Frazer and others believed that primitive magic would give way to science. So far, their expectations have not been realized.

The evolution of religion has become another platitude that we can set alongside its greatly exaggerated social, psychological and aetiological benefits. Today anodyne books purport to describe some sort of evolution of the idea of God and the ways in which people worship him in the hope that such information will make the whole notion of religion more palatable. In the shadow of Darwin it is difficult to write a history of religion without implying some notions of evolution. Those from a Christian background inevitably do so as they invoke 'progressive revelation', the idea that God revealed himself progressively, though partially, in the Old Testament and then fully in the New. For instance, David Robertson, the pastor of St Peter's Free Church of Scotland, writes of progressive revelation:

This is the idea that the Bible, written over a period of more than 1,000 years, progressively reveals God to us. Little by little, the curtain is opened and the light comes in. Therefore, some aspects of earlier revelation are superseded by the later.[17]

Many questions arise. Why did God not reveal the absolutely fundamental virtue of compassion, even towards enemies, right at the beginning? Was the story of Lot and so many other Old Testament tales an 'earlier revelation' or was it just plain barbarism? Can later 'revelations' be said to 'supersede' earlier ones when they flatly contradict them? (Is the nature of God not eternal?) In any event, why was 'progressive revelation' necessary? Why did not the loving God reveal himself at the beginning instead of consigning millions to the fate of Lot's daughters? We can go on. Is the New Testament the final revelation, or are we to expect further 'superseding', probably from the clergy and theologians? Is the following a final word? 'If any man come to me, and hate not his father, and mother, and wife, and children, and brethren, and sisters, yea, and his own life also, he cannot be my disciple' (LUKE 14:26).[18] If the New Testament is the full and final revelation are we to take the Virgin Birth and all the miracles as facts (as did the writers themselves and their early readers) or has 'progressive revelation' opened up the possibility that all these difficult passages may be metaphors? 'Progressive revelation' sounds like making the best of an embarrassing past.

But let us return to the grander theme of the supposed evolution of religion in general. What could any evolutionary sequence of religious development mean, be it Tylor's, Frazer's or anyone else's? Does it imply that religion was getting 'better' as time went by, that superstition was being left behind and 'real' religion was taking its place? If we think this, we have to remember that the nineteenth-century writers who proposed the developmental sequences were themselves believers in, or associated with, a major, monotheistic religion. The British Empire, which facilitated scholars' access to animistic and polytheistic 'primitive' religions, was built on a proselytizing monotheistic religion – Christianity. It is no coincidence that the rising scale of animism, totemism, polytheism and monotheism was devised by people in a highly hierarchical, monotheistic society. We must also ask what we mean by 'better' and 'real'. Religion largely concerns finding out what the spirit world is all about and deriving supposed social, psychological and explanatory benefits from that realm. If we mean that religion was getting closer and closer to the 'truth' about the spirit world and that that truth would set humankind free from superstition, we are clearly mistaken. It has done nothing of the kind. Indeed, it would be hard to identify areas of life in which monotheistic religions are, in this sense, 'better' than polytheistic ones. Certainly, major religious conflicts and internecine wars are more characteristic of monotheism than of polytheism or indeed of any of the so-called 'primitive' religions. Provocatively, we could say that monotheism means conflict and murder; polytheism means tolerance.

In any event, polytheism lingers on the fringes of monotheistic religions. The Christian Trinity is the example that immediately springs to mind: God has three persons, but they are really only one God. But we should also understand that the Virgin Mary and the saints are, in effect if not in name, lesser gods – supernatural beings to whom human beings can appeal and who have the ability to intervene in human lives. Declared monotheists themselves believe in more than one kind of supernatural being; these include angels, devils, saints, evil spirits, and, in mainstream Christianity, the Trinitarian Godhead. In other religions too these kinds of beings are arranged in a hierarchy, with a chief god at the top. Monotheism is thus polytheism onto which the social hierarchy of complex human societies has been projected. There is really no major difference between polytheism and monotheism, and the supposed evolution of religion is an illusion.

Origin and persistence

Beliefs in a supernatural realm persist and, however diverse they may be, they persist worldwide. I argue that the persistence of religion into our modern, generally materialistic Western milieu in fact points to the answer to the problem of the origin of religion. Instead of religion as an answer to social and psychological needs, and in place of supposed evolutionary stages of religion, I prefer to think of *origin-as-process*. The two cannot be separated. The reasons why religion persists today are, in some fundamental ways, the same as those that explain why religion came into being in the first place.

Modern research on the ways in which the human brain functions to produce the complex experience we call consciousness provides a foundation for an understanding of religion that unites its social, psychological and aetiological elements. The dreams that various earlier writers, such as Tylor, emphasized in their explanations for the origin of religion are but a part of human consciousness. Isolated from the rest of consciousness, they do indeed seem a trivial explanation for so complex a phenomenon as religion. There is, of course, much more to religion than merely dreams. Still, as we shall see in the next chapter, features of dreams and some altered states of consciousness that are neurologically wired and generated by the electrochemical functioning of the brain occur in most religions, though they do not inescapably lead to religious beliefs. These common features suggest that religion is closely related to mental states *daily* and *necessarily* generated by the electrochemical functioning of the human brain.

So rather than adopt a straightforward functionalist account of religion – people have religion because it fulfils this or that need – I identify what I call three domains of religion. They are religious experience, religious belief and religious practice.

In his book *Why Gods Persist*, Robert Hinde, the Cambridge biologist who has studied animal behaviour and the psychology of human relationships, reaches a similar conclusion, though he identifies four domains: 'Religion involves feeling, thinking, acting, and relating'.[19] He goes on to add that 'there are tremendous individual differences in their relative importance', something that our overview of the development of science in the West brought out. Some people emphasize feeling and pay little heed to doctrine. Others, like Aquinas, cogitate and set less store by feeling. Still others, like Orthodox Jews, emphasize behaviour and ritual observance more than religious experience. But all are recognizably religious.

In brief, this is what I argue in the following three chapters:

Religion is one possible explanation, not for natural phenomena, but for highly complex experiences that the human brain generates. It does so in such a way that a whole range of further explanations (for natural events, death and so forth) becomes available. Moreover, religion makes possible powerful social and political hierarchies not based on sex or brute strength. The persistence of the neurology of the brain through time ensures that the 'origin' of religion is always with us.

In his *Ode: Intimations of Immortality from Recollections of Early Childhood* Wordsworth thinks of human beings as entering this world with built-in insights into a divine realm from which their souls came:

> Our birth is but a sleep and a forgetting:
> The Soul that rises with us, our life's Star,
> Hath had elsewhere its setting,
> And cometh from afar:
> Not in entire forgetfulness,
> And not in utter nakedness,
> But trailing clouds of glory do we come
> From God, who is our home:
> Heaven lies about us in our infancy!

In fact, we come trailing clouds of neural activity in our brains that long ago, in the emerging social and intellectual dispensation of *Homo sapiens*, grew 'naturally' into religion.

CHAPTER 5

Religious Experience

This chapter deals with the first of the three domains of religion that I identify. My approach amounts to an analysis of a complex phenomenon that we, by general consent, call religion.

An analysis, a word often loosely used to mean simply discussion, breaks a phenomenon down into constituent parts. Further, an analysis of religion will propose a model, a framework, to help us understand complex and diverse data. A model, in this sense, shows how parts of an analysed whole fit together and influence one another. All models are like hypotheses: if they are useful, we retain them; if they do not seem to make sense of our observations, we discard them and try to devise others. By 'useful' I mean that the model furthers the research aims of the person who devises it. Other research aims may require other models to make sense of the same (or nearly the same) data. If a model throws light on and organizes observable data and, moreover, directs a researcher to further data that would have been otherwise missed, it will be deemed useful. More than that cannot be said. Certainly, I do not claim that the account I give in this and the next two chapters is definitive in the sense that it is the only one possible. I merely contend that understanding the three domains that I describe will help us to comprehend the origin and persistence of religion and also the conflict between science and religion as it is being played out in our world today.

Durkheim wrote of 'imagining gods'. He believed that this 'imagining' grew out of social and psychological factors. Life and personal experiences virtually forced people to invent a supernatural realm. By contrast, I argue that the social and psychological functions of religion are better thought of as consequent upon the 'imagining' of gods. I do not mean that there was a time gap between the 'imagining' and the social and psychological functions. Those functions probably went hand in hand with the elaboration of beliefs in supernatural realms. The real questions are: what does 'imagining' mean? What leads people everywhere to suspect the existence of an unseen realm?

Westerners and others continue to seek in religions a blissful mental state that will lift them above the turmoil of daily life. Like all mental states, their bliss must be generated by the functioning of the human brain, whether they realize that or not. We therefore need to examine not so much psychology as neurology: that is, part of the hard, observable anatomical make-up and electrochemical functioning of human beings. This is not a new idea. The Harvard psychologist William James, brother of the novelist Henry, began his 1901–1902 Gifford Lectures at the University of Edinburgh by addressing the relationship between religion and neurology. He concluded that religious experience, formalized by creeds, has the worldwide impact it does because it is 'amongst the most important biological functions of mankind'.[1] He saw that religious experience was a 'biological function', not an invasion of our brains by an invisible, supernatural, divine presence. James was being bolder than we may at first realize. Those who deliver the Gifford Lectures are required to deal with 'natural theology'. Given this remit and what was by all accounts James's humane generosity of spirit, we do not encounter in his Gifford Lectures the kind of incisive reasoning and (I believe) ultimate, inescapable rejection of the supernatural that we find in many modern discussions of religion. But it was nevertheless there as a subtext.

As a result of vastly improved techniques for studying brain functioning we are getting closer to knowing what happens when we think and feel than James could manage at the very beginning of the twentieth century. We now know that *all* our inner experiences have a neurological foundation, even if we are still in the dark about the details of those foundations. Some evidence suggests that experiences that we may label religious originate in the right temporal lobe of the brain and may be neurologically akin, but not identical, to the experiences of temporal lobe epilepsy and schizophrenia.[2] Despite major advances in neurology, we still know more about what we experience than we do about what happens neurologically in the brain. I shall therefore focus on experiences that are widely and cross-culturally reported rather than on neurology – experiences that we ourselves have had.[3] But, even if we do not go into details, we need to note that, as neuroscience has progressed by leaps and bounds in recent years, it has uncovered no suggestion of any immaterial, supernatural goings-on in the brain. There is no evidence for a ghost in the machine.[4]

Human consciousness: a spectrum

Human beings are not either conscious or unconscious, as may be popularly supposed. Consciousness should rather be thought of as a spectrum (Fig. 19).[5]

Archaeologists who today seek the origin of fully human consciousness fail because they wrongly imagine that consciousness is a single, consolidated state. It is impossible to study consciousness without taking its various forms into account.

The idea of a consciousness spectrum is, of course, a metaphor drawn from the study of light. As is well known, light can be broken down into a range of colours. But these named colours are in fact arbitrary divisions of a spectrum in which one colour grades imperceptibly into another. Not all cultures and languages recognize the seven colours that English speakers do (red, orange, yellow, green, blue, indigo, violet). The traditional Welsh language, for instance, names only four colours. This does not mean that the Welsh are congenitally unable to distinguish finer grades of colours. Merely that their social context named only four colours.[6] Like all reasonably good metaphors, the notion of a spectrum has the power to arrest attention and to help people to grasp ideas. A metaphor does not imply total identity – if it did, it would not be a metaphor. Rather, it highlights and focuses the mind on aspects of what we are trying to understand. So the spectrum of consciousness that I now describe is an *aid to understanding*, not an empirical description. A metaphor can thus be a model, a framework into which we can fit our observations.

19 *Human consciousness is constantly shifting. It consists of a normal, daily trajectory and an intensified trajectory that leads to overwhelming hallucinations.*

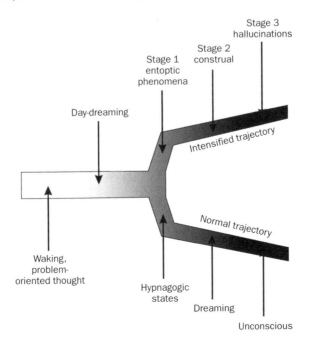

I certainly do not imply that all people at all times have recognized the divisions of the consciousness spectrum that I identify, though some major areas of the spectrum must be part of everyone's life.

At one end of the spectrum is alert consciousness – the kind that we daily use to relate rationally to our environment and to solve the problems that it presents. This is an outward-directed perspective. A little farther along the spectrum are more introverted states in which we solve problems by inward thought. In these states, sensory input, while not eliminated, is diminished. Relax a little bit more, and we are day-dreaming: mental images come and go at will, unfettered by the material world around us. This state is sometimes known as reverie. Gradually, we slip into sleep. On the threshold between sleep and waking is the hypnagogic state. In it, people experience vivid hallucinations that are principally visual but may also have an aural component. From there, sleepers drift into normal dreaming, a world of changing forms and impossible circumstances. Although the specifics of dreams are cultural and personal, the structure of dreams and the sort of events, transformations, blurring of time, and so forth that recur in them are neurologically engendered. Dreaming is a human universal created by the electrochemical functioning of the brain.

Even what we think of as normal, waking consciousness is not a discrete, consolidated state. Evidence suggests that our normal day is comprised of cycles of 90 to 120 minutes of moving between outward-directed attention and inward-directed, meditative states.[7] Waking consciousness is, through its inward phases, thus closely linked to dreaming with its fragments of remembrance, bizarre concatenations of events, and transformations. Sometimes memories of people, events and even dreams simply pop up in our minds without any obvious trigger: where do they come from?

Altered states of consciousness

So far, I have spoken about the ever-changing states of consciousness that *everyone* experiences in the natural course of a day. But there is more to human consciousness. In all religions there is an ecstatic component, whether of the introverted (meditative) or the extroverted (frenzied) kind. This type of consciousness differs from what we experience in a normal day.

Knee-jerk negative reactions to this observation from academics and others who are committed to social and psychological explanations of religion are common. They fail to acknowledge what seems to me to be an undeniable fact: universally, many human beings are intrigued by the 'autistic' (inward-directed, but not pathological) end of the consciousness spectrum that they inevitably glimpse in dreams. Dreams are evanescent, fleeting glimpses of what seems to be another, sometimes frightening but always intriguing, realm. Even memory of dreams usually fades quickly. People can remember what they did yesterday but (often) not what they dreamed last night. They wonder why there are these two kinds of memory. Can it be that the events of dreams are special in some supernatural way and that is why memories of them fade more quickly? In dealing with this sort of dilemma, people try to intensify 'autistic' consciousness by a wide variety of means that include:

- ingestion of psychotropic substances,
- intense, rhythmic dancing,
- auditory driving (e.g., chanting, clapping, drumming),
- electrical stimulation,
- flickering light,
- fatigue,
- hunger,
- sensory deprivation,
- stress,
- extreme pain,
- intense concentration (meditation).

In medieval times some of these factors were intentionally combined to trigger hallucinatory episodes: they included malnutrition, fasting, self-flagellation, prolonged prayer and meditation. Because medieval acceptance of a spirit realm was so pervasive such episodes were both moulded and accepted by sufferers and religious authorities as messages from God. After all, God himself had said: 'If there be a prophet among you, I the Lord will make myself known unto him in a vision, and will speak unto him in a dream' (NUMBERS 12:6).

In addition, intensification may be brought about by pathological conditions, such as migraine, temporal lobe epilepsy, schizophrenia and other pathologies, and near-death experiences.

Pathological conditions are sometimes difficult to define because they are culturally situated. What is identified as schizophrenia in a modern Western community (hearing voices, for instance) may be regarded in a more benign way in a non-industrial society.[8] Schizophrenia, with its sense of passivity, voices and visions, could in the Middle Ages be accepted as a transcendental experience. In the medieval context, there does not seem to have been any distinction between dreams and the visions of altered states of consciousness. Acceptance of the validity of a revelation seems to have been based on social standing: the higher clergy were more likely to have their visions accepted at face value than laity or children.[9] The same continues to be the case today in some Christian churches where all ecstatic experiences, some of which may be pathological, are taken to be possession by the Holy Spirit or, less fortunately, the Devil (Fig. 20). These beliefs persist in close-knit devout communities that are, in effect, islands in present-day Western secularism.

All the means of inducing altered states, either singly or in combinations, lead to three intergrading stages of what most people think of as 'altered states of consciousness'; these states are an extension of the normal, daily spectrum of consciousness, not something bizarre and discrete.[10]

STAGE ONE

Altered states of consciousness often start with geometric mental images. Researchers have given these percepts various names: form constants,[11] phosphenes,[12] endogenous percepts[13] and entoptic phenomena.[14] By a short head, I prefer the term 'entoptic', which means generated anywhere in the optic system, not necessarily within the eye itself. By and large, six frequently repeated forms can be identified:

1 A basic grid and its development into a lattice and expanding hexagon pattern;
2 Sets of parallel lines;
3 Bright dots and short flecks;
4 Zigzag lines, reported by some subjects as angular, by others as undulating;
5 Nested catenary curves, the outer arc of which comprises flickering zigzags (well known to migraine sufferers as the 'fortification illusion'; see Chapter 9);
6 Filigrees, or thin meandering lines.

Because these forms are unstable, the six categories are not as rigid as this list seems to imply. Moreover, the percepts pulsate, or vibrate, with bright light that is independent of any source in the subjects' environment. They rotate, expand, contract, combine and change one into another. They may be projected onto one's environment when the eyes are open (this is an important point, as we shall see in Chapter 8). Another point is that the 'fortification

20 *At the Church of Jesus Christ is Love, Buenos Aires, ecstatic, swaying worshippers raise their hands to receive the Holy Spirit. For them, religious experience is all-important.*

illusion', or scotoma (type 5), has a blind spot in the centre of the scintillating arc. When the scotoma is projected onto, say, another human being, the head of that person becomes invisible. The ubiquity of these entoptic forms suggests that they are not culturally determined. Rather, they are wired into the neurology of the human brain.

Nonetheless, stage one is not ineluctable. Sometimes people are catapulted directly into the third stage that I describe in a moment. Further, some cultural contexts focus so sharply on stage three that, when experienced, the shimmering geometrics of stage one are considered 'noise' and ignored.

Although all people have the potential to see all the geometric forms, cultural emphases may lead them to value some and to ignore others. People in societies that accord altered states of consciousness important religious status therefore sometimes watch for and try to cultivate a restricted range of forms, the ones to which their religion ascribes emotionally charged spiritual meanings. The Tukano people of South America, for instance, take undulating parallel lines (type 4) to represent 'the thought of the Sun-Father'.

An arc of several multicoloured parallel lines (type 5) is taken, understandably enough, to represent a rainbow, but in some mythological contexts that would be impossible for an outsider to guess, it is said to be the Sun-Father's penis.[15]

The southern African San, on the other hand, concentrate on brilliant lines (types 2 and 6) that they believe to be 'threads of light' that healers climb, or along which they float, to the Great God in the sky.[16]

When you dance and get hot, you will see a rope hanging from the sky. As you stretch and become taller, you can climb this rope.... It is white and thin as a blade of grass. If you dance and see this rope, you don't have to grab or touch it. You just float away with that rope.... It is a big day when you first get the big power and see the ropes of light.[17]

Similar funicular visions have been reported from Australia. A boy taking part in an initiation ceremony that involved looking at 'a large, bright crystal that stole the light from the dawn and dazzled their eyes' sank 'into a state of repose that was almost sleep'. Then he saw cords that 'seemed to rise into the air, and the old fellows climbed hand over hand up them to treetop height'.[18] The Tukano, the San and the Australians were all making sense of entoptic phenomena.

In sum, entoptic phenomena are not deterministic; they are always culturally framed, and the meanings that people ascribe to them may therefore change. The important point is that the *forms* are universal, though their selection and meaning are culturally determined.

Stage Two

When people move into the second and deeper stage of altered consciousness, they begin to try to make sense of the entoptic forms they are seeing by construing them as objects with emotional or religious significance,[19] a tendency we have already seen in the examples of stage one visions I have described. In a normal state of consciousness the brain receives a constant stream of sense impressions. A visual image reaching the brain is decoded (as, of course, are communications from the other senses) by being matched against a store of experience. If a 'fit' can be effected, the image is 'recognized'. In altered states of consciousness the nervous system itself becomes a 'sixth sense'[20] that produces a variety of images including entoptic phenomena. The brain attempts to recognize, or decode, these forms as it does impressions supplied by the nervous system in a normal state of consciousness. Mardi Horowitz links this process of making sense to the disposition of the subject:

Thus the same ambiguous round shape on initial perceptual representation can be 'illusioned' into an orange (if the subject is hungry), a breast (if he is in a state of heightened sexual drive), a cup of water (if he is thirsty), or an anarchist's bomb (if he is hostile or fearful).[21]

As a further example, I argue in Chapter 9 that the fortification scotoma (type 5) has been interpreted as the battlements of Augustine's City of God.

STAGE THREE

In the third (still 'deeper') stage, subjects find themselves in a bizarre, ever-changing world of enveloping hallucinations that is comparable to dreaming but that is more intense, more memorable and more prolonged. It is in this stage that people see animals, monsters, emotionally charged sacred things and persons, and all the other experiences that we associate with visions. All the senses hallucinate, not just vision. People report somatic hallucinations, such as attenuation of limbs and bodies, intense awareness of one's body, polymelia (the sensation of having extra digits and limbs), changing into animals, and other transformations. They also report aural hallucinations, voices speaking to them about the future. The hallucinations of Paul, Constantine, Augustine, Joan of Arc and the young girls at Lourdes in France all fall into this category.

Another feature of stage-three experiences like these is that the entoptic forms of stage one persist, peripherally or integrated with iconic hallucinations. Benny Shanon, a professor of psychology at the Hebrew University of Jerusalem, who experimented with the South American hallucinogen ayahuasca, saw 'lizards popping out of arabesques. The lizard images were repetitive and embedded within the geometric pattern.'[22]

Another Western subject found that the lattice, or grid form (type 1), merged with his body:

[H]e saw fretwork before his eyes, that his arms, hands, and fingers turned into fretwork. There was no difference between the fretwork and himself, between inside and outside. All objects in the room and the walls changed into fretwork and thus became identical with him. While writing, the words turned into fretwork and there was, therefore, an identity of fretwork and handwriting. 'The fretwork is I.' He also felt, saw, tasted, and smelled tones that became fretwork. He himself was the tone. On the day following the experiment, there was Nissl (whom he had known in 1914) sitting somewhere in the air, and Nissl was fretwork.[23]

Here we see further points of interest:

– Hallucinations are experienced in all senses.
– The senses become confused so that one may smell a sound (synesthesia).
– Entoptic forms can become all pervasive.
– Subjects can themselves become an entoptic form.
– So-called after-images may recur unexpectedly some time after a hallucinatory experience.

As long ago as 1890, William James recorded the experiences of a friend who had ingested hashish; my observations are in square brackets:

Directly I lay down upon a sofa there appeared before my eyes several rows of human hands [polyopsia], which oscillated for a moment [pulsating], revolved and then changed into spoons [transformation]. The same motions were repeated, the objects changing to wheels, tin soldiers, lamp-posts, brooms, and countless other absurdities [some of these transformations appear to derive from entoptic forms 2 and 3]... I became aware of the fact that my pulse was beating rapidly.... I could feel each pulsation through my whole system [somatic intensity].... There were moments of apparent lucidity, when it seemed as if I could see within myself, and watch the pumping of my heart [preternatural sight]. A strange fear came over me, a certainty that I should never recover from the effects [heightened emotions].... Suddenly there was a roar and a blast of sound and the word 'Ismara!' [aural hallucination].... I thought of a fox, and instantly I was transformed into that animal. I could distinctly feel myself a fox, could see my long ears and bushy tail, and by a sort of introversion felt that my complete anatomy was that of a fox [transformation into an animal]. Suddenly, the point of vision changed. My eyes seemed to be located at the back of my mouth; I looked out between parted lips [somatic transformation].[24]

The reports of altered states of consciousness that I have quoted show that there is an interaction between neurologically wired experiences that are activated in altered states and the culturally specific content that is incorporated into those experiences. Human brains exist in diverse societies, but they remain human brains. Human neurology is inescapable. At the same time, it is true that most people do not experience profoundly altered consciousness – visions and hallucinations – but only inwardly oriented states and dreams.

All in all, we can be sure that the structure and functioning of the human nervous system means that consciousness is, inescapably, a shifting state. It gives rise to two kinds of experiences: those that relate to the material environment and that can be checked by normal sight, touch, sound and taste, and those that cannot be thus checked because, though vivid, they are illusory. We are on the threshold of faith.

Religion in the body: universality and diversity

Religious experience emerges from the human body. Or rather I should say that the body provides raw material for what, in a variety of social contexts, is accepted as some sort of trafficking with supernatural forces or beings. I now give examples of people dealing with neurologically generated experiences. Although the experiences are, overall, different, few would contest that they may be labelled 'religious'.

Gopi Krishna, a Kundalini practitioner, described his religious experience as follows. As he sat upright and unmoving 'for hours at a time' in the lotus posture, he breathed 'slowly and rhythmically' and focused his attention exclusively on 'the crown of [his] head, contemplating an imaginary lotus in full bloom, radiating light'. Gradually his breathing slowed down until 'it was barely perceptible'.[25] Then he experienced 'a strange sensation below the base of the spine'; it was extraordinary and pleasing. The feeling extended upwards and grew in intensity. Then came the climax:

Suddenly, with a roar like that of a waterfall, I felt a stream of liquid light entering my brain through the spinal cord.... The illumination grew brighter and brighter, the roaring louder, I experienced a rocking sensation and then felt myself slipping out of my body, entirely enveloped in a halo of light.... I felt the point of consciousness that was myself growing wider, surrounded by waves of light.... I was now all consciousness, without any outline, without any idea of corporeal appendage...immersed in a sea of light.[26]

Now let us travel to southern Africa, first to the central parts of what is today South Africa. In the nineteenth century, /Xam San *!gi:ten* (singular, *!gi:xa*; translatable as shaman) spoke of leaving their bodies and going on journeys to the spirit world and to other parts of the land. The first syllable of *!gi:ten* (*!gi:*) means supernatural potency. The second syllable of the singular form, *-xa*, means 'full of'. (The plural is formed by the suffix *-ten*.) San *!gi:ten* were thus people who were filled with potency. One of these men, Diä!kwain, described the rising sensations that he experienced. He said it was as if his 'vertebral artery has risen up' and 'would break'.[27] There is some doubt as to what he meant by 'vertebral artery', but the word translated 'vertebral artery' (*!khaua*) can, as a verb, mean 'to boil'.[28] Diä!kwain experienced a hot, boiling sensation up his spine. When his 'vertebral artery' rose up, people danced, sang powerful songs and gave him aromatic herbs to smell so that it would 'lie down'. It was believed that if they did not do this he would grow hairs on his back and turn into a lion. In his feline frenzy he would bite people.

In the twenty-first century Kalahari Desert, some 1,300 km (800 miles) to the north of where the /Xam lived, the same experience is still described. The San people living here today speak different languages from those who lived to the south in the nineteenth century. The Kalahari informants in the next three quotations speak of *n/om* (formerly transcribed as *n/um*), a supernatural energy that has been likened to electricity;[29] *n/om* and the southern San's *!gi:* are one and the same thing. In the northern Ju/'hoan (!Kung) San language, a person who 'owns' *n/om* is a *n/om k"xau.*

Bushman medicine is put into the body through the backbone. It boils in my belly and boils up to my head like beer.... You feel your blood become very hot just like blood boiling on a fire and then you start healing.[30]

You dance, dance, dance, dance. New *n/um* lifts you in your belly and lifts you in your back, and then you start to shiver. *N/um* makes you tremble; it's hot.[31]

In your backbone you feel a pointed something and it works its way up. The base of your spine is tingling, tingling, tingling, tingling. Then num makes your thoughts nothing in your head.[32]

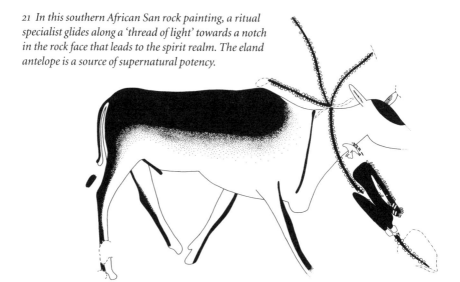

21 *In this southern African San rock painting, a ritual specialist glides along a 'thread of light' towards a notch in the rock face that leads to the spirit realm. The eland antelope is a source of supernatural potency.*

When the *n/om* reaches the head, a dancer enters *!kia*, a word translated as 'trance' or 'altered state of consciousness'.[33] In this state the spirit is believed to leave the body and to visit the spirit realm and also other places in the vast Kalahari Desert (Fig. 21). *!Kia* is experienced as an 'opening up'. One informant put it like this: 'When I pick up num, it explodes and throws me up in the air, and I enter heaven and then fall down.' Others expressed the climactic sensation thus: 'Bursting open, like a ripe pod.'[34] Another said: 'You become so light that you simply fly away. You don't know whether your feet are on the ground or not. I feel like the wind when this happens.'[35] One man said:

The power makes my head vibrate very fast. Sometimes it feels like a liquid is inside me, being poured over my brain and flowing down the rest of my body from inside. It is a magical feeling. It is the best feeling.[36]

Like the nineteenth-century southern people, the Kalahari San also speak of their transforming experience in terms of lions. After his first shattering transformation, a man retains his leonine alter ego. A San man explained:

Once you turn into a lion, you always have the lion inside you. Just thinking about the time that you turned into a lion brings back its power.... The transformation has to take place only once and then you have that power for life.[37]

The huge separation in space and time between the Kundalini exponent and the San shamans can be explained only by their possessing the same human nervous system. Their different religious practices – meditation and ecstatic dancing together with hyperventilation – activate the nervous system in such a way as to produce similar religious experiences.

Together, the Kundalini and the San reports point to yet another way of understanding the same neurologically generated experiences. It is the 'oceanic' or 'one with the universe' experience, sometimes known as Absolute Unitary Being (AUB). In Buddhism, it is related to Nirvana, a state of bliss, happiness and Ultimate Reality.[38] In that state, the individual 'seeks to extinguish himself in God' – at least in some Buddhist traditions (Fig. 22).

> In the higher realm of true Being,
> There is neither 'self' nor 'other'.
> When direct identification is sought,
> We can only say, 'Not two'.
> One in all;
> ALL in one.
> *Lankavatara Sutra* (Zen Buddhism)[40]

22 *The Great Buddha of Kamakura, Japan, sits in the lotus position. Devotees adopt this posture as they empty their minds of worldly things. Repeating the name of the Buddha, suddenly, stopping and then gazing into the emptiness is said to induce 'liberation'.*

Meditation, either by emptying the mind of all thoughts or by focusing on a single object or mantra, seems to do the trick: the process stalls the normal functioning of the brain and the discriminations it makes. It more or less collapses in upon itself and all becomes one. People in this state feel that boundaries between themselves and others (including the world itself) break down. This is what James described as 'a reconciliation', an experience that 'soaks up and absorbs its opposite into itself'.[41]

Lest it be thought that I argue that radically altered states of consciousness, as popularly conceived, are the *only* basis for religion, I emphasize other, milder, mental states that many people would not consider 'altered'. In some secular circumstances, people understand these experiences, not as supernatural in origin, but as some sort of aesthetic effulgence – Wordsworth's response to beauty, 'Ne'er saw I, never felt, a calm so deep!' AUB can be triggered by views of natural beauty as in Wordsworth's 'Tintern Abbey':

> Nor less, I trust,
> To them I may have owed another gift,
> Of aspect more sublime; that blessed mood,
> In which the burthen of the mystery,
> In which the heavy and the weary weight
> Of all this unintelligible world,
> Is lightened: – that serene and blessed mood,
> In which the affections gently lead us on, –
> Until, the breath of this corporeal frame
> And even the motion of our human blood
> Almost suspended, we are laid asleep
> In body, and become a living soul:
> While with an eye made quiet by the power
> Of harmony, and the deep power of joy,
> We see into the life of things.

Wordsworth felt no need to invoke an anthropomorphic god or any other routine religious concept to explain his feelings.

There is nothing mysterious about this feeling of seeing 'into the life of things'. Indeed, the neurological foundation of AUB has been studied.[42] The sensations of religious exaltation and calm are wired into the brain. Why is it that some people seem more susceptible to the surge of AUB than others, whether they interpret their feelings as divine or not? The answer probably lies in the physical make up and functioning of individual human brains: some are more labile than others. Each and every member of a community does not experience the full gamut of altered states.

It is clear that some who experience AUB and other potentially religious experiences may interpret them in a secular way. Robert Hinde writes:

[T]here seems to be no evidence for differences between the concomitants of religious and similar secular experiences.... It is probable that the two are to be distinguished primarily by the interpretation put on the experience, and this is influenced by the cultural and social experience of the individual in question.

Ominously, he adds: '[S]uch links do, however, suggest that religious experience is less special than has sometimes been thought.'[43]

To sum up this discussion, I give examples of three kinds of AUB that the psychologist Basil Douglas-Smith identified.[44] They are theistic, nature and monist (unitary) mysticism. He obtained the statements in response to questionnaires.

THEISTIC MYSTICISM: 'It was as if I were for the first time truly myself.... Perception and comprehension were now one.... The Presence on which I "gazed" was utterly *sui generis*.... All that had ever allured or attracted me in my whole life was but a feeble symbolic sparkle or reflection of This...I became aware that I was dealing with a Person, although personality in Him was far other than in me.... Although loving kindness itself, He was inexorable justice.'

NATURE MYSTICISM: 'I seemed to become one with the life that ran through the earth.... The trees, the grass, the flowers, and even the earth itself, inanimate things, like the stones which had come from the earth, were full of the power of the Creator.'

MONIST MYSTICISM: 'I suddenly *knew* (with the same certitude with which I was and am aware of my own identity and my own past) that "I was there when the stars were born." It was a clear and unquestioning awareness with the cosmic "I".'

The genesis of religion

We can now see that what we have learned about the shifting nature of human consciousness helps us to tackle the difficult matter of the origin of religion.

Some writers have argued that altered states induced by hallucinogens were the origin of religion.[45] While not wishing to underestimate the role of hallucinogens and the part they still play in many religions, it seems to me that we should take the *whole* spectrum of consciousness into account. Even without the impact of hallucinogens, complex human consciousness opens up vistas onto another realm.

We can, with a comfortable degree of certainty, accept that the type of shifting human consciousness I have described developed as the human brain was evolving. The brain is a complex, integrated organ that requires certain neural wiring and electrochemical processes for it to function. Complex consciousness evolved as a 'package deal'.[46]

I do not argue that the spectrum of consciousness (daily and altered) became part of human experience because shifting mental states were in themselves adaptive. Rather, the electrochemical functioning of the brain that, *as a by-product*, produces the spectrum was adaptive. Shifting consciousness permitted the brain to function, to renew itself day in and day out, to develop a reliable memory and to generate the kind of symbolic thought that facilitated a creature's survival in a real world that was stocked with predators, drought, impending starvation and other people who were antagonistic and were competing for resources. Individuals who had dependable memories and symbolic thought (the foundation of language) were more likely to survive and pass their genes on to subsequent generations. To sustain this kind of functioning the brain had to manufacture proteins – to recuperate from a day's labour – in the partially shut-down state we call sleep.

Neuroscientists today propose various functions for dreaming, but there is wide agreement that dreaming is random: it has no mystical purpose. In sleep, the body, especially the brain, is simply rejuvenating itself in a physical, anatomical sense. Freud, Jung and others who sought meaning in dreams were no better than people in earlier generations who believed that God spoke to them in dreams.[47] Today an idea may come to a scientist by chance in a dream; the cross-firing of neural pathways in the brain during sleep may, by chance, connect ideas in a way that is not readily apparent in normal waking states. But, if something like this happens, it has to be substantiated by scientific method. So too an early hunter who dreamed of a new type of weapon would have to test it in practice in the material world. Dreams, I must emphasize, are not consistently useful and cannot be relied on.

I therefore argue that beliefs about a supernatural realm are in themselves neither adaptive nor maladaptive. Certainly, if a human being were to rely too heavily on supernatural beliefs, the result would be maladaptive because they would get in the way of that person's relationship with the real world. Examples of belief in a supernatural realm getting in the way of useful dealing with the real world abound. Today the Catholic Church's response to the AIDS pandemic, based as it is on ancient revealed knowledge, is a tragic example.

What does a neurologically integrated spectrum of consciousness mean for our understanding of the evolution of the human brain/mind? Just where does the notion of adaptation fit in?

Let us set dreaming aside for a moment and consider states closer to alert waking, yet not fully alert. Withdrawing into a reverie (I use 'reverie' to mean free-flowing, dreamlike fantasy) does not *in itself* aid practical engagement with the material environment. On the other hand, withdrawal into contemplative symbolic thought of the kind to which I suspect the Blombos carvings point, does facilitate coping with and exploiting the environment – that is, it has an adaptive value. A person indulging in this kind of consciousness is able to manipulate ideas and words that are in effect symbols of things in the environment and, importantly, of different kinds of relationships between those things. Undetected feline + hunter = Bad Thing. Green vegetation = water, a Good Thing. By this sort of mental manipulation a person is able to solve problems in the present and to decide on future behaviour. This 'thought world' is an experimental simulacrum of the real world that, while it is being experienced, diminishes attention to sensory input from the material environment. Moreover, for it to be effective some of the elements derived from dreams and reverie (e.g., divine interventions, suspension of normal causality) must be excluded, or at any rate marginalized, if the end result of the thought process is to be of any *practical* value.

If this way of symbolic thinking and mental problem solving did not arrive suddenly and miraculously fully formed, then its roots must have been in still older, pre-Blombos, pre-*Homo sapiens* animal behaviour. Here I am thinking of the specific sounds that certain species of monkeys emit to warn others when they spot a raptor circling above them. The monkeys' call does not *resemble* a raptor. If the call and its significance ('Take cover!') were programmed into the inherited brains of the species, as seems likely, we have a possible origin of symbolic thinking in natural evolution. With increased brain capacity and more complex neural wiring, a subsequent species (us) was able to break free from its neurologically inherited links between sounds and raptors and to create its own limitless series of symbolic links – language. Fully modern language enables people to utter a limitless range of never-before-heard sentences and expect them to be understood – something monkeys and apes cannot manage. Some of these utterances may be lies: not only in Denmark, 'one may smile, and smile, and be a villain.'

The realm of reverie, dreams and visions has elements derived from the environment (people, animals and so forth) and also emotions like fear and joy. But these 'real-world' elements behave in distorted, unrealistic ways; they are not bound by the restraints of the material world. All this is a product of the electrochemical functioning of the human brain. It was, and still is, common to all people, though perhaps to varying degrees from individual to individual. We may therefore accept that it was present in fully human experience in its nascent stages. So right from the outset, emerging human consciousness was, by the very anatomical nature and functioning of the human brain, of different kinds. The spectrum of consciousness was there in our brains from the beginnings of *Homo sapiens*, and experiences generated by inescapable introverted states had to be accommodated. They had to be explained in a way that was different from practical thinking in the material world.

That being so, I argue that, even as other components of human consciousness were emerging (such as a sense of self), people were aware of two realms: they did not have to invent a realm of inner experience, dreams and visions. They merely had to deal with the weird, non-real experiences that their brains sometimes generated. And they did so by accepting the existence of a parallel realm where all this was happening, what today we call the supernatural. Belief in a supernatural realm thus formed in parallel with conscious perception of the material world, symbolic thought and the notion of a person as an individual distinct from other individuals.

This distinction between natural and supernatural realms, the foundation of religion, was useful. All sorts of aberrant experiences, such as bizarre dreams in which people cavorted peacefully with lions, could be relegated to the supernatural realm and not allowed to influence behaviour in the material environment in tragic ways. There was thus a ready-made framework for sorting out the different kinds of experiences that everyone has. This is why the ancient Greeks (and indeed most peoples) distinguished between significant dreams in which some higher power was communicating vital knowledge and other dreams that were of no significance. It could be argued that the foundation of Paul's difference of opinion with the Athenian philosophers was founded on different ways of interpreting the spectrum of human consciousness: he believed that visions and dreams imparted truth and he had experienced this on the road to Damascus; the philosophers, by and large, believed that truth could be achieved only by rational thought.

The evolutionary advantage of discriminating between real and supernatural experiences was that perception of the material environment was more effective, not that belief in a supernatural realm had in itself adaptive advantages. As we saw in the previous chapter, the supposed adaptive advantages of religion are questionable and, even if they are advantageous, they are consequent upon belief in the supernatural, not reasons for its invention.

At any point in human history, the acceptance of certain mental precepts as supernatural was contingent upon the intellectual and social milieu of the time. Belief in supernatural realms is not ineluctable. Indeed, acceptance of the supernatural as an essential and true part of human perception has fluctuated through history, as Chapters 1, 2 and 3 showed. Belief in the supernatural has been and still is persistent because other factors (such as social acceptance and political power) enter into the equation; they can make recognition of the supernatural as a non-real product of the functioning of the human brain difficult for people embedded in a religious milieu.

Making sense of consciousness

How does this understanding of human neurology and the experiences it generates illuminate the worldwide phenomenon of religion? How do we get from the individual to communally understood experiences and beliefs?

Once people realize that the mind can open up to realms of experience other than those of everyday life, they begin to ask questions. What sort of world, or two worlds, are we living in? How do we harmonize such diverse experiences? Where do the beings that we see in dreams and visions live? Where do all these strange transformations take place?

These are key questions. People have no option but to ask them and to think up answers to them. Many of the mental states that I have described are inevitably produced daily by the functioning of human neurology even without conscious attempts to create them. They are as integral to the human body as, say, the digestive system: fundamentally, they are neurological. What people make of them, how they react to them are matters for psychology.

Throughout the history that I outlined in the first two chapters, people have concluded, logically enough, that there are two realms: daily life and an immaterial spirit realm. Today, among people of a scientific turn of mind who know something about the anatomy and workings of the brain, there is another consensus: the experiences in question are the product of the human brain, not the result of invasion by a spirit realm.

Here we must remember that people do not confront questions about the significance of the inner world as individuals. Inevitably, they discuss their mental experiences with other people and thus discover that they are not alone. The explanations for the diverse mental states that come to be generally accepted in a community are, by and large, consensual conclusions. The spectrum metaphor is again useful. As I have pointed out, human communities divide up the colour spectrum in various ways: some identify fewer colours than the seven that English-speaking Westerners name. Similarly, with the consciousness spectrum, communities identify different mental states. Some, like Buddhists, are fascinated by mental states and therefore identify more of them than, say, a materialistically minded Westerner.

But consensus is never complete. As communities negotiate their way towards generally accepted ideas about what shifting consciousness means, they find that some individuals seem to have (or at any rate to speak about) more vivid and elaborate mental experiences than others. Communities also find that, as in other areas of social interaction, some individuals have more influence – more charisma – than others. So, as a community moves towards consensus about the nature, content and significance of mental experiences there remains a degree of fluidity, debate and argument. Conflict remains – indeed, must remain – endemic. In the nature of things, agreement on something that does not exist – like a supernatural realm – can never be settled to everyone's satisfaction. So mercurial a thing as mental experiences of a supposedly supernatural realm guarantee discord.

Nevertheless, at the foundation of this sort of contestation, there is a major agreement: overwhelmingly, most people accept that *something* is going on in an adjacent, normally invisible realm. After all, their own dreams and reveries repeatedly confirm this 'fact' by giving them glimpses of that realm. Here we see what lay at the base of the historical trajectory that we traversed in Chapters 1 and 2. *Something* is going on both 'out there' and 'in my head'. But exactly what?

We are now close to discussing religious belief and religious practice, the topics of the next two chapters. If certain individuals can manoeuvre themselves into a position of special insight into, and revelations about, this other realm, they can formulate doctrine and rise to positions of influence and power in their communities. People feel that these 'gifted' individuals have some sort of insight into their own minds. The 'gifted' need not necessarily do this arrogantly. When we are dealing with an ineffable spiritual realm, feigned humility can be a source of great power, as the wealthy medieval monastic orders found.

An important point needs to be emphasized, especially because the foundational nature of religious experience is today so often overlooked. The experiences that we have identified can come without being consciously generated by openly religious practices. They are often spontaneous and unsought because they are wired into the human brain and can be triggered by the variety of causes that I have listed. The experiences are therefore independent of religious beliefs and practices. They can come whether one believes in God or not. They are thus both prior and fundamental to the complex that we term 'religion'. Religion is only one way of dealing with them, though history and anthropology show that, in the absence of commitment to rational thought and science, people find that the existence of a spirit realm is the most obvious explanation.

Dealing with religious experiences

Religious experience is generated by the human brain, either spontaneously or with intent. As we have seen, these states may be mild and peaceful, like those experienced by temperate Christian believers at prayer. They are encouraged by the dim light, soaring architecture, chanting and incense in a medieval cathedral. But they may also be more overwhelming, like those experienced through intense meditation and ecstatic dancing. Ultimately, they all come down to tinkering with the neurology of the brain in one of many possible ways. Neurological activity in the brain, not sociality, comfort in distress or a need for explanation, is the foundation on which the whole edifice of religion is built. But we must remember that religion cannot be reduced to mental states. There is much more to it. Moreover, because the range of experiences we call religious originates naturally in the brain, their designation as religious is post hoc and not inescapable. Why some people do not interpret their mental states as contact with some supernatural entity while so many others do is a separate question.

In the next chapter I consider how shifting consciousness leads to religious doctrine, sometimes fairly directly, sometimes circuitously.

CHAPTER 6

Religious Belief

A significant part of all religions is devoted to explaining mysteries of their own making. These 'mysteries' are not ordinary puzzles. In a more specific sense, the word denotes divinely revealed religious truths that are beyond human comprehension. In Christianity these truths are, principally, the Trinity, Original Sin and the Incarnation – hence the medieval 'mystery plays' that dealt with such matters and brought them before an illiterate public. Once people take the route of accepting certain kinds of neurological activity as evidence for a supernatural realm their chosen path leads them to a cornucopia of mysteries, enough to keep religious pundits and their followers busy for millennia. This is the enterprise that we call theology, a facet of the second of the three domains of religion that I distinguish, religious belief.

The neurologically generated experiences that I have considered have to be understood in a shared way for them to become the foundation of a religion. Belief is therefore in some ways similar to, and yet different from, religious experience. It derives, in the first instance, from attempts to codify religious experiences in specific social circumstances.

I say 'in the first instance', because a religious belief system need not refer back to religious experiences in every respect. Once the existence of a supernatural realm is accepted, religious devotees, whether they themselves have had religious experiences or not, begin to build theological superstructures. These are elaborate, often convoluted systems of thought that purport to clarify what is, when all is said and done, a false understanding of the workings of the human brain. There is no supernatural realm apart from the one that people create inside their heads.

In such a system, there will be core beliefs founded on supposedly revealed unequivocal knowledge and a host of peripheral beliefs that admittedly derive from the cogitations of theologians. Deciding into which category a specific belief falls is often a disputed matter and demands firm rulings from the Church. Hence the concept of *credo* – a carefully worded, parroted formula that is in some ways comparable to the recitation of magical spells, despite the best efforts of the clergy in confirmation classes.

Most religions spend time hammering out a formulation of what they believe. They do this in specific economic and political circumstances that inevitably influence their conclusions. Before deciding on a particular tenet, they have to consider the economic and political consequences of making it mandatory. This is especially true of literate societies; their beliefs have to be written down – there is no alternative. Writing dictates precision, and precision inevitably leads to intolerance. On the other hand, small-scale, pre-literate societies do not have a mechanism for pinning down and passing on an approved formulation of beliefs; as a result, they are more humane.

Since the time of Constantine and his conclave at Nicea, Christian congregations have repeated a statement known as the Nicene Creed. But what do people mean when they say, 'I believe'? Are they drawing a distinction between what they believe and what they *know*? Certainly, the recitation of a creed, parts of which are incomprehensible and contradictory to many people in the congregation, suggests that *credo* means 'I accept without reasons'. For some devout believers *credo* means 'I am saved simply because I believe this list of unlikely things'. Deep personal commitment to a static creed becomes inextricably bound up with inner experiences. Recitation of a creed creates emotional and mental states that seem to confirm the content of the creed.

I start this discussion of religious belief with an ancient concept that is widely held throughout the world in one form or another. I argue that its widespread, though not absolutely universal, nature is best explained in neurological terms. This is why it is part of many religions.

The religious cosmos

As debate about a spirit realm ebbs and flows within a community, there is a tendency to think about this 'other world' in geographical, or topographical, terms. We know the material world by its mountains and valleys, its rivers, lakes and seas, its towns and villages, and, importantly, the places where different kinds of people live; some are friends, others are enemies. Is the supernatural world like this too? Does it, like this world, have 'many mansions'? Do different kinds of supernatural beings (God, saints, angels, the dead in general, evil spirits) enjoy their own residences?

A cosmology is a culturally constructed view of the universe. The shape of a community's cosmology is governed by the degree of credence that people give to the introverted end of the consciousness spectrum. Those who emphasize the revelatory nature of dreams and visions have a different cosmology from those who insist on alert observation as the only means to understanding the material world – hence all the trouble with Copernicus and Galileo.

Here, I am principally concerned with societies that take altered consciousness seriously. A feature of the religions of such societies virtually worldwide is that people believe in cosmological levels. At its simplest, they believe the cosmos to have three levels:

> A realm above
> inhabited by beings and spirit animals.
> _____
>
> The daily world
> in which people live in amity and strife.
> _____
>
> An underworld
> in which other beings and spirit animals dwell.

Some societies, especially more complex ones, believe in more than these three tiers, but the generally layered nature is retained. Subdivisions of the upper and lower levels seem to reflect social divisions that divide up life in the material world. One thing has to be said for a tiered cosmos: it provides a framework not only for the experiences of dreams and visions and the way in which mythology apparently explains natural phenomena but also for daily life. It co-ordinates the diverse experiences of the consciousness spectrum. Everything has its place somewhere in a tiered cosmos.

This is true of traditional Christianity. The geography of the spirit realm greatly intrigued early and medieval mystics. Their visions often included trips to Heaven and Hell. The visionary Hildegard of Bingen (more of whom in Chapter 9) was but one who strove to produce diagrams of the cosmos that allotted spaces to grades of spirit beings: her cosmology included not only the material world of medieval Germany but also spiritual dimensions that she believed God had revealed to her in her visions.[1] In the centre of Hildegard's cosmology is a spherical earth. It is surrounded by concentric zones. She took over the then-common notion of concentric spheres, an idea that chimed well with the mystical music that she composed and that has become so popular today. She was somewhat vague about the location of Hell and Purgatory (obligatory components of a medieval Christian cosmology), settling for spaces shaped like truncated cones in the centre of the earth.

Probably, it was accounts of visionary grand tours of Heaven and Hell that inspired Dante to write in his *Inferno* about his trip through multiple levels of depravity and punishment.

A tiered cosmos is embodied in and provides a framework for the Christian Apostles' Creed, shorter and better known than the Nicene Creed. This is how it appears in the Anglican 1662 *Book of Common Prayer*:

I believe in God the Father Almighty, Maker of heaven and earth: And in Jesus Christ his only Son our Lord, who was conceived by the Holy Ghost, Born of the Virgin Mary, Suffered under Pontius Pilate, Was crucified, dead and buried: He descended into hell; The third day he rose again from the dead; He ascended into heaven, And sitteth on the right hand of God the father Almighty; From thence he shall come to judge the quick and the dead. I believe in the Holy Ghost; The holy Catholick Church; The Communion of Saints; The Forgiveness of sins; The Resurrection of the body, And the life everlasting. Amen.

For centuries it was impossible for people to believe the central Christian doctrines without a tiered framework into which they could slot their otherwise incredible beliefs: Heaven, earth, Hell; rose again, ascended, resurrection. It is all there in the Creed. Now, post-Copernicus, theologians feel a need to explain the concept away. Radical theologians are, understandably enough, embarrassed by notions of a tiered cosmos. They are at pains to argue that Christianity, at least and at last, has outgrown and put aside such childish things (Chapter 10). But despite what theologians say, vast swathes of Christianity, as in the American Bible Belt, still prefer to stay with literalness. It is they, not liberal theologians, who influence persons in high political office. It is they who plot to have their people in influential positions so that they can control American foreign and internal policies. They want the USA to be, in effect if not in legislation, a theocratic state. We should not allow more liberal theologians to divert our attention from the existence of vast numbers of influential fundamentalists. The theologians tell us that it is they who are the 'true' Christianity of today, but we have heard that claim for centuries.

How can we account for the general ubiquity of tiered cosmological beliefs? They are not arrived at by rational inferences from observation of the daily world around us. On the contrary, I believe that this near-universally believed cosmology derives from mental experiences that are wired into the human brain. Two neurologically generated experiences that I have not yet considered are key parts of the consciousness spectrum.

Vortex

As subjects move towards stage three of the spectrum of altered consciousness, they often experience a vortex or tunnel, at the end of which is bright light. On the internal surface of the vortex there is sometimes a grid, in the compartments of which appear the first hallucinatory images of people, animals, monsters, and so forth.[2]

In the Prolegomena to this book, I briefly considered the Maya. They believed that 'all natural openings into the earth, whether caves or cenotes (sunken waterholes), were portals to the Otherworld'.[3] The notion is indeed widespread.

Today, the vortex is especially associated with near-death experiences and is frequently discussed.[4] Examples abound. A child who was grievously ill and was rushed to hospital later recalled: 'I was moving through this...long dark place. It seemed like a sewer or something.'[5] Another person reported: 'It was like being in a cylinder which had no air in it.'[6] More beatific experiences are common (Figs 23, 24). Another person who had been near to death recounted his passing down a passageway: 'I floated on down the hall and out of the door onto the screened-in porch.... I floated right straight on through the screen...and up into this pure crystal light.'[7]

I wish to dispel the possibility that these experiences may be unique to the Western culture from which these examples come. Ritual specialists around the world frequently use similar imagery to describe their out-of-body travels to the spirit realm, regardless of how they alter their consciousness.

In the 1870s, a /Xam San man named //Kabbo ('Dream') explained to Lucy Lloyd that everyone who dies goes along a path, 'a Bushman's path',[8] that leads to a large hole in the ground. There they live on, and old men become attractive again. More recently, a southern African Ju/'hoan San healer spoke of 'a big hole in the spirit world' and of following a line 'that goes underground'.[9]

God wanted me to be a doctor. That's why he made me sick years ago. In my sickness, God came to me and took me to visit other places. He took me inside a deep hole in the ground. I went into it and there God taught me how to be a good doctor.[10]

This sensation of entering the earth is linked to the 'threads of light' that I described in the previous chapter. Kxao Giraffe, a Ju/'hoan man, told how he became a healer:

Then my protector told me that I would enter the earth. That I would travel far through the earth and then emerge at another place. When we emerged, we began to climb the thread – it was the thread of the sky![11]

Similarly, a Sora shaman in India climbs down a huge tree that leads to the underworld: 'The path includes dizzying precipices on the descent to the "murky-sun country, cock-crowlight country".'[12] Descent to an underworld is indeed a common shamanistic experience, and the historian of religions Mircea Eliade collated a number of accounts of such journeys in his book *Shamanism: Archaic Techniques of Ecstasy*. For instance, Siberian Yakut shamans' costumes had attached to them symbols known as 'Opening into the Earth' or 'Hole of the Spirits'; these enabled them to travel to the nether realms.[13] A Tungus shaman, also in Central Asia, sometimes descends to the underworld, a dangerous undertaking during which 'he goes into ecstasy' and 'goes down through a narrow hole and crosses three streams before he comes upon the spirits of the infernal regions'.[14]

Another Siberian shaman told of his initiation. His spirit guide, who was associated with a tree that the young man had cut down, took him to a hole in the earth.

My companion asked: 'What hole is this? If your destiny is to make a [shaman's] drum of this tree, find it out!' I replied: 'It is through this hole that the shaman receives the spirit of his voice.' The hole became larger and larger. We descended through it and arrived at a river with two streams flowing in opposite directions. 'Well, find out this one too!' said my companion, 'one stream goes from the centre to the north, the other to the south – the sunny side. If you are destined to fall into a trance, find it out!'[15]

23, 24 LEFT *An engulfing vortex with a bright light at the end leads to the hallucinations of deeply altered consciousness;* OPPOSITE *Gustave Doré's depiction of Dante's vision of the route to the Empyrean translates the neurological vortex into a swirl of angels.*

Such experiences are also associated with pathological conditions. A Western patient suffering from schizophrenia described his visions:

There were small suns and strange twilight worlds of lakes and islands.... An ancient cave, passage, or hollow ladder, seemed to connect new earths; perhaps this was such as Jacob saw, for it was an image of remote antiquity.[16]

A corollary of the vortex needs to be mentioned. One of the near-death experiences I cited recorded the sensation of being in a cylinder 'which had no air in it,'[17] and the Siberian shaman spoke of 'two streams flowing in opposite directions'.[18] Similarly, a southern African San shaman said that he entered a wide river: 'My feet were behind, and my head was in front...I travelled like this. My sides were pressed by pieces of metal. Metal things fastened my sides.'[19] The sensations of passing through a constraining vortex, difficulty in breathing, affected vision, a sense of being in another world, and weightlessness are frequently interpreted as being underwater. '[S]ubmersion in pools, springs, whirlpools, and rivers provides access to the underworld. The process by which one travels there is akin to drowning.'[20] Indeed, many shamans speak of diving into water. Lapp shamans, for instance, refer to altered states as 'immersion', and Inuit shamans situate the beyond in the sea.[21]

In Western reports we read of corridors, trains in tunnels, sewers and so forth; the ethnographic reports express the same sensations in their own imagery of holes in the ground, trees that link the sky (their foliage) to the underworld (their roots), subaquatic travel and so forth.

Together, all the descriptions I have given show that there must be some common human proclivity to experience passage through a vortex. That commonality is wired into the human nervous system and manifests itself in certain conditions of altered consciousness. The wiring is principally in the functional architecture of the striate cortex.[22] The human brain generates a worldwide (potentially) religious experience.

Flight

The vortex is one means of access to stage-three hallucinations. Flight is a sensation frequently experienced in that stage.

Sometimes writers argue that this aerial sensation derived from people's observations of birds soaring above. They claim that people simply think themselves into what it must be like to fly, as Leonardo da Vinci did when he set about designing ways by means of which human beings could fly. This is one of those simple, initially attractive explanations that warrant closer attention. Before we accept it, we must notice that flight is universally

25 A stonecut print by Mary Pitseciak entitled In the Night Sky *depicts an Inuit shaman's flight through the heavens to the spirit realm. Reported worldwide, the sensations of weightlessness and flight are induced by the human nervous system as it enters certain altered states.*

reported *as part of altered consciousness* and that it is people who experience altered states who talk about their own experience of flight. In some instances, birds provide a way of 'rationalizing' the neurologically generated sensation of flight, not the source of it. People who simply try to imagine what it must be like to be a bird do not tell others about their flying experiences as if they actually happened.

There are many examples of flight being associated with altered states. For instance, South American Tapirapé shamans speak of travelling through the cosmos in an aerial canoe. Like the Siberian shaman's horse, a feature of the open steppe, the canoe is clearly related to the riverine life of the Amazon Basin. The Tapirapé also speak of changing into birds and flying through the cosmos, a more widespread trope.[23] Other South American people, the Tupinamba of Brazil and the Caribs of Guyana, ingest a hallucinogen which helps a shaman's 'soul to leave his body and fly'.[24] A recurring shamanistic dream among the South American Bororo is one of soaring very high above the earth like a vulture, accompanied by the soul of some living person, often but not always a shaman. The dreamer sees 'a curiously altered but perceptually vivid world, in which "things are very little and close to one another".'[25]

North American Inuit similarly speak of shamanistic transformation from human to bird and carve ivory shamanistic bears in a flying posture (Fig. 25). In Siberia, Khanty drum rhythms facilitated shamanistic flight. Shamans 'were able to fly faster than a speeding arrow and pierced the sky on drums, flying to the golden residence of the sky god Torum'.[27] Then, too, in southern Africa, nineteenth-century /Xam San spoke of their shamans transforming themselves into birds:

At some other time, when we are liable to forget him [a shaman], he turns into a little bird, he comes to see us where we live and flies about our heads. Sometimes he sits on our heads, he sits peeping at us to see if we are still as we were when we left him.[28]

Eliade recognized the ubiquity of spiritual flight in his survey of world shamanism: 'All over the world, indeed, shamans and sorcerers are credited with the power to fly, to cover immense distances in a twinkling, and to become invisible.'[29] The anthropologist Piers Vitebsky comments: '[T]here are astonishing similarities, which are not easy to explain, between shamanistic ideas and practices as far apart as the Arctic, Amazonia and Borneo, even though these societies have probably never had any contact with each other.'[30] The widespread nature of such flights is indeed striking.

In summing up, we can say that, worldwide, there are beliefs in tiered cosmological levels with passages between them and flight. How can this virtually pan-human situation be explained? I argue that there is only one explanation. Both descent into a tunnel and flight to a realm above (or through the air above the real world) are sensations wired into the human brain and are activated in certain altered states of consciousness.[31] Beliefs in vortex travel and magical flight seem to lead inevitably to beliefs about a tiered cosmos.

The psychologist William James, as usual with this perspicacious writer, expressed the matter memorably:

We may now lay it down as certain that in the distinctively religious sphere of experience, many persons (how many we cannot tell) possess the objects of their belief, not in the form of mere conceptions which their intellect accepts as true, but rather in the form of quasi-sensible realities directly apprehended.[32]

'[Q]uasi-sensible realities directly apprehended' lead from experience to belief. Neurological processes explain why there has been, through the millennia, a steady stream of religious belief in supernatural dimensions and, in addition, the ways in which beliefs of this kind are co-ordinated with daily life in a coherent tiered cosmos. Religion and a tiered cosmology have long been inextricably related.

Transcendence and immanence

The notion of a tiered cosmos brings its own problems – its own 'mysteries'. On the one hand, it seems to situate supernatural beings in a distant realm of their own. On the other, religious experience suggests the tiers interdigitate: gods leave their realm and visit the hearts and material surroundings of human beings.

I suggested that the Blombos people of 75,000 years ago may have believed that something powerful but intangible was inside the pieces of ochre on which they engraved motifs of crosses. Ochre itself was at that time probably a symbolic substance, not a meaningless *tabula rasa* on which people could engrave anything they wished. We may have here an early hint of an important component of religious thought: immanence. Gods and supernatural powers can be *inside* statues, mountains, lakes, seas, nature itself, and of course people.

Although shamans and other mystics talk of travelling between layers of the cosmos, they also describe the way in which spirit beings and influences are present in this world as well as above and below. Spirits are *here*, not exclusively *there*. This widespread belief leads Vitebsky to an interesting idea. He suggests that space is a metaphor for

the otherness of the spirit realm. If we see spirits around us, then this realm is not geographically removed.... Space is a way of expressing difference and separation, but the shaman's journey expresses the possibility of coming together again.[33]

In some sense he is probably right. But people in many societies, from medieval Christian Europe to some present-day religions, take separation of 'the other' by space as literal, not as a metaphor. It is progressive theologians who tell these people that it would be more prudent – indeed, modern and sophisticated – to think of traditional beliefs as metaphors.

The problem of *how* the spirit realms can be spatially removed and, at the same time, immanent in the material world is resolved by a further understanding of human neurology. Flight and movement through a vortex, to be sure, suggest space, but the sensations of flight and the tunnel are experienced within the brain, that is, inside one's head. Then, too, hallucinations, as we have seen, are sometimes projected onto a person's immediate environment – outside of one's head, they become part of the material world. Hallucinated spirit beings and animals are therefore also part of this world.

That at least some people who experience visions are aware of this duality of transcendence and immanence is evident in an explanation that a Huichol shaman gave to the medical anthropologist Joan Halifax:

There is a doorway within our minds that usually remains hidden and secret until the time of death. The Huichol word for it is nieríka. Nieríka is a cosmic portway or interface between so-called ordinary and nonordinary realities. It is a passageway and at the same time a barrier between worlds.[34]

The Huichol notion of doorways within our minds recalls Aldous Huxley's *The Doors of Perception*, and Huston Smith's *Cleansing the Doors of Perception*.[35] These writers take the idea from William Blake: 'If the doors of perception were cleansed, everything would appear to man as it is, infinite.' A comparable take on perception comes from Coleridge: 'The primary imagination I hold to be the living Power and prime Agent of all human Perception, and as a repetition in the finite mind of the eternal act of creation in the infinite I am.'[36]

Writers of this persuasion take experiences at the far end of the consciousness spectrum as authentic, not hallucinatory. Some argue that hallucinogens, whether botanical or chemical, should rather be termed 'entheogens' because they bring God and profound insights to those who ingest them. In inviting agreement, they use phrases like 'open mindedness': who does not wish to be thought 'open minded'? Beware of rhetoric. The question is: does the far end of the consciousness spectrum vouchsafe genuine insights or merely trivia on which the ingestion of psychotropic substances and any of the many other triggering factors bestow a kind of sacred reality? When these seekers after Truth wake up do they have something that we sceptics do not have? William James wrote of these 'insights':

Depth beyond depth of truth seems revealed to the inhaler [of nitrous oxide]. This truth fades out, however, or escapes, at the moment of coming to; and if any words remain over in which to clothe itself, they prove to be the veriest nonsense. Nevertheless the sense of a profound meaning having been there persists.[37]

Though James thought it wise to be 'open minded', but in a very restricted way, I think he sums up the matter well: 'veriest nonsense' indeed. Despite all the protestations of 'entheogen' users, I can detect nothing of interest in what they claim to have learned. Their insights are trivial. They have brought nothing to the world that has been of any use. Interest in passing through doors in our minds seems to me to be self-indulgent, if not addictive in the worst sense of the word. Indeed, there is no evidence that there is any sort of 'other realm' filled with wisdom, be it tucked away in our brains or 'out there' in another dimension. The interplay between transcendence and immanence is indeed complex.

Doctrine

The beliefs with which I have so far dealt are suggested, though not ineluctably, by the functioning of the human nervous system. That is why they are so widespread. But it is only the basic notions of a tiered cosmos and immanence that are widespread: their details are filled in, or elaborated, in specific cultures. Mere belief in a supernatural realm is insufficient to secure human unanimity on what that realm is like and the demands it makes on people. It is out of this open-ended situation that belief grows into doctrine. People try to reach agreement on not only the fundamentals but also the details of their beliefs. Because disputes about religion cannot be settled empirically, theology will continue to flourish.

Inevitably, as human society becomes more hierarchical, doctrine becomes divisive. In largely egalitarian communities, like those of the San, there is no argument about, say, what //Gauwa, God, is like. Some say he is a little fellow covered with yellow hairs. That is how they saw him in a vision. Others respond that they believe he is like a Land Rover. No one minds. They simply say that these matters are of another world and different people see things differently. Unlike Christians and Muslims they do not kill for their beliefs. When we come to literate, hierarchical societies, we encounter a more fraught situation. Those in power try to formulate and impose doctrine on those who have less power. Doctrine is now equivalent to revealed knowledge in the eyes of those who formulate it, and it becomes an instrument of power. Some doctrines, such as Augustine's 'no salvation outside the Church', have immense economic implications. Eventually certain points of belief achieve such social significance that they become worth killing and torturing for. God would have it so.

The birth of the Anglican Church is a case in point. Henry VIII wanted to separate himself from the Roman Catholic Church so that he could divorce an unsatisfactory wife and marry one who would bear him a male heir. This was no sentimental desire: political power and inheritance were at stake, and the King wanted to divert the flow of money going to Rome into his own coffers so that war with Catholic France would become a viable option. To characterize his new Church of England, the king selected some doctrines and rejected others. By and large he tried to negotiate a middle road between the highly sacramental Church of Rome and what he saw as the excesses of Luther and other Continental reformers. As a result, people on both sides, 'extreme' reformers and resolute Catholics, were tortured and executed by being disembowelled and burned alive. Books were burned and possession of the 'wrong' books could lead to indescribable physical suffering. While all this was going on, Archbishop Thomas Cranmer and other divines approved of such horrific acts. Reformation theological issues aside, they saw power for themselves. Cranmer later became a revered Protestant martyr, his cruelty forgotten. In this economic and political milieu, the doctrines that people chose to believe became matters of life and terrible death. The comfortable, easy-going modern Church of England was born in horror, not piety.

Once a community achieves writing and doctrine comes to be written down, leaders have an immensely powerful tool. Written scriptures become a canon (established by the most powerful social groups) to which people appeal for support for their own views.[38] Far from consolidating religious communities, as is generally believed, sacred scriptures become the foundation for struggles far more bitter than any between communities that do not have scriptures. The written codification of verbally shared beliefs is an outcome of, and a contributor to, dissent and power struggles within a society.

Knowledge is something that other people don't have. 'Common knowledge', a dismissive phrase if ever there was one, is of little use to anyone except in the practical day-to-day matters in which everyone more or less automatically uses it. By contrast, there is, within all religious traditions, a ladder of knowledge that adherents can climb. Within these traditions, 'common knowledge' (e.g., God exists) earns little if any respect within the community. But to be able to explain (if not comprehend), for example, the Trinity is an intellectual status that commands respect. All the religions that are founded on sacred scriptures admire 'learning' of this kind – and it bestows immense prestige and influence. Catholic Jesuits are intimidating in their manipulation of doctrine and their casuistry. Protestant pastors earn respect as they leave their university departments of theology and take up a post as the spiritual (and often economic) leader of a local church. Famous Jewish rabbis of the Kabbalah or legalistic traditions are revered. Muslim clerics run whole nations.

Written beliefs are there to facilitate the identification and exclusion of heretics and to provide a platform for power struggles. This was clearly brought out in Chapters 1 and 2.

Complaisant doctrine

The first two chapters of this book also showed that doctrine does change, despite the presence of supposedly anchoring scriptures. Christianity in the modern age is (fortunately) rather different from Christianity in the Middle Ages. Fourteenth-century wall paintings depicting the agonies and demons of Hell are not part of modern churches, except as antique curiosities. What was canonized as the definitive Word of God in one epoch becomes an embarrassment in another. At the same time, religions have to maintain the impression that they are *not* changing. God (however defined) remains the same yesterday, today and forever – world without end. 'I am Alpha and Omega, the beginning and the ending, saith the Lord, which is, and which was, and which is to come, the Almighty' (REVELATION 1:8). Christianity and other major religions are faced with what seems to be an insoluble 'mystery' – unless of course they get around to acknowledging that there is no such thing as an eternal, unchanging divine being. But that way out seems too radical for most believers.

To illustrate some of the difficulties that changing religious belief encounters, let us return to the biblical narrative of Lot and his family. After all, Jesus himself said, 'Remember Lot's wife' (LUKE 17:32).

Today, many people of all three patriarchal religions – Judaism, Christianity and Islam – take the extended narratives of which the tale of Lot is a small part to be *generally* true and that the patriarchs were real, historical figures. Some archaeologists, such as William Albright,[39] believed that the stories of Genesis and Holy Land archaeology could be convincingly woven together and that they suggest a date for the patriarchs between 2000 and 1500 BC, though the texts were probably written as recently as the seventh century BC. Today it is widely agreed that those early hopes were too sanguine: there is no persuasive and sufficiently precise archaeological evidence for the historicity of the patriarchal narratives. As Israel Finkelstein and Neil Silberman put it, 'Perhaps we should see [Genesis as] an attempt to present the patriarchal traditions as a sort of pious history.'[40]

The fundamentals of Lot's story appear to have been widely told in ancient times. Like myths everywhere, there are other versions of the tale. Indeed, the existence of different versions sometimes points to the mythical status of an ancient narrative, whatever its historical content may be. In considering the Lot narrative, one thinks, for instance, of Orpheus looking back to see if poor Eurydice was still following him out of Hades. His 'sin' of looking back meant that he lost his beloved. Most unfairly, she suffered because of his disobedience. Like Lot's daughters, her view of things did not matter. In the Book of Judges there is a close but (understandably, given its content) little-known parallel. An old man of the town Gibeah took in a mysterious 'wayfaring man', who may be akin to Wagner's Wanderer, Wotan, chief of the gods in disguise. In Judges, men of the town besieged the Gibeah house, demanding that the visitor be given to them so 'that we may know him' (that phrase again). Instead, the old man offered them his concubine: 'And they knew her, and abused her all the night until the morning: and when the day began to spring, they let her go' (JUDGES 19:15–28). The old man found her in the morning on his doorstep, violated and broken: 'The woman his concubine was fallen down at the door of the house and her hands were upon the threshold' (JUDGES 19:27). This is a touching image. But the old man brusquely ordered her to get up. Not surprisingly, she was unable to speak. Instead of caring for the shocked and tormented woman, 'he took a knife, and laid hold on his concubine, and divided her, together with her bones, into twelve pieces, and sent her into all the coasts of Israel' (JUDGES 19:29). This version of the Lot narrative certainly has a more grisly ending than the one in Genesis. What could it possibly mean? Why is it preserved in divinely inspired, inerrant scriptures?

Today, people tend to avoid such disturbing questions. Instead they adopt a more practical – one may say, diversionary – viewpoint and wonder if there really were two towns called Sodom and Gomorrah – the stuff of TV documentaries. Other biblical sites have been found – one need mention only Jerusalem, Jericho and Galilee – so why not Sodom and Gomorrah? Searchers have concentrated on the southern end of the Dead Sea, the 'Salt Sea' of the Bible, where there are the ruins of a number of ancient towns. Writers have argued that some of these towns were destroyed by an earthquake and submerged in the Dead Sea. Eruptions along the faultlines that created the below-sea-level valley and also volcanoes farther to the north could, they say, account for the rain of fire described in the Bible. Moreover, the resulting saltiness of the neighbouring landscape may have led to the naïve notion that some natural features were solidified people – all in all, a convenient and comprehensive natural explanation.[41] Unfortunately for those who accept this naturalistic reading of the biblical tale, geologists can find no evidence for volcanic activity in the region during the last 4,000 years, though some movement along faultlines may have occurred. Faced with this limitation, others have suggested that the fire and brimstone came from a meteor.

It is curious how even ardent believers look for natural explanations of biblical miracles and ignore the moral implications of the events. Why? Perhaps natural (though, they would add, divinely occasioned) explanations provide a halfway house between belief in untrue narratives and scepticism that would banish 'moderate believers' from their conservative religious communities.

Even if archaeologists do one day manage to find sites that are persuasively identifiable as Sodom and Gomorrah, the truth of the towns' destruction as a God-given punishment for sin will not be confirmed. Some writers naively imply that if they can establish the general historical accuracy of biblical narratives (there really was a man named Moses, the Israelites really did wander in the Sinai desert, there really was a great flood, there really was a prophet named Jesus) they will have, in some measure, vindicated the general spiritual truths of the Bible. Not so. These are two separate issues. The possible historical status of places, people and events in the Bible does not tell us anything about the spiritual implications of the narratives or of the book as a whole. If it can be shown that there really were two towns called Sodom and Gomorrah and that they were destroyed by fire, this archaeological success will have no bearing on the spiritual, or moral, meaning of the story of Lot and his wife. The believers' frequent cry 'Archaeology once again proves the truth of the Bible!' is overly optimistic and comprehensive. Biblical archaeology often misses the point.

Instead of rejecting outright the Sodom and Gomorrah story as the ancient mythical and morally repugnant nonsense it is, less extreme forms of Judaism, Christianity and Islam struggle to find a moral in it. Simply because the tale is in the Bible (compiled over a millennium and a half ago, let us remember), believers feel that they must make some sense of it. This is especially so today, when non-judgmental evaluations of homosexuality are being accepted by the Western public and the Anglican Church faces a major schism as a result of different views on the issue. The story of Lot and other explicit condemnations of homosexuality[42] are now an embarrassment, so the story must be bowdlerized in some way.

Some believers get around the problem by arguing that the sins of the destroyed towns were multiple and not merely sodomy. The value of this vague position escapes me. Other believers are more focused. They claim that the sin of inhospitality, not homosexuality, was the overriding cause of the problem. This view seems equally insupportable, at least for Christians who believe in the divine inspiration of the whole Bible. Jude, in his New Testament epistle, relishes this tale of sexual depravity and its spectacular punishment. He wrote, 'Even as Sodom and Gomorrah, and the cities about them in like manner, giving themselves over to fornication, and going after strange flesh, are set forth for an example, suffering the vengeance of eternal fire' (JUDE 7; see also 2 PETER 2:6–10). Sexual sins and their savage punishment have long fascinated Christianity. Paradoxically, sex has simultaneously supplied Christians with a rich source of metaphor for religious experience and representation. For instance, Christ is said to have been born of a virgin (the sexual act passes on inherited sin), and the Church is said to be the bride of Christ. There was no sex in the Garden of Eden, despite God having created naked Adam and Eve with all the necessary anatomical equipment fully functional. It was only after God had driven them from Eden that 'Adam knew Eve his wife' (GENESIS 4:1). Then, too, sexual and blood imagery blend: 'Come hither, I will shew thee the bride, the [sacrificial] Lamb's wife' (REVELATION 21:9; brackets added). The knifing of a sacrificial animal parallels the deflowering of a bride. For most churchgoers this thought is best not explored.

Yet another escape route from the trap of sacred scriptures is even more abstruse. The story of Lot's daughters, like Abraham's decision to gather firewood and take his (silent?) son up the mountain to his death, is so beyond any acceptable morality that some believers have postulated the suspension of normal ethics in the interests of a 'higher good'. We should not, so we are told, judge Old Testament passages by the standards of today's morality. Does it follow that we should judge today by today's moral standards and not by some revealed ethics?

Another way around such problems is to treat 'difficult' tales as allegories. During the early centuries of Christianity and in medieval times this was a favourite, if arcane, way of interpreting the Bible – when the Church Fathers were not attacking one or other perceived heresy, which was most of the time. For instance, Irenaeus (c. 130–200), an influential bishop of Lyons, suggested that 'Lot stands for the Father; his seed for the Holy Spirit; his two daughters for the "two congregations", the Jewish and the Christian Church; and then his wife for the Church".[43] Nowadays few would go along with this sort of reading.

Today the continuing dilemma of what is to be taken literally and what is metaphor is sometimes confronted in a more esoteric way. Some 'advanced' theologians deny any dichotomy between reality and metaphor. The Bible, they tell us, should be read without asking questions about what actually happened and what should be seen as metaphor. For them, the question of whether Old Testament Jonah was *really* swallowed by a 'great fish' (JONAH 1:17) is meaningless, and we simply should not ask it. (The fish is identified in the New Testament as a whale: MATTHEW 12:40.)

These theologians are more concerned with the spirit than the letter of the Bible. Too much in the Good Book is simply unbelievable in any historical way for modern educated minds. But distilling the spirit from the embarrassingly bizarre and sometimes barbaric biblical narratives, not to mention vindictive divine injunctions, is no easy task. A similar situation exists within Church tradition. Theologians work hard to distance themselves from old ideas of, for instance, a long-bearded Old Man in the Sky, a cloven-hoofed, snarling Devil, and the burning of heretics. Yet these concepts, as medieval wall-paintings graphically show, served the faithful well enough for centuries.

This process of etherealizing solid concepts continued fitfully through the years to the twentieth-century theological movement of 'demythologizing' the Bible and beyond. For example, instead of believing in a literal Virgin Birth, some Christian theologians today offer the notion that that miraculous event should be seen as a metaphor, merely a way of speaking about the uniqueness of Jesus. That the Church got it wrong for 2,000 years does not seem to trouble them.

This approach does not, of course, really demystify or demythologize the Bible. Rather, it re-mythologizes it in terms of metaphor rather than narrative. Where does it lead? Are we, we begin to ask, to 'demythologize' all seemingly miraculous and supernatural events? Gradually, some theologians have pulled their demarcating line back to what they consider the absolute fundamental Christian beliefs, the last bastions of dogma. The Resurrection of Jesus is often taken to be one of these, though some of the more 'advanced' Christian thinkers insert the word 'bodily' to distinguish the old, literalist narrative (the one Christians believed for two millennia) from a more metaphorical, more spiritual, Resurrection that will be palatable to modern sceptics. Even fundamental dogmas can be circumvented. No wonder that many who stand outside the Church find all this ducking and diving unconvincing and self-discrediting.

But the problem today is not whether certain events really took place or not – whether Sodom and Gomorrah were really destroyed by fire and brimstone. We should come back to what I see as the bedrock of religion. We need to ask a more fundamental question: can we continue to have faith in supernatural realms and beings? The role of metaphor in belief systems should be considered only *after* this basic question has been settled. Of course, theologians will argue that the notion of distinct material and supernatural realms is itself misleading, something made up by excessively scientifically minded modern people. Twentieth-century theologians such as Rudolf Bultmann and Paul Tillich saw that supernaturalism is a major stumbling block to modern belief. To get around the problem they argued that modern, educated people should not think in terms of a supernatural realm 'out there', but rather in terms of 'depth'. God is 'the ground of all our being'. The supposed reality of a supernatural sphere independent of the material world is traded in for the metaphor of depth (Chapter 10).[44] A comparable phrase comes from a philosophical point of view: 'Ground of Information, or an Ambience of Information, otherwise known as God.'[45] Should we pray, 'O Ground of all our Being, please intervene and save my child'?

It is hard to see how essential Christian notions of prayer and divine intervention (surely among the absolutely fundamental beliefs) can survive so radical a reformulation of the God concept. Augustine did not take this route. In all the disputes in which he found himself, he could not and did not abandon his belief in inviolate, supernatural revelations that could not be questioned by any human being. They could be explained and explicated by human beings (as he himself was doing) but not challenged. If he, or the Church, let go its foundation in specially revealed supernatural knowledge, all would eventually be lost.

But if God is 'the ground of all our being', what are we to say about evil? The problem of how an omnipotent God could allow evil to flourish in the world has exercised the minds of believers from the earliest days of the Church.[46] Taking up an Old Testament belief, Jesus thought of the Devil as a supernatural person, a fallen angel: 'I beheld Satan as lightning fall from heaven' (LUKE 10:18) – the idea that Milton vividly developed in *Paradise Lost.* Then Satan tempted Jesus by taking him up 'an exceeding high mountain' and showing him 'all the kingdoms of the world, and the glory of them' (MATTHEW 4:8). The theology-obsessed early Church had to weave these and the many other biblical references to Satan and the Devil into a coherent doctrine that squared evil with the existence of a benign God.[47] According to Augustine, it was all fairly simple: Satan was indeed a disobedient and fallen angel, the model by which believers could understand their own lives. For Satan and for believers, the problem was pride, Aristotle's hubris.

Theology, standardized belief, has a life of its own, and the Church Fathers developed a complex demonology. Augustine, for instance, accepted the existence of lesser demons who were able to have sexual intercourse with human beings.[48] This idea of the Devil as Incubus had dire consequences for women suspected of witchcraft; given the contemporary attitude to dreams, they allowed their own natural sexual dreams to convince themselves of their guilt. Aquinas took the matter further and developed the idea of a Succubus that could bring to life the body of a dead woman, have sex with a man and then, transforming himself into an Incubus, have sex with a woman and thus impregnate her by depositing the man's semen into her.[49]

When we read this sort of horrific, indeed wicked, nonsense, we wonder why the Church continues to admire people like Augustine and Aquinas. They were not merely 'of their time'. Their obsessed, twisted minds verged on madness. We also wonder how the Church of today deals with the problem of evil. Significantly, Satan does not appear in the creeds. Nowadays Satan is, generally speaking, yet another embarrassment to a Church that is struggling to slough off its mythologically laden past, though Satan is of course alive and well among fundamentalists. On the other hand, he is usually banished from more liberal sermons, though his fall from heaven, his temptation of Eve and his presence as an explanation for the Church's many failures is useful background material.[50]

But, the faithful may say, Jesus himself believed in the Devil. An answer to them: when Christ came to earth he emptied himself of omniscience so that he could be fully human; he therefore believed the superstitions of his time. Doctrine is indeed complaisant.

Religion without doctrine

Disaffected by theological disputes, many people in the West today are seeking religious experience – the bliss that Christianity and Judaism claim to provide but all too often fail to do so. These folk have lost confidence in all the ins and outs and disputes of doctrine. Increasingly, they hope to find happiness in a less theologically structured Eastern mysticism. As far as they are concerned, the vaguer the better. They rightly recognize that theological precision leads only to dissent and unpleasantness. Buddhism seems to be the answer.

The Buddha (Sanskrit for 'the Enlightened') lived in sixth-century BC India. The religion named after him now takes many forms. As with numerous other religious founders and leaders, he is said to have abandoned a life of luxury for more austere circumstances (Moses, Augustine and, in a different sense, Jesus come to mind). Always, he claimed to be only a poor, humble man. Indeed, peace of mind, spiritual insights and Nirvana (the final deliverance of the soul) came to him through asceticism. Buddhism does not have a creed or codified beliefs. Rather, it emphasizes ethics: right speech, right action and right livelihood. The practical difficulty lies in what constitutes 'rightness'. Fundamentally, Buddhist ethics is thus little different from

Judaism's or Christianity's, though it has a much greater emphasis on emptying the mind. It disengages the believer from the material world. In doing so, it reduces the immediacy of real-world moral dilemmas.

A modern seeker after a Buddhist enlightenment is Stephen Batchelor. Having been educated in a number of Buddhist monasteries, he advocates an agnostic Buddhism, what he calls a 'culture of awakening'. He says that the world does not need another Buddhist Church, another sect, but rather 'creative imagination and social engagement'.[51] But readers of his work will notice that, when he wrote *Buddhism without Beliefs*, he was himself Director of Studies at Sharpham College for Buddhist Studies and Contemporary Enquiry in Devon, England. Despite his protestations, all this sounds like another school of Buddhism, just what he says he wants to avoid – and it affords him a position of influence. At least he seems to back off from reincarnation and prayer wheels, seemingly matters integral to Eastern Buddhism. Emphasis on such beliefs in present-day secular England would be suicidal.

Buddhism, like all religions, has many branches and its own disputes. Identifying 'true' Buddhism is a difficult matter. The escapist claim that *real* Buddhism has no beliefs is, of course, a belief in itself, one that sets the stage for a set of further beliefs. Of these, Buddhism's emphasis on compassion, serenity and joy is attractive, though not unique. It is humanism by another name. Every book on Buddhism seems to claim that most other writers have misunderstood the fundamentals, but now this writer will tell it like it is. Like modern Christianity, Buddhism never turns out to be what you think it is.[52] One reason for this parallel is that in a tradition like Buddhism much respect and wealth are given to teachers who are considered to speak the 'Truth'. They become gurus. Despite its other-worldliness, Buddhism involves serious economics and politics, as the Dalai Lama daily illustrates but his followers fail to notice or choose to ignore. Buddhism has become an industry. (The same could be said of the Christian Church through the centuries.[53])

All in all, Eastern religion without embarrassing doctrines is no more than consciousness fiddling. Devotees seek bliss through shifting their consciousness to the introverted end of the spectrum and enjoying the results. This is 'awakening'. As it stands, it seems innocuous enough. For the individual, it provides a harmless and effective way of coping with modern-day stress. There seems to be little doubt that meditation can be beneficial.[54] But it would be well for devotees to separate meditation from belief in supernatural intervention and, moreover, enquire who is making money out of their altered states of consciousness.

Degrees of belief

To end this chapter I point out that it is useful to distinguish between *what* people say they believe and *how* they believe it. What religious people believe is comparatively easy to ascertain: simply ask them and, for the most part, they will tell you. Then difficulties arise. In what sense do they 'believe' a particular doctrine? How do they square contradictory beliefs? Here the notion of a spectrum is again useful: acknowledge, accept, believe metaphorically, *really* believe.

Many people, including clergy, are content to remain within the Church without being *real* believers. They *acknowledge* the teachings of the Church and experience no difficulty in reciting the Creed. Some of the ancient Greek philosophers took this position: they took religion into their vocabulary without really believing that the Gods were doing what the myths related. Other people *accept* that, say, the doctrine of the Trinity is true without bothering to try to understand it. More intellectually minded Christians take the *metaphorical* or philosophical route; they are the ones who talk about 'the ground of all our being' or 'a supreme objective reality'. Then there are those who *truly believe* that God has revealed to them (through their understanding of the Bible and the teachings of their particular branch of the Church) the Truth about the cosmos, sin, salvation, Heaven and Hell. These are the ones who place themselves in indefensible positions of belief in a scientific world and are, as a result, dangerous.

CHAPTER 7

Religious Practice

When people experience what they believe to be a supernatural realm and formulate beliefs about it, they inevitably feel that they must do something in response to it. This is what I call religious practice. On both sides of the divide between what the anthropologist Harvey Whitehouse designates imagistic and doctrinal religions (those without and those with sacred scriptures) we find people performing acts that are related, closely or more loosely, to their religious beliefs.[1]

This is a highly complex field and sorting out different kinds of religious practice is no easy matter. Some of these practices, private or public, are formulaic and repeated: we call them rituals, though a watertight definition of that word is hard to come by. Of course, not all rituals are religious: some, like shaking hands, are purely secular, though they may be roped in for some religious occasions. Other practices are less obviously ritualistic and we may miss the fact that they are religious. A requiem Mass in St Peter's, Rome, is not representative of all rituals since it is highly formalized; in many societies outsiders hardly notice some rituals as the functionaries are not easily distinguished by accoutrements, and the place where the ritual is performed is not prepared in any way. It is a matter of degree.

Some religious practices are expressly designed to create experiences of the kind that I described in Chapter 5. In these rituals, religion loops back on itself; it becomes self-perpetuating. Other practices seem designed to support the whole religious structure together with its social, economic and political implications; these may or may not have an experiential component.

Inducing religious experiences

I have made the point that religious belief as embodied in an oft-repeated creed can impact upon religious experience. To a greater degree, some religious practices expressly generate religious experience. Rituals, such as the Mass, are more than symbolic representations of belief or enactments of myths. Participation in them creates experiences that reinforce beliefs.

More than that, they show that beliefs embodied in the Apostles' Creed
and other creeds are true because participants can experience them in themselves: their own bodies and emotions react in a validating way. For the true
believer, consumption of the body and blood of Christ can have an exalting
effect. Cannibalism is for most people a revolting concept, and this aspect
of the Mass brings devout participants to a knife-edge of contradiction that
leaves them emotionally churned up. But people's reactions to rituals vary
greatly. Most participants do not, I suspect, dwell much on what lies in the
words of the Mass: for them, there is only comfort and a sense of safety
derived from repetition of familiar words and actions in a specific, clearly
defined social setting.

This sort of thing can be encountered to a greater or lesser extent in any
religious service (here I include Christianity, Judaism, Islam and other
religions). Liturgies, music, prominently placed symbols, and architecture
combine to induce soaring emotions and a type of inward consciousness
that is essentially pacific. The historian Norman Cantor has described how
potent combined sounds, spaces and images must have been for medieval
Christian congregations:

All arts were employed, all senses were stirred, to render a momentary insight into the
indescribable glory of the heavenly life. The stained glass refracted the Divine Light and
bathed the altar in a myriad of miraculous colours.... As the choir intoned the complex
harmonies of its hymns and chants, as the bishop or his adjutant stood before the altar
in his golden vestments, as the Saviour, Virgin, and saints blazed forth from the glass
mosaics in the clerestory and were made to stand out below in sculptured roundness
from the surrounding gloom by the falling light, it was easy to imagine the angelic host
as the supporters of this divine temple.[2]

'[I]t was easy to imagine': we recall Durkheim's imagining of gods, now
intentionally contrived. The complexly induced, uplifting experience that
Cantor describes is of the 'Absolute Unitary Being' (AUB) type that I discussed
in Chapter 5: participants feel absorbed into the universe or the godhead.
In some other traditions, the same basic factors, though with more highly
rhythmic and louder musical accompaniment, induce ecstatic reactions that
may include catalepsy and glossalalia (speaking in tongues).

The power of images and chanting to induce religious experience can be
seen in the famous saint St Teresa of Avila's (1512–1582) vision of the Virgin.
She reported that she was in church and that the chanting of the *Salve Regina*
had just begun. Then she saw 'the Mother of God surrounded by a host of

angels descending from the heavens and placing herself at the Prioress's stall, just where the picture of Our Lady is situated'. Then the painting was superseded by the vision. Teresa at once 'fell into a state of great ecstasy'.[3] The picture of a vision combined with a highly charged context to induce a vision. 'Imagining' was taken a step further. These trance experiences are much further along the consciousness spectrum than the peace of mild meditation that most mainstream, church-going folk seek.

We see ways of inducing religious experience among the Maya that entail more intense assaults on the body than mere music, chanting and clapping. Maya researchers have described frenzied dances that, in a way comparable to Teresa's visions, induced 'mystical transformation of human beings into supernaturals by means of visionary trance'.[4] Indeed, the ultimate aim of Maya ritual was for the rulers and other participants to see visions, often of one of the great Vision Serpents that they believed inhabited the upper and lower levels of their cosmos. Direct, visionary contact with the gods was a result of the pain and shock of preliminary fasting and excessive bleeding. Bloodletting was a religious ritual that altered consciousness and thus

26, 27 RIGHT *A Maya panel at Yaxchilan shows Lady Xoc pulling a thorn-trimmed cord through her tongue, while the king, Shield Jaguar, holds aloft a flaming torch.* OPPOSITE *Lady Xoc looks up at her visionary double-headed serpent. In her right hand and on the floor are shallow bowls to receive her blood. A warrior emerges from the mouth of the writhing Vision Serpent.*

enabled participants to 'see' the spirit realm and to acquire a special type of knowledge unavailable to ordinary people. They saw what ordinary people could not see. They were, literally, 'visionaries' and 'seers'.

Many Maya carvings show such rituals. In some instances, the intricate Maya calendar has enabled archaeologists to establish the date of the gruesome events depicted. One of these at Yaxchilan (known as Lintel 24) is dated to *c.* AD 725.[5] It shows the ruler, Shield Jaguar, standing before his wife, Lady Xoc. He holds a torch, which suggests that the ritual is being performed in the dark, probably in a temple on top of a mountain-pyramid. The shrunken head of an earlier sacrificial victim is attached to his head. His kneeling wife draws a thorn-fringed cord through her lacerated tongue (Fig. 26). An adjacent carving shows Lady Xoc's ritually induced hallucination: it is a huge, double-headed rearing serpent that rises from a plate bearing bloodied paper and lancets (probably stingray spines) (Fig. 27). A warrior emerges from the serpent's mouth, as does a mask that was associated with war, sacrifice and bloodletting.

Bloodletting with its attendant pain was probably not the only ritual practice that the Maya used to induce visions. The ingestion of psychotropic drugs also led sufferers to experience hallucinations (Chapter 5).[6] In the 1970s Marlene Dobkin de Rios took cognizance of the widespread Mesoamerican use of hallucinogens in religious contexts and argued that the Maya may well have ingested psychotropic mushrooms. She also suggested the use of hallucinogenic frogs and water lilies.[7] Her arguments, based on images of these things in Maya carvings, received a mixed reception. There has, however, recently been more openness to the idea that the Maya ritually induced trance by means of hallucinogens.

Once we get into the realm of interacting religious practice, imagery and belief, the waters become murky indeed. Questions arise that have far-reaching implications for our overall evaluation of religion in its present-day relationship with science. Did the sixteenth-century Spaniards sense any parallels between Maya religion and their own? Did they see the bloody flagellation and crucifixion of a thorn-crowned Jesus reflected in some of the cruel Maya carvings? Did they appreciate the importance of blood, the letting of blood, and, in their daily celebration of Mass, the drinking of blood in their own Christianity? Did they see a parallel between their own murderous conquest of Central America and the bloody genocide that, according to the Bible, the Israelites inflicted on the Canaanites when they entered the Promised Land? Did they realize that the killing of their own martyrs, together with the flow of their blood and pain that is so often exalted in Christian art, pleased their own God and bought salvation? Does not the Judeo-Christian tradition also hold that there is a reciprocal relationship between humans and God, a relation forged by the sacrificial blood of Christ? As St Paul wrote, '[W]e have redemption through his blood' (EPHESIANS 1:7). St John the Divine explains this point by saying that Christ 'washed us from our sins in his own blood' (REVELATION 1:5). And the writer of Hebrews puts it succinctly: '[W]ithout shedding of blood is no remission' (HEBREWS 9:22). The Maya gods too had to be fed and placated with blood. Some of those shocked Spaniards did see parallels; some even tried to find points of contact with the indigenous religions. But the overall impression is that those parallels were deeply disturbing.[8]

So we may ask: is the Christian God any different from the Maya deities? Anyone who doubts the unpleasant realities of ritual and blood in the Christian and Jewish traditions should consult a biblical concordance and count the number of entries listed under the word and its permutations. We reject

Maya rituals as abhorrent. Why do we not do the same with much of the Bible, the New as well as the Old Testaments, where blood and death are supposed to secure salvation?[9] Does the God who created this whole universe with its spiralling galaxies really demand that people must shed blood and kill in order to obtain salvation? To identify his faithful, does the all-powerful, all-knowing God really need to see the blood of sacrificial lambs on their doorposts? Elements of bloody sacrificial rituals are disturbingly widespread; they are not confined to the Maya. They speak to us not about the reality of a spirit realm but rather about the workings of the human brain and body and their manifestations in various social settings. That being so, we come to an inevitable question. Did the same kind of electrochemical events in the human brain produce both Maya hallucinations of Vision Serpents and the sometimes (though not always) beatific visions of Christian saints? I argue that they did.

The use of hallucinogens in religious rituals is still today widespread.[10] The ingestion of certain substances is a sure way of shifting consciousness towards altered states in which visions are assured. The use of hallucinogens in mainstream Christian contexts should therefore come as no surprise, though it is little known. Detailed studies of Christian medieval imagery have revealed what appear to be depictions of hallucinogenic mushrooms.[11] Present-day Christianity is, of course, reluctant to countenance or even to admit to the ingestion of psychotropic drugs.

Prayer is another matter: it is central to Christianity. Mainstream churches of today play down any suggestion of altered states of consciousness being induced by intense prayer and are generally silent on the connection between prayer and meditation. However, inspection of the activities that take place in the less inhibited charismatic churches leaves no doubt that altered states are sought in twentieth-century Christian contexts: huge crowds sway, wave their raised arms, allow their facial expressions to express extreme ecstasy, fall to the ground in a trance and gabble as they 'speak in tongues'. How much of this is real and how much is faked? Simulation of ecstasy is nothing new: it was recognized and condemned in the sixteenth and seventeenth centuries.[12] More honest people in these intense situations genuinely believe that prayer has brought them into contact, or even union, with the divine – an interpretation of the AUB experience.

In medieval times the situation was different: the established Catholic Church openly advocated types of prayer that led to ecstasy. Illustrated man-uals taught devout readers a series of bodily positions that they could adopt and advocated flagellation (shades of the Maya) to help things along. One such book demonstrates 'how to pray so that the soul possesses the body and its parts or so that the soul is possessed by the body to enable it occasionally to reach a state of contemplative ecstasy as though it had left the body'.[13] Again, it is the AUB experience, but here combined with the out-of-body experience.

The art historian Victor Stoichita points out that a novice could 'acquire a physical programme to induce ecstasy'.[14] In his book *Liber de Oratione et Specibus Illius*, late twelfth-century Peter the Chanter commended suitable postures. One was standing with outstretched arms in imitation of Christ on the cross. At this moment, it was 'sometimes possible to witness levitation'.[15] Note the use of 'sometimes': ecstasy is not inevitable; it is something for which to strive. Mystical flight, so often depicted in religious art, was an acceptable, normal part of Christian experience. In 1585 Diego Pérez de Valdivia wrote: 'Some souls lose consciousness...others levitate, while others remain standing, kneeling, or virtually lying down'.[16] In another recommended posture, standing with arms raised above the head, the body 'at last becomes like an arrow shot into the sky...and ecstatic access to the "other world" takes place'.[17] In other ethnic and social contexts, we would not hesitate to use the word 'shamanism'. Christianity is but one way of understanding neurologically induced experiences.

Modern-day readers will wonder if sex played any role in all this ecstasy. Inevitably, it did. St Teresa's vision illustrates this point; it was depicted by Bernini in the Cornaro Chapel, Rome, and by a painter of the school of Murillo (Fig. 28).[18] In 1574, she wrote:

I saw the physical manifestation of an angel standing next to me on my left.... He was not tall, but small and very beautiful; his radiant features revealed that he was one of those spirits from the top of the hierarchy who seemed to be all love and fervour.... In the hands of this angel I could see a long gold spear, tipped with a glowing iron arrowhead. It seemed to me that now and then he plunged this through my heart and into my entrails, appearing to take them with him as he withdrew it, leaving me wholly encompassed in God's love. The pain of this wound was so acute that it drew from me the groans of which I spoke earlier: but so great was the sweetness of this terrible pain, that I was incapable of wanting it to cease, any more than I was capable of finding happiness outside God. It was not a physical pain but a thoroughly spiritual one, though the body was not above feeling it a little, acutely even.[19]

No present-day reader will require any exegesis of this sexually charged passage. The ecclesiastical authorities of the time, however, accepted it and it

28 In Gianlorenzo Bernini's altarpiece from 1647-1652 an ecstatic St Teresa awaits penetration by a spear. It is wielded by a beautiful young male angel.

played a role in St Teresa's canonization. In stark contrast, a Franciscan nun, Juana Asensi, was executed for speaking of a similar vision: 'the whole of Our Lord Jesus Christ's body had confronted her and united with her, face to face, eye to eye, mouth to mouth and other parts of the body also.'[20] As Stoichita remarks of Francisco Ribalta's depiction of St Francis ecstatically embracing Christ on the cross, 'Without a doubt, an indiscreet painting.'

The relationship between prayer (an apparently innocuous religious practice), ecstasy and the nature of that ecstasy is a little-publicized part of the Christian tradition. Today theologians avoid it as much as possible: apart from charismatic congregations, religion has become more cerebral than emotional. Still, seen in the context of worldwide experiences and neurological studies, Christian religious experience is no different from any other. The inference to be drawn from this is obvious: religious experience is not created by divine visitations.

Constructed spaces

One of the implications of religious practice is that it has to happen somewhere. Rituals can be conducted in natural surroundings, but people often want spaces that are more precise and controlled. They therefore construct buildings and delineate spaces that they consider appropriate for their religious practices. These constructions involve not only spaces imbued with special significances. They also implicate people who, in the performance of religious practices, move between those spaces. Inevitably, a religious, space-demarcating structure becomes a template for shifting consciousness, the dramatization of belief, movements of people, and social discrimination.

The second 'window' in the Prolegomena to this book showed that the Maya constructed pyramids so that they would have a meaningful series of spaces, each of which was relevant to a part of their cosmology and to their system of beliefs. These beliefs included practices that led to approved ways of mystically traversing the cosmos and making contact with divine beings. We can go a step further. Visually, the spaces gave meaning to the movement of people through them, and, at the same time, beliefs gave meaning to the spaces. The whole ensemble of constructed symbolic spaces and their traversal by people had an extraordinarily powerful effect comparable to what took place in medieval cathedrals. First, those who immersed themselves in the ritual ensemble found their consciousness altered, some to extreme ecstasy, some to more moderate awe. Secondly, the combination of space, movement and meaning became political: the power structure of Maya society was dramatized – made visual – and thereby reproduced and strengthened. Everyone stood somewhere and thus became physically a part of the political structure. Seemingly inevitably, religious practice is political. *Pace* the American Founding Fathers, can religion and politics ever be prised apart?

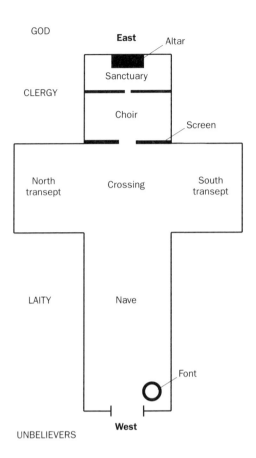

GOD

East — Altar

Sanctuary

CLERGY

Choir

Screen

North transept — Crossing — South transept

LAITY — Nave

Font

West

UNBELIEVERS

29 *A cruciform Christian church is a template in which social divisions are dramatized. The laity wait in the nave, while the degrees of clergy approach the sanctuary and altar.*

A more familiar example of the religious use of space and meaning than the Maya is the plan of a cruciform Christian cathedral or, in lesser kind, a parish church (Fig. 29). The overall layout of the cathedral dramatizes not just the central event of Christian belief – the death on the cross and resurrection of Jesus – but also distinctions between clergy of various ranks, the choir, the laity and, of course, non-believers who remain conspicuously outside the building. Although these major divisions are in many churches the foundation for much more intricate, indeed arcane, symbolism, we shall confine our attention to them.

Concepts immanent in the spaces of a cruciform church are brought to life in rituals. As a procession, clad in a hierarchy of distinguishing vestments, enters at the beginning of a service and moves through the church, it makes social distinctions plain for all to see. Rank is paramount. Lowly choristers and lesser clergy lead as they all walk in reverse order of precedence. The most senior clergyman brings up the rear. He (or she in some churches) passes up the central aisle, through the nave and past the laity standing in their pews, through the empty crossing (a cordon sanitaire that emphasizes the distinction between laity and clergy?) and into the choir. He then takes his assigned seat in the chancel. In many medieval churches there was a carved screen that prevented the congregation in the nave from seeing what was going on in the choir and the chancel, though they could glimpse some of it through the interstices of the screen and hear (if imperfectly) what was being said and sung. Their partial view was comparable to ephemeral dreams, while the clergy within the chancel could see more clearly and participate more directly: they enjoyed the most immediate 'vision', the most intimate contact with revelation. In some medieval churches there was also a screened off seating area for the nobility. In this way, the nobility created a visual parallel between themselves and the clergy. Their separated presence was clearly signalled by the screened area, but the hoi polloi could not see them at their prayers.

Later in the service, the senior clergyman and his attendants pass through a demarcating rail to enter the extreme east end of the church, the sanctuary, where the altar is located and, in Catholic churches, the host (bread conse-crated in the Eucharist) is kept. This is the most sacred space in the church. And the most exalted clergy are now as far away from the laity as it is possible to be.

Because the altar is placed at the east end, the climactic ritual, the enactment of the sacrificial Mass, takes place where the rising sun daily symbolizes (thanks in part to Constantine) the resurrection of Jesus on the first Easter *Sun*day. The cruciform church structure is thus designed to fit in with a natural phenomenon. If Christians think about it, the rising sun daily reminds them of the Resurrection. At the same time, a hierarchy of priests and attendants is also seen to be automatically in accord with the very nature of the cosmos. The building and the ritual drama enacted in it together show, first, that the Christian life demands the intercession of ordained priests and, secondly, that the complex of belief and ritual is directly related to the functioning of the solar system.

Imagine the cruciform church raised vertically like the ancient instrument of torture and execution it represents. The congregation in the nave clusters around the foot of the cross, as did those followers of Jesus who were brave enough to be there. The clergy are in the part of the cross where the body of Jesus would have been. The upper clergy are at the top of the cross where Jesus' head would have been. And all this is positioned so that it accords with the rotation of the earth. Doctrinally, the Resurrection of both Jesus and believers in him ties in with nature and the world as God created it. Creation and salvation are cut from the same cloth.

The whole ritual practice, conducted as it is in a meaningfully fashioned and subdivided cruciform space, thus not only symbolizes Christian beliefs about how sinners may approach the ineffable deity. It also dramatizes and visibly *reproduces* the power structure of the church. Salvation and the hierarchy of the Church are virtually indivisible. Augustine's complex, abstract ideas of 'the Church of God' and 'the City of God' become immanent in a human-made structure.

In today's anti-authoritarian climate, and somewhat embarrassed by its inherited hierarchical structure, the Christian Church in some of its manifestations consciously adopts novel architecture to reflect a more egalitarian view. The circular form of Liverpool Roman Catholic Cathedral in England is an example: the altar is highly visible and surrounded by the clergy and congregation. In some other churches, saddled with an ancient building that may not be changed for reasons of 'heritage', the clergy decide to move the altar from the extreme east end of the church and situate it in the crossing – nearer to the now sparse laity and very visible. The local congregation is using the traditional lay-out of the cruciform church to contest tradition, to make a statement that contradicts the exclusivity of medieval times. But we should not take all this at face value. On closer examination it turns out that these spatial ploys merely disguise an ecclesiastical hierarchy that has no real wish to relinquish its power: equality and egalitarianism have no place in religion, especially not in Christianity.

Charismatic televangelists, who whip up their congregations into an emotional frenzy of altered states, seem to prefer a building closely resembling a theatre, but which nevertheless leaves no doubt in the minds of the congregation who is in charge. The architecture and seating-plan focus on the preacher.

At this point, it is not too fanciful to think of Stonehenge and again to recall the Maya. At Stonehenge, a long avenue led up to where the circle of standing stones and their lintels demarcated a space for the performance of sacred rites, while ordinary people remained without. The whole structure was oriented to the solar cycle, which annually confirmed the status and revealed knowledge of the 'priests' (Chapter 4). Similarly, the Maya built pyramids that not only partly symbolized the cosmos but also associated the king and priests with its higher reaches, far removed from the ordinary people who crowded the plaza below. At Stonehenge and at Maya temples natural phenomena were not explained. Rather they were harnessed in a complex nexus of beliefs, practices, symbols and social divisions. The same is true of a cruciform Christian church.

Religions that devotees believe to be uniquely provided by God turn out to be in so many ways mere variations on widely encountered themes. The claims that God revealed himself to only one Chosen People and that one particular religion possesses a divinely and uniquely revealed path to salvation begin to look suspiciously chauvinistic – or, to put it another way, plainly wrong: a delusion.

The act of making

People objectify (or reify) religious beliefs not only in natural phenomena, such as the rising sun, but also in things that they themselves manufacture – statues, icons, altars and buildings. In doing so, they externalize what would otherwise be merely ideas existing in human heads. The *notion* of God is all very well, but how much more impressive is an image of God, or a manifestation of God, or a house in which God dwells? The immanence of ideas in objects is, as we have seen, a central religious idea. Once turned into objects, ideas attain a new kind of reality. They start to have influence (or as some theorists put it, agency) of their own.

In this regard, the Blombos ochre raises many points of interest. I'll mention only one. Today we automatically assume that images, geometric (for instance, the Star of David) or representational (say, pictures of the Crucifixion), are made primarily to be seen and to be contemplated. If we break out of our Western tradition, we shall be able to allow for another way of considering religious objects, whether large or small. Some may have been made because the *act of making them* was more important than the result. So we may ask the question: was the Blombos motif made as an *act of communication* with the power inside (immanent in) the piece of ochre rather than as an image, a 'picture', to be contemplated by human beings?

In *The Mind in the Cave* I argued that the handprints of Upper Palaeolithic cave art were the result of complex ritual procedures that linked people to a supernatural realm and were not primarily 'pictures' of hands – or, for that matter, prehistoric equivalents of 'I was here!' The ritual placing of a person's paint-smeared hand on the rock surface (a membrane between this world and the deep spirit realm) to absorb supernatural power or to enter the spirit world was what really mattered.[22] Further, some Upper Palaeolithic images of animals may have been made to be seen not by human beings, as we unthinkingly assume, but by spirit beings inhabiting a lower level of a tiered cosmos. Many images deep in the caves were apparently made and never looked at again. In this way, spells and instructions on ancient Egyptian sarcophagi were intended to advise the dead person, not to be admired by the living. In many medieval churches beautiful stained glass windows were placed so high as to render their images invisible to people below. They were placed there 'for the glory of God' rather than for human admiration. The making and placing of the windows in the positions they occupy was a religious act.

Building the City of God on earth

The building of large religious monuments is of great antiquity. Palaeolithic people took the caves of France and Spain and adapted them by embellishing them with images and by performing rituals in them. By Neolithic times, people had moved the focus of their rituals above ground, away from the subterranean passages and chambers; now they built custom-designed structures for the performance of religious practices and the induction of religious experiences. One step away from nature, Neolithic people had greater control of religious experience, belief and practice, together with their social implications. At the very beginning of the Neolithic period, even before they started planting crops and herding animals, this trend can be discerned. I have already referred to Göbekli Tepe. At this site in what is now south-eastern Turkey people came together seasonally to construct circular 'crypts' with huge, beautifully carved pillars.[23] Here, as long ago as 9,600 BC, we can see religious architectural practices that were to continue up to the present. People designed and demarcated spaces that would dramatize their beliefs and social structure. The making of these structures was in itself important.

Today we tend to think of the construction of a building as a necessary and purely technical phase through which we must go to reach our aim: a 'finished' building. The completed building is seen as an end-product. It was not always so. The Maya, we have seen, built pyramid-mountains and then added to them, encasing the old pyramid within new ones. In this way a pyramid accumulated power and importance. More than that, the protracted act of building was a religious, economic and political process that engaged many people and kept society ticking. The act of building was as important as, if not more so than, the finished product.

Certainly, that appears to have been the case at Göbekli Tepe many millennia before the Maya. The crypts with their striking carvings of animals and peripheral benches seem to have been made to be used: some sort of ritual took place within them. But then the people filled them in with rubble and started to build another one close by. The striking carvings of animals and birds that make the pillars so impressive were buried; people could no longer see them. When they began a new cycle of building, they cut cupules in the tops of the now-buried but still just visible pillars, presumably to receive offerings or libations of some kind that would filter down through the pillars to the realm beneath. They thus changed the use of the old structure: old 'crypts' and pillars were used for different but no doubt related purposes when new ones were built. The continual construction and re-construction of impressive monuments was an on-going religious practice.

Let us now fast-forward to medieval Europe. In the twelfth century, the theological school at Chartres was growing in fame and influence as it became a conduit for the teachings of Aristotle and other Greek philosophers.[24] As we saw in Chapter 2, a stream of knowledge flowed into Europe via the Muslim world. At the end of the twelfth century, when the old Carolingian (though much altered) cathedral burnt down, Bishop Regnault de Mouçon, perhaps feeling a bit challenged by the learned clerics in the adjacent school, decided that a new, massive, modern cathedral was called for. The Catholic Church had to be *seen* to overshadow any secular institution and, of course, any rival bishopric. The existing structure housed two famous relics: the tunic that Mary was said to have worn when she gave birth to Jesus (some say she wore it at the Annunciation) and, grotesquely, the head of Mary's mother, Anne. These dubious items were believed to possess a supernatural power comparable to the southern African San's *n/om* (Chapter 5). Years before, in 911, when a Viking named Rollon besieged Chartres, the bishop ordered Mary's garment to be displayed on the city ramparts. Rollon fled, was converted and then made Duke of Normandy. A grand new cathedral seemed a good idea.

But religious practice in this cathedral-building case was not without its problems. The local population had been heavily taxed by the wealthy bishop and, twenty years after building started, they rioted. The clergy had a way of dealing with this sort of recalcitrance: they excommunicated the whole town.

30 The labyrinth in the nave in Chartres Cathedral probably symbolizes the pilgrims' arduous route to spiritual transformation. Their ultimate goal, the high altar, is in the distance.

The builders were not the pious folk who feature in the medieval myth of a beneficent Church with its cared-for and dutifully devoted flock. They wanted better pay, they went on strike, they skimped with their work and even attacked the dean's house. They saw through the public piety of the luxury-loving priesthood to the grasping, power-hungry men they were. Like most European cathedrals, the one at Chartres was added to and altered in succeeding centuries. Chartres was never 'finished', and today's endless ecclesiastical restorations may be seen as part of a centuries-old architectural religious practice.

The tale of Chartres shows that religious practices can be devoid of religious experiences. Was that also true at Göbekli Tepe and the Maya pyramids? Probably. But once built (in a surprisingly short time, given all the difficulties), Chartres emerged as a marvellous structure that filtered divine light through its famously blue windows to pilgrims seeking to alter their consciousness, have religious experiences and obtain remission for their sins.

After entering one of the great portals, pilgrims found themselves facing the so-called labyrinth (Fig. 30). Overall, its four segments recall the cross, and the rosette at the centre symbolizes enlightenment. Was this pattern set in the floor intended as a convoluted route for pilgrims to follow as a symbol of the road to God? If they traversed the labyrinth on their knees, the pilgrims

would not easily have discerned its trick: it contains no dead-ends. Those who traverse it double back through each of the quadrants before inevitably landing up at the centre. Perhaps it was telling pilgrims that their journey to Chartres *inevitably* led them to God and his blessings, no matter what vicissitudes and disappointments they encountered on the way. It was a visual guarantee that the convoluted theology of the time could be trusted, if not understood.

In the previous chapter, I pointed out that religious belief is never a rigid, static, ineluctable formulation. Beliefs and theology change, though not in response to the sort of evidence that leads to change in science. Change in doctrine has implications for large-scale religious practices. Material structures, like Göbekli Tepe, Maya pyramids or Christian cathedrals, are not imposed inexorably on societies and individuals. It is people who create structures in the first place, even if unanimity in this process was elusive. In the daily flow of life, people manipulate their notions of spatial divisions and mediators and constantly (and often consciously) employ them to shape and reshape society. This is especially clear in changes to buildings and the ways in which they are used. The labour and economics entailed in the construction and re-construction of higher and higher pyramids helped to build Maya society and power. Material manifestations thus created, entrenched and modified abstract beliefs and social relations. So it is with cathedrals, always in need of repair or alteration. Perhaps baking cakes for a bazaar to pay for a new church roof is an innocuous present-day religious practice.

'To ferne halwes, couthe in sundry londes'[25]

Pilgrimages are an integral part not only of Christianity but also of many other religions as well. Journeying from one place to another, from a mundane to a sacred place and back, seems to have been a universal way of achieving spiritual development and change. In some ways the procession through the spaces of a cruciform church is analogous to a pilgrimage to the Holy Land. But pilgrimages should not be taken at face value. Inevitably, much more than the faith of the devout is at stake.

Today no one doubts the economic benefits of pilgrimages. Where would Lourdes or Santiago de Compostela be without all those money-bringing people? It was not pure piety that, in 2007, led the Vatican to establish its own airline to ferry pilgrims to various shrines. Its toe-curling motto is 'I'm searching for your face, Lord.'

Lourdes, the most famous of Catholic pilgrimage destinations, receives as many as six million pilgrims per year, despite the secular trend in Europe and despite the separation of Church and State in the French constitution. Some who visit Lourdes are put off by the vast number of tacky souvenir shops and the tawdry items they sell, but sell they do. For 2008, the 150th anniversary of the supposed appearance of the Virgin Mary to an adolescent girl, the Pope offered special indulgences to those who went to Lourdes. The recipients believed that their time in Purgatory, the half-way house of rather uncomfortable purification en route to heaven, would be shortened. This was an offer many believers considered unwise to refuse.

But it is of course promises of miracles that draw the elderly, the frail, the maimed and the terminally ill. Today this emphasis on miracles is hard to sustain. The Church itself has so far managed to authenticate very few so-called miracles. The clergy therefore shift the meaning of the word 'miracles' and claim that the real miracle is not something as trivial as restored sight (though that would be nice) but rather the sense of belonging and the social conviviality that pilgrims enjoy. The pilgrimage as a whole, they say, bolsters faith and that is a 'miracle'.[26] Of course it is nothing of the sort. Again, we see the Church caught in the trap of history: *pace* the Pope and other hardline, reactionary theologians, old beliefs become impediments to the acceptance of the Church in modern society, and new slants on those beliefs have to be devised. The enormous gulf between miracles in the normal sense of the word and what today is passed off as miracles is ignored: the faithful are deceived by sophistry. Indeed, when should we stop talking about 'the faithful' and instead refer to 'the gullible'? Were there not a religious foundation to all this, we should have spoken frankly long ago.

The secular component of religious pilgrimages can be clearly seen in Chaucer's *Canterbury Tales*. In the late fourteenth century, the taverns, not just the now-famous Tabard in Southwark, were making a healthy profit. Who undertook these pilgrimages? Some were lay folk; others were either in holy orders or were outwardly pious. Let us remember the Prioress, whose greatest oath was by St Loy (the Bishop of Noyon who scrupled to take an oath), the Monk, a manly man who loved hunting, the Friar who knew every barmaid, the Franklin who was Epicurus' own son, the gap-toothed, somewhat deaf Wife of Bath who had been through five husbands and who was out of all charity if someone in the parish went to the offering before her, the Pardoner, newly come from Rome with a bag of relics (including pigs' bones) that he used to cheat the poor out of what little money they had.

Despite all this worldliness, there was nevertheless a poor Parson who was rich in holy thought and work, who did first and only then taught. Chaucer knew that there were many reasons why people performed religious practices; he did not succumb to the pious glosses of the Church.[27] But, in the end, both the worldly majority and the poor Parson supported the shrine of Thomas Becket in Canterbury Cathedral – and, of course, all the attendant inns and shops. The jollity of the journey with its entertaining tales held the pious and the secular components of society together: everyone benefited, even the corrupt Pardoner who was selling fake relics and who probably saw the relics of Thomas Becket in the same light.

Opposition to the economic exploitation of pilgrimage and the moral laxity that often accompanied pilgrimages built up from the tolerance of Chaucer's time to the outright rejection of the Reformation. In the 1530s, Henry VIII put an end to the Canterbury pilgrimage and to the flourishing ones at other places, such as the Marian shrine at Walsingham that was believed to contain a crystal phial of Mary's milk. It is hard to understand why, even today in the twenty-first century, there are still people who believe in the efficacy of such palpable nonsense. They believe that the relics at the end of their pilgrimage contain healing properties and that their self-abnegation, often proceeding for considerable distances on their knees, will please God and he will consequently forgive at least some of their sins. Penance was one of the reasons for going on a pilgrimage, and could also be imposed by secular authorities in the Middle Ages.

Chaucer's motley but merry band is emblematic of all religion, not just the Christian Church. However indulgently Chaucer may view the foibles of humanity, there is little to be said for most of his pilgrims. Yet there was that poor Parson. The Church, cleric and lay, is not wholly bad, utterly evil. Still, the poor Parson could have been just as good without the Church. A less attractive thought is that, by choosing to be part of the Church, the Parson was unwittingly leading people into a frame of mind in which the Pardoner could fleece them. Does one poor Parson redeem the Church?

The anthropologist Victor Turner and his wife Edith analysed pilgrimage in terms of a schema of stages and liminality that was constructed by an earlier anthropologist, Arnold van Gennep.[28] Van Gennep divided rites of passage (such as from childhood to adulthood) into phases. First, novices separate themselves from their past. Then, in a liminal condition, they allow images and doctrines to impact on themselves in highly charged situations. Finally, they return to society remade, in a new form. This, argue the Turners, is what pilgrimage is all about. People return internally changed. Today researchers are wary of such broad generalizations.[29] To get around them, they try to 'theorize' pilgrimage in new ways. I contend that, in cases like this, a fundamental issue must be brought into the foreground. If we do not believe in a supernatural realm, let alone relics and miraculous cures, we can see what is going on in pilgrimages more clearly. Like other religious practices, they are implicated in exploitative and intellectually crippling strategies.

There is a much darker, yet seemingly inevitable, religious practice that is closely related to pilgrimages. The idea of pilgrimage provided the *raison d'être* and set the stage for the waging of crusades and genocidal war. For some, this comprehensive view is uncomfortable: certainly, political and economic factors loom much larger in crusades than they do in pilgrimages. People therefore often try to separate religious belief from religious practice by claiming that religious wars and persecutions (often triggered by individuals' religious experiences) are not true to the fundamental beliefs and experiences of a given religion. Those who wage the wars, of course, claim that *they* are the genuine fundamentalists, in the strict sense of the word. Today few Christians would speak openly in favour of the medieval Crusades. Yet they were initiated and supported by the Church whose God is the same yesterday, today and forever. Have we today suddenly (or gradually) received a new divine revelation, an injunction against taking up arms against people of a different religion?

Pilgrimage has become emblematic of the religious (mental or emotional) journey that people like to think they undertake in the management of their lives. When we sing John Bunyan's stirring hymn 'Who would true valour see, Let him come hither.... To be a pilgrim', let us remember Chaucer's motley band. There is more to pilgrimage than piety.

More mutilation

I have drawn attention to the prominence of blood and mutilation in numerous religions, but I have not so far properly commented on a bloodletting religious practice that still flourishes today and that is widely countenanced. It is genital mutilation, euphemistically called circumcision. The shedding of blood is a key part of circumcision rituals.[30]

That the God who created this whole universe could possibly want people to mutilate the genitals of their male babies is surely one of the most bizarre of religious practices. Yet that is precisely what many people of numerous religions believe. According to the *Book of Common Prayer*, the Anglican Church has a day set aside to celebrate the 'Circumcision of our Lord'. But where did the idea of circumcision originate? According to Genesis 17:11, God said to Abraham, 'And ye shall circumcise the flesh of your foreskin; and it shall be a token of the covenant betwixt me and you.' So Abraham, at the age of 99, circumcised not only himself and his son (Ishmael, poor fellow, was thirteen at the time) but also his slaves and all the men born in his house (GENESIS 17:27). The picture is revolting. Thereafter, the circumcision of enemies was a bloody Israelite custom: David 'slew of the Philistines two hundred men; and David brought their foreskins, and gave them in full tale to the king, that he might be the king's son in law. And Saul gave him Michal his daughter to wife' (1 SAMUEL 18:27). Was all this part of divine and benign 'progressive revelation'? Later Paul, trying to make Christianity acceptable to gentiles and at the same time not offending circumcised Jews, spiritualized the ritual. He absolved Christians from the obligation to mutilate their children. Instead of rejecting the whole custom as cruel, he spoke of circumcision as a spiritual matter: 'circumcision is that of the heart, in the spirit, and not in the letter' (ROMANS 2:25–29; 3:1). Paul seems to have got the idea from Deuteronomy (10:16): 'Circumcise therefore the foreskin of your heart, and be no more stiffnecked.' Was Paul making the best of a bad job?

The imposition of family religion on children even to the extent of mutilating them is, of course, widespread. A present-day Jewish thinker, Shalom Auslander, writes of his unhappy childhood in an Orthodox family.[31] Apart from a thoroughly unpleasant father, Auslander had to endure debilitating terror lest he, a young boy, wittingly or unwittingly offended against some dietary prohibitions and a vengeful God would punish him. In later life, he worries about whether he should have his new-born son circumcised. His book is satirically entitled *Foreskin's Lament*. According to the Old Testament, God wanted his Chosen People to be circumcised so that they (well, it was only the men who mattered) could be distinguished from the uncircumcised gentiles. With Islam and secular people also adopting circumcision, the practice no longer fulfils that function. It is easy for more moderate and humane Jews to say that they and Judaism itself are not as rigid as Auslander portrays them, but that is just too convenient an escape from responsibility. What is Judaism doing to rid itself of crushing – and mutilating – Orthodoxy?

For believers in divine creation (whether by miracle or evolution), circumcision poses a problem. We sometimes hear that God gave the Israelites circumcision not as an identification mark as the Bible says (it is, after all, seldom displayed) but for (spurious) 'reasons of health'. This, we are told, is why present-day fundamentalist Christians and non-believers alike still practise the custom. If we adopt this view, we must allow that, for utterly inexplicable reasons, God lied to the Israelites. If foreskins are indeed a health hazard, we must grant that they were a glitch in Intelligent Design. Or did God create foreskins simply so that they could be cut off?

In a moral milieu that defends the rights of children, circumcision is a shocking anachronism. To cut off a child's foreskin is not only a gross invasion of its human rights (think of the rights that some believers ascribe even to the earliest foetuses); it is a form of child abuse. Yet, because the word 'religion' is attached to the practice, governments dare not outlaw it.

Today there is rightly growing opposition to female genital mutilation wherever it is practised. Few readers of this book will need any persuasion that female genital mutilation is barbaric. But what of male circumcision? In any event, why this obsession with genitalia? As I suggested earlier in this chapter, more than one writer has noted how close religion and sex are to one another in a number of ways. For the present, simply to note religion's obsession with sex and the genital mutilation to which it often leads is enough to dissuade us from accepting religion as God-given.

A tripartite unity within religion

Religious practices raise many questions. I suggest that what we today may see as ethically 'acceptable' and 'unacceptable' practices can *simultaneously* flow from religious experience and belief. A religion should be judged (I see no reason why religions should not be judged) not only by what its sages or ambiguous scriptures say its tenets are but also by what adherents *practise*. The diverse social practice of a religion is inseparable from its systematized beliefs. Experience, belief and practice form an integrated whole. Increasingly, people today are finding that whole unattractive.

Many religious people therefore respond that violent and sectarian religion is not religion at all, and that religion in general should be judged by what 'we' find acceptable. This is the line that the Anglican theologian Keith Ward advocates in his book *Is Religion Dangerous?* Responding to Richard Dawkins's trenchant denunciation of religion, he allows that religion *can* be dangerous but that it is 'also one of the most powerful forces in the world for good'.[32] In writing this book, I have had to struggle with this matter again and again. Much as I should like to agree with Ward, I find that, in the end, I cannot.

Who is to say what constitutes 'good' practice and what is 'bad' practice? The Christian Crusaders (the original jihadists) believed that they were doing God's will and waging a 'holy war'. Today, Muslim suicide bombers believe they are practising 'good' religion. From where do these appalling practices come? Reluctantly, and *pace* Ward, I conclude that practitioners of 'good' religion are, by teaching belief in supernaturally vouchsafed knowledge and the supposed reality of divine interventions, laying the foundations for fanaticism – unwittingly maybe, but also inevitably. If 'good' religious practice is as believers say it is, why do we not see these believers in united, forthright, unequivocal denunciations of 'bad' religion? If Islam, Christianity and Judaism are to be cleansed of 'bad' religion, it is the apparent moderates who must undertake the task. The rejection of extremism must come forcefully and explicitly from within religion. It cannot be done by outsiders.

In the next two chapters I argue that the three interlocking domains of religion that I have identified help us to understand very diverse manifestations of religion. We shall see religious experience, belief and practice being negotiated in contrasting contexts, yet in recognizable ways.

CHAPTER 8

Stone Age Religion

In the previous three chapters I set out a way of looking at religion. Experiences, beliefs and practices together add up to what we may call a religious package-deal. In a nutshell, I argue that:

- Human beings have no option but to make sense of their shifting consciousness.
- One way of doing this is to interpret certain brain functions as evidence for a supernatural realm.
- Then, in a never-ending process, people hammer out a set of socially shared, yet contested, beliefs about their and others' experiences and their implications.
- They also have to *do* things that those experiences and (often contested) beliefs entail.
- The people who do these things do not necessarily all experience religious mental states.

If these conclusions are valid, we should be able to find traces of ancient peoples' religious practices in the archaeological record – as has so famously been the case with the early Neolithic and Bronze Age communities of the Near East, the Egyptians, the Maya and many other societies. Indeed, evidence for ancient religions is ubiquitous. But can we gain some idea, no matter how sketchy, of what the *earliest* religion looked like? What forms did Stone Age religious experience, belief and practice take?

Apart from the pieces of ochre themselves, evidence for religion at the time of the southern African Blombos finds (75,000 BC) is indeed slight. We shall therefore have to set Blombos aside for a while, at least until we have a better idea of what was happening in the early periods of the southern African Stone Age. For the present, it is better for us to turn to another Stone Age context, one for which there is much, albeit sometimes puzzling, evidence.

We need to see if the points I have made about religion in general help us to understand the most ancient evidence that we have for activities that we may call 'religious'. Does this evidence help us to piece together very early interacting religious experiences, beliefs and practices? This chapter therefore deals with the Upper Palaeolithic caves of France and Spain (approximately 50,000 to 12,000 BC). These deep limestone caverns contain the well-known cave paintings that have for long been considered to be one of the greatest achievements – and mysteries – of the past. Caves like Lascaux, discovered by schoolboys in 1940, and Cosquer, the entrance to which is now under the waters of the Mediterranean, have fired the popular imagination. Why, everyone asks, did Stone Age people climb, crawl and squeeze sometimes for a kilometre and more underground in order to make pictures of horses, bison, woolly mammoths and other animals? Are the endlessly convoluted underground passages and chambers and the scatters of pictures in them simply one great meaningless muddle? We find ourselves in a very intriguing realm indeed. How do we start to answer questions about the 'religion of the caves'?

I answered some of these questions in *The Mind in the Cave*,[1] so I shall not repeat everything here. Instead, I take a set of caves that I did not consider in any detail in that book and see how they can be understood in terms of religious experience, belief and practice.

Activity areas

Let us begin with a non-controversial statement. The practices and products of a sane human mind are ordered, and, because minds are situated in communities, those practices and products are socially meaningful, even though some individuals may harbour idiosyncratic ideas. It follows that, no matter how randomly images and other evidence may seem to be scattered through the passages and chambers, Upper Palaeolithic caves meant something to the people of the time: the images and the underground spaces were intelligible, and moving through them must have been a meaningful experience, if not exactly the same experience throughout the period. In the previous chapter we saw that Christian beliefs and rituals make sense of the layout of medieval cathedrals. Upper Palaeolithic people must have had some comparable mental template that made sense of the caves and the activities that they performed in them. So let us keep the Christian example of designated spaces in mind as we proceed. The details of Upper Palaeolithic cave use may remain forever unknown, but the broad lineaments are waiting to be discovered. We don't have to know everything in order to know something.

When we look closely at the caves, we find that concentrations of images and other evidence, along with apparently unused spaces between concentrations, suggest that people often (certainly not always) tended to group their underground activities in specific chambers, passages and niches that they considered appropriate in accordance with their beliefs and what they hoped to achieve. We should therefore attend to what I call 'activity areas' and their distributions in the caves if we wish to get some idea of Upper Palaeolithic religious experience, belief and practice. Because finds that archaeologists have made in activity areas imply different kinds of rituals and emotionally charged attitudes to the caves, we need to explore the ways in which the human brain functions to produce experiences of that kind.

Ritual and the human brain

As we saw in previous chapters, neuropsychology constitutes an evidential strand that can be profitably intertwined with others in discussions of ancient religions. Nevertheless, I must emphasize that neuropsychology is no magic wand. It constitutes only one of a number of evidential strands, and its exploitation goes hand-in-hand with, for example, broad ethnographic precedents and theory concerning the social construction of space. As a discipline, neuropsychology differs theoretically and methodologically from these other strands. The philosopher Alison Wylie[2] has pointed out that this kind of theoretical and methodological *dis*unity is a highly positive and strengthening characteristic of the type of argument that I develop here: '[T]o establish evidential claims of any kind, practitioners must exploit a range of inter-field and inter-theory connections.'[3] This is what I try to do.

A need for a neuropsychological component in the explanation of certain social behaviours has in fact been widely recognized. For instance, the anthropologist Roy Rappaport aptly summed up the frequent importance of altered states in the creation and maintenance of differentiated societies:

Trance and less profound alterations of consciousness are frequent concomitants of ritual participation.... The relationship between alterations of the social condition and alterations of consciousness is not a simple one, but it is safe to say that they augment and abet each other.[4]

We need to ask how this augmenting and abetting may have taken place in the embellished caves and how neuropsychology aids our enquiry. All researchers rightly assume that, whatever may have been the case with earlier hominids, such as the earlier Neanderthals, Upper Palaeolithic people had the same brain and cognitive abilities as we do.[5] This means that they experienced, or had the potential to experience, the full consciousness spectrum which, as we saw in Chapter 5, ranges from alert consciousness through profound introspection to sleep and dreaming, as well as deeply altered states that can be induced by a wide variety of means.[6]

In Chapter 6, I argued that two neuronally generated experiences – passage through a vortex and flight – contributed substantially to near-universal beliefs in a tiered cosmos. Upper Palaeolithic myths of transcosmological journeys would probably have taken place within and served to reproduce and explain a tiered universe.[7] Here is a link between the Upper Palaeolithic and the Maya, who constructed 'caves' in their mountain-pyramids, and many other societies throughout history that have believed that caves and tunnels led to an underworld and then built these ideas into their myths.

Given the universality of the human brain, it is reasonable to assume that Upper Palaeolithic people probably took entry into the caves as equivalent to entry into an underworld. Then, bearing in mind the hard-wired vortex experience, we can suggest that they probably saw the passages of the caves in terms of the vortex – the tunnel – that leads underground *and* into deep, hallucinatory states of consciousness. The cave passages were the 'entrails' of the underworld, and the walls, floors and ceilings were thin 'membranes' that could be penetrated for access to what lay beyond them.[8] Activity areas were therefore subdivisions of a nether realm.

Do these broad conclusions help us to make sense of *specific* Upper Palae-olithic caves? I find that it is not very satisfying to talk in generalities about 'Upper Palaeolithic religion'. I therefore move on to a discussion of a single cave system. We need to see if my explanation makes sense of actual caves.

The Volp Caves

In *The Mind in the Cave* I compared and contrasted the caves of Lascaux and Gabillou, both in the French Dordogne.[9] Here I consider a cave complex in the Ariège Department (south-western France) known as the Volp caves. Unlike so many other diversely embellished sites, these caves have been miraculously preserved by their owners, the Bégouën family;[10] the caves have not been trampled and destroyed by generations of visitors.

During the Magdalenian period (approximately 16,000–8,000 BC), two of the Volp caves, Enlène and Les Trois Frères, were joined by a narrow passage

(Fig. 31); there was no other access to Les Trois Frères.[11] There is no known connection between these two linked caves and the adjacent third Volp cave, Tuc d'Audoubert. Social and image-making relationships between the unconnected caves are not clear and must await discussion at a later time. At present, Enlène and Les Trois Frères together provide the largest samples of portable and parietal (cave wall) images made by a single community. At the same time, the care with which the caves have been conserved has ensured the survival of a variety of other traces.[12] Robert Bégouën, the present owner, writes of 'the thousand and one features of Magdalenian expression.'[13]

Fascinating contrasts between activity areas in Enlène and those in Les Trois Frères have become evident. Upper Palaeolithic people did not wander aimlessly through the caves making pictures whenever the fancy took them. Following the Upper Palaeolithic route from the outside world into Enlène and then on to Les Trois Frères, we now explore the Volp caves.

ENLÈNE

Except for the Entrance Chamber, the first Magdalenian area of notable dimensions in Enlène is some 160 m (525 ft) from the entrance. It is in total darkness. Known as the Salle des Morts, it was used in the much more recent Bronze Age as a cemetery. When the Bégouën family first explored Enlène at the beginning of the twentieth century, the Bronze Age deposits had been excavated, and debris from that work covered the entrance to the passage leading to Les Trois Frères. There are also Gravettian and some slight Magdalenian deposits near the entrance to Enlène. The deeper Magdalenian parts of the cave have been radiocarbon dated by means of charcoal from hearths to 12,000–11,500 BC.[14] Bégouën and the French authority on cave art Jean Clottes believe that the Magdalenians may have used the deeper parts of Enlène for a period of about 1,000 years at most.[15]

The dated hearths are evidence for one kind of activity. It appears that people started fires with wood, but then used bone as the chief combustible material. Vast quantities of bone (bison, reindeer, horse, etc.) were brought into the cave; the meat was eaten and the bones were split to obtain the marrow.[16] The presence of all this bone suggests that people must have remained underground for extended periods or that they visited the Salle des Morts frequently, but this does not mean that the chamber was a living area in the generally accepted sense. Certainly, it is extremely unusual to find evidence so deep underground for prolonged Magdalenian habitation. Why, then, were the occupants of the cave burning bone so deep underground?

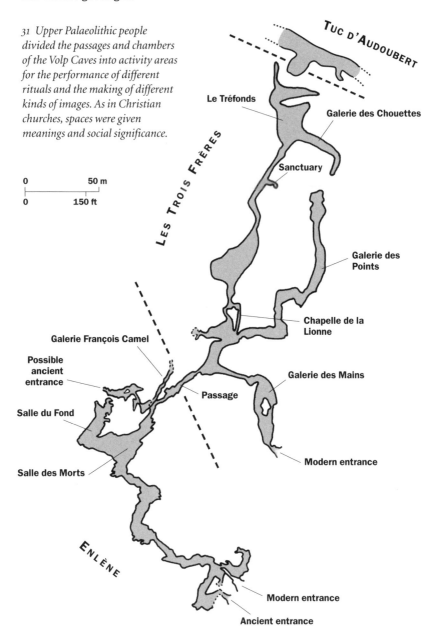

31 *Upper Palaeolithic people divided the passages and chambers of the Volp Caves into activity areas for the performance of different rituals and the making of different kinds of images. As in Christian churches, spaces were given meanings and social significance.*

TUC D'AUDOUBERT

Le Tréfonds

Galerie des Chouettes

LES TROIS FRÈRES

Sanctuary

Galerie des Points

Chapelle de la Lionne

Galerie François Camel

Possible ancient entrance

Galerie des Mains

Passage

Salle du Fond

Salle des Morts

Modern entrance

ENLÈNE

Modern entrance

Ancient entrance

One is led to suspect that their reasons were not the demands of daily life. On occasions when the fires were lit the stench of burning bone must, in that confined space, have been overpowering – though not necessarily as distasteful as we may have found it. Indeed, the odour may have had its own positive significance, deriving as it did from animal bone. We need only recall the use of sickly incense in Christian cathedrals. It may be that we have here the earliest evidence for a nascent notion of sacrifice, many thousands of years before Abraham believed he was called upon to kill his son Isaac, Jesus died on the cross, and the Maya slaughtered their captives. As we have seen, killing and religion often go hand in hand. The burning of vast quantities of bone was probably an Upper Palaeolithic practice that created a religious ambience.

There is a further clue in the Salle du Fond. When the chamber was first occupied in the Middle Magdalenian, about 60 bone artefacts, some engraved, were thrust vertically into the ground. They seem to have been conceptually directed at the floor of the cave. A centrally perforated bone disc engraved with an image of a bison and with peripheral decorations, as well as a more fragmentary bone disc with perforations around its circumference, was also found in the Salle du Fond. In addition, there were many other engraved bones and antlers.[17] One of these is engraved with two images of fish and seven encircling lines; it was found in the Salle des Morts.[18]

Animal bone received another kind of treatment in the Salle du Fond. Hundreds of small pieces, some burnt, others not, were thrust into fissures and cracks in the rock walls.[19] Apart from the tip of a reindeer antler, these pieces are too small to be identified as to species.[20] Why were bone fragments pushed into cracks? Their small size, vast numbers and diversity of angle and height above the floor preclude a practical explanation: they could not, for instance, have been used to hang bags or other equipment. Rather, people carefully and meticulously pushed bone fragments into the cracks for reasons that are not immediately clear. But we can make a suggestion.

I argue that we are again encountering the notion of immanence: something inside something apparently solid. The pieces of bone may be evidence for some sort of communication, or mediation, between the subterranean space in which the people were and a spirit realm beyond the rock face. A belief along these lines would have heightened the emotional impact of what we see merely as a rock surface: for Upper Palaeolithic people, touching the rock was touching a powerful, perhaps dangerous, cosmological interface. We must also notice that it was fragments of dead animals that people were passing from one cosmological space to another. Was there, we may again ask, some notion here of sacrifice? Were dead animals seen as conduits between cosmological and spiritual areas? Were Upper Palaeolithic people 'feeding' supernatural beings behind the rock face?

There is also evidence for Magdalenian flint-knapping in the Salle des Morts and the deeper Salle du Fond.[21] This activity may, like the hearths, be taken at face value to suggest mundane occupation of the cave. But we should not accept unquestioningly that tool-making was a task that had no social and symbolic connotations.[22] Many flint tools, it should be remembered, related in various ways to animals: hunting for meat and for supernatural animal power are frequently related activities.[23] Separation of the utilitarian from the symbolic is often impossible. Moreover, the power of that realm may have been 'activated' in some way by the sound of the knapping of imported flint. Rhythmic sound, such as drumming and chanting, is widely considered a way of contacting the spirit world,[24] and researchers have studied caves and other archaeological sites for the effects of resonance.[25]

Another non-utilitarian activity conducted in both the Salle des Morts and the Salle du Fond was the making of engraved stone plaquettes.[26] Indeed, few Upper Palaeolithic sites can compete with the number of plaquettes discovered in Enlène. Over 1,100 have been found; most await study. It seems that large quantities of local stone were brought into the cave for the purpose of engraving them. So far, there is no evidence that these plaquettes were engraved outside the cave and then carried into the depths. People went underground to think about animals, to burn their bones and to make images of them.

Usually, only one animal image appears on each plaquette, though some have images on both sides. Among the animals depicted, bison and horses are numerically dominant. Uniquely for Upper Palaeolithic image-making, anthropomorphic figures constitute a common theme.[27] Both kinds of images are overlain by what seems to be random scribbling that makes decipherment a taxing task (Fig. 32). As in parietal imagery, natural contours of the stone were occasionally used to complete animal forms (9 out of the 85 thus far studied).[28]

To Western eyes, the images are often ill-proportioned and appear to be sketched rather than executed with a sure line. It would, however, be wrong to infer that they were necessarily made by novices in the course of practising in order to become 'perfect' before tackling more sacred cave walls. The sketched, sometimes vague, 'style' of the plaquettes may have been specifically desired and may not have resulted simply from inchoate skills. The mode of sketching probably had a different significance from drawing with a sure line. The plaquettes are therefore evidence for a particular class and context of image-making that differed from sure-line parietal image-making. On the other hand, the plaquettes were associated with the underworld – like parietal imagery but in a different way.

The large number of fractured plaquettes suggests that they were intentionally broken; they are fairly substantial, not fragile. The breaking of images was part of the overall ritual. Broken or not, many were used to pave the floor – though perhaps not for the sort of mundane reasons that 'paved' calls to mind today. Turning them over or picking them up to reveal images would probably have generated powerful emotions. We should always remember that images were not as abundant in Upper Palaeolithic communities as they are in Western and other societies today. In those ancient times images were rare; they were something special.

In the depths of the Salle du Fond, the deepest chamber of Enlène and where most of the engraved plaquettes were found, there are a number of marks made with red pigment on the walls and on stalactites. They are not recognizable as representational images. In one place there are four parallel red lines made with a finger. The placement of these marks in out-of-the-way locales within the deepest chamber suggests that they were not accidental. It seems likely that it was the mere presence of paint on the surfaces that was meaningful. Paint was probably not simply a technical substance, such as we might

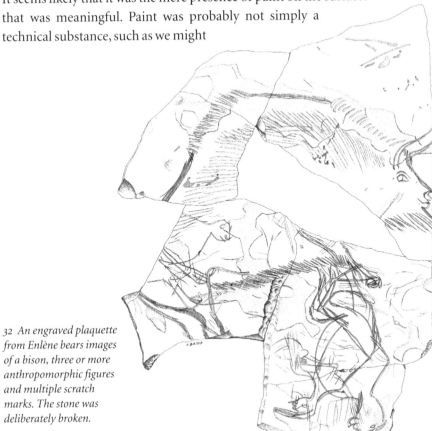

32 An engraved plaquette from Enlène bears images of a bison, three or more anthropomorphic figures and multiple scratch marks. The stone was deliberately broken.

purchase in a hardware shop; it probably had its own power and significance. Even a small smear of it meant something important because it interacted in some way with the rock face and thus established contact with what lay behind the rock.[29] The act of placing a paint-loaded finger on the rock was both rare and daring.

All in all, the most striking feature of Enlène is the absence of representational wall paintings. The cave was used for other activities, some of which related to the walls, floors and ceilings of the chambers. The cave was also used for the making and reusing of engraved plaquettes. We must not forget that all these practices were conducted in two selected areas deep underground and not uniformly throughout the passages and chambers of the cave. The larger spaces afforded by the Salle du Fond and the Salle des Morts were the only areas deemed appropriate for the activities that took place within them. The choice of larger spaces suggests that those religious practices involved a small number of people simultaneously, but not a great many.

The activities conducted in Enlène, though apparently diverse, were parts of a coherent whole, and that whole, I argue, derived from a notion that animals were mediators of cosmological levels, as indeed they are believed to be in many parts of the world.[30] Enveloping smoke from their bones, pieces of them thrust into cracks or set vertically in the floor, and images of them made and overscored on plaquettes – all these transformations of animals were situated in and interacted with a defined area within the nether realm of a tiered cosmos. That area was a place for complex interactions between people, animals and the beings and powers of the underground realm.

Exactly what those interactions were we do not know, but, as worldwide ethnography and mythology suggests, it may well have involved the absorption or activation of animal essence or power,[31] the complex processes whereby shamans and other ritual specialists learn to acquire, engage with and manipulate mental imagery,[32] and perhaps some notion of sacrifice.[33]

THE PASSAGE

Leading off from the Salle des Morts is the low, 60-m (200-ft) Passage to Les Trois Frères, a much larger cave (Fig. 31). At this point of traversal a different kind of movement is necessary. One can no longer walk from chamber to chamber: one must now crawl on hands and knees. Middle Magdalenian artefacts found on the floor show that the Passage was used during that period. Indeed, researchers who know the caves well believe that the Passage afforded the only Magdalenian access from Enlène to Les Trois Frères.[34]

There is, however, evidence for an earlier (Gravettian) entrance from the outside to Les Trois Frères,[35] but it is with the Magdalenian parts of the cave only that I am concerned.

The archaeological deposit in the Passage ends at the Les Trois Frères cave, and it is here that parietal engravings begin (though there is also a smaller gallery, the Galerie François Camel, that branches off the Passage; it has a few engravings). As Bégouën remarks, 'This cannot be due to chance.'[36] At the juncture between the Passage and Les Trois Frères, again standing upright, Magdalenians entered a new realm within the underworld, complete with its own distinctive religious experiences and practices.

LES TROIS FRÈRES

In Les Trois Frères, as in Enlène, human activities were concentrated in distinct areas. To the right of the point where the connecting Passage enters Les Trois Frères is evidence for earlier Gravettian occupation, and there are hand-stencils in the Galerie des Mains. In this section of Les Trois Frères, it is difficult to be sure which parts were used exclusively during the Gravettian or the Magdalenian.

After one has passed the entrance to the Galerie des Mains, the Galerie des Points branches off to the right. This large but little studied gallery is characterized by black and red dots[37] and a variety of parietal engravings and other features that have not yet been published. I therefore omit the Galerie des Points from this discussion and concentrate on the Chapelle de la Lionne and the Sanctuary.

CHAPELLE DE LA LIONNE

About 50 m (165 ft) from the crawling Passage and down the high-ceilinged main gallery of Les Trois Frères there is a small chamber on the right. The eminent French prehistorian of the first half of the twentieth century Henri Breuil named this area the Chapelle de la Lionne. The only engraved stone plaquette found in Les Trois Frères lay just outside it. It bears a depiction of a bison and other scratch marks.[38] The entrance to the chamber is, compared with that to the Galerie des Points, hidden, and the narrow Chapelle can accommodate only a few people at one time.

In addition to the large eponymous lion engraving, there are images of a horse, a bear and another small lion.[39] As Breuil recorded it, the principal lion has two tails and two or more feline heads placed around it.[40] The second tail is, however, a human arm and hand reaching out to the rear of the animal.[41]

This arm exemplifies another non-real relationship between human beings and animals: manual contact with animals and animal images was ritually important. As we shall see, a particular facet of this relationship was developed further by the most prominent image in the Sanctuary.

In the Chapelle de la Lionne, large animal teeth, including a bear's lower molar,[42] not the small fragments of bone characteristic of Enlène, were placed in, rather than thrust into, niches. One niche contains five items: four pieces of worked stone and the bear's tooth already mentioned.[43] As the knapping in Enlène suggested, worked stone clearly had more than utilitarian significance and that significance was related to animal parts and the subterranean realm. A large fossil shell was also found in a fissure near the entrance to the Chapelle de la Lionne; both faces had been treated with red ochre. The shell may suggest that the subterranean realm was also conceived of as subaquatic, ethnographically a common enough belief.

The items 'given' to, or placed in, the walls are more diverse and consistently larger than those in the Salle du Fond. They thus represent an 'advance' beyond the Enlène practices: they suggest more direct, more substantial, contact with the mediating 'membrane' of the rock face.[44] This closer contact may have been associated with beliefs about felines. Worldwide, shamans are commonly associated with felines.[45] We may therefore entertain the possibility that the Chapelle de la Lionne was a confined place where initiates and others conducted solitary, perhaps terrifying, vision quests and made important offerings to what lay behind the wall. It was a distinct activity area, possibly a necessary stage for anyone en route to proceed to what lay deeper in the cave.

THE SANCTUARY

After one leaves the entrance to the Chapelle de la Lionne, the floor of Les Trois Frères slopes steeply down. Deeper and farther into the cave is the Sanctuary. On the left, high above this chamber, a narrow ledge leads to the ultimate parts of the cave, Le Tréfonds. Here there are paintings and engravings, including a red horse, fine engravings of bison, finger, stone or bone scrapings in the form of a grid, and a claviform. In the nearby Galerie des Chouettes are two owls with what appear to be their young,[46] and a mammoth; these representational images were made by scraping the clay surface with fingers. Le Tréfonds and the Galerie des Chouettes lie beyond the scope of our investigation.

On the right-hand wall of the Sanctuary is the famous image of the therianthropic 'sorcerer', or *Le Dieu cornu* (Fig. 33).[47] Engraved and painted, this

33 The Abbé Breuil's rendering of the Les Trois Frères 'sorcerer'. Situated on a high ledge, it has owl-like eyes, antlers, human legs and a horse's tail. Seeing it was probably the climax of a person's visit to the cave.

fusion of human and animal forms is on the wall behind a ledge some 4 m (13 ft) above the floor of the Sanctuary. Breuil describes climbing with great difficulty up the cave wall to the ledge.[48]

In prehistoric times the 'sorcerer' could be reached more easily through a narrow 'tube' that curves up from a small, richly engraved hollow in which one must crouch or lie; it can accommodate no more than two persons at a time. Although it was necessary to lie prone in order to traverse the narrow ascending tunnel, its sides were embellished with images that could have been seen by only one person at a time and then only when the viewer was edging his or her way up towards the narrow ledge in front of the 'sorcerer.' These images have not been recorded for fear of damaging them. This 'tube' with its swirling images is even more reminiscent of the neurological vortex than the Passage from Enlène to Les Trois Frères, which has no parietal images. Movement through the Sanctuary 'tube' may point to another and even deeper level of altered consciousness and consequently more overwhelming mental imagery associated with the close proximity to the 'sorcerer' that the narrow ledge necessitates.

The 'sorcerer' itself demands close attention. It has antlers, a horse's tail, seemingly human male genitals but set back as on a feline, bear or feline paws, large owl-like eyes, what appears to be a beard, and human legs and feet.

Almost certainly, this striking being will have featured in myths that conveyed his nature and function.

The part-human, part-animal 'sorcerer' is unique in the Volp Caves in a number of ways. It is both painted and engraved. Its elevated position well above eye level is not duplicated elsewhere. Its size exceeds that of all other images. The manner in which it is both painted and engraved suggests repeated renewals and modifications that were meaningful to those who effected them. The sequence of production remains to be determined, but it seems likely that the different techniques of painting and engraving had different meanings. Today, the 'sorcerer' is not well preserved, but in prehistoric times it would have dominated the chamber even more strikingly than it does now and, potentially, it could, unlike the images in the Chapelle de la Lionne, have been simultaneously visible to a fairly large number of people.

The raised, isolated 'sorcerer' may thus have played a role comparable to (but not identical with) that played by the large, brilliant images in the Hall of the Bulls in Lascaux or the 'spotted horses' in Pech Merle.[49] Contemplation of large, probably communally produced and dramatically revealed images, would have informed the sort of hallucinations that people would experience in deep areas of sensory deprivation, or by ingestion of psychotropic substances and other means. At least in part, people hallucinate what they expect to hallucinate, and a blending of human and animal features is, in any event, a common characteristic of deeply altered states of consciousness.[50] The single, elevated Les Trois Frères therianthrope suggests belief in a powerful chthonic being whose transcendence of the human:animal dichotomy people emulated in altered states. It is also possible that this multi-species being may have functioned as a Lord of the Animals who created all creatures in the underworld and, suitably approached and propitiated, released them to hunters.[51]

On the wall below the 'sorcerer' and to the right of the entrance to the ascending tunnel is a complex entanglement of engraved images with lines scored across them. It seems that in some parts of Upper Palaeolithic caves, people strung out their images, each standing, as it were, by itself and asserting its independence.[52] In other underground spaces, as here, they chose to concentrate images in a comparatively small area.[53]

As many as 307 parietal animal images have been located in the Sanctuary, but far fewer elsewhere in Les Trois Frères.[54] They include horse, bison, reindeer, ibex, bear, mammoth and rhinoceros. Unlike the images on the plaquettes in Enlène, they are of 'high quality' and 'are nearly always made with a sure hand'.[55] The sketchiness of the plaquette images is absent from the

Sanctuary. As I have suggested, the 'sure hand' should not be attributed simply to practice having led to perfection; these are a different kind of image from those on the plaquettes and in a different context. They may have been associated with deeper, more vivid experiences of altered consciousness. A few images have lines emanating from their mouths.[56] Noel Smith, a professor of psychology, takes these lines to represent the spirit, or power, of animals harnessed in shamanic rituals.[57] He is probably correct.

Among the images below the multi-species 'sorcerer' are two therianthropic figures: they combine human and bison features. Some writers believe that one of these therianthropes (Fig. 34) may be playing a musical bow, though I doubt this interpretation. Like the 'sorcerer', this figure has also been identified as a shaman.[58]

Therianthropic figures like the 'sorcerer' are rare, though persistent, in Upper Palaeolithic imagery. They date back to the striking Aurignacian (30,000–22,000 BC) ivory carvings of human figures with lions' heads at Hohlenstein-Stadel and Hohle-Fels in southern Germany[59] and the 'sorcerer' in the Aurignacian Chauvet Cave that seems to be a bison with a human arm and hand associated with a female lower body.[60]

Part of the extraordinary density of the engraved Sanctuary panel is created by overlying lines. They are not like the apparently random scribbling on the Enlène plaquettes: many are straight 'cut marks' that sometimes appear in rows that seem to have resulted from a single episode of scoring. Sometimes they converge in chevrons or three-part acute angles that have

34 In a densely engraved panel below the Les Trois Frères 'sorcerer' a half-human, half-bison figure appears to be related to two animals. Visitors to the cave repeatedly scratched this panel, possibly in attempts to derive power from it.

been taken to represent spears. Whether or not some of them depict spears, I suggest that these lines resulted from a distinct way of relating to the rock interface with its existing images and the spirit realm that lay behind it, a way that went beyond mere looking. Whereas the bone splinters in Enlène passed from the activity area into the realm beyond, the 'cut marks' may have released power so that it flowed out from the beyond, via animal images, into the activity area of image-making and, very possibly, other activities that have left no traces.

The religious practices that took place in the Sanctuary were clearly complex. People viewed the 'sorcerer' from afar; perhaps it appeared to float above them in the darkness. Some crouched below it and engaged with the wall of the cave by making images and by scoring across them. A few wormed their way up the tunnel to come into close proximity with the 'sorcerer.' (The undamaged state of the tunnel suggests that it was seldom used.) Some people may have participated in dancing and chanting that was related to the images but without themselves making images. There was thus a distinction between communal and personal experiences of images.

Contrasts

In comparing Enlène and Les Trois Frères, we can detect an overall difference that may be summed up as a *narrowing down*, or *intensification*, from Enlène to Les Trois Frères, not just topographically through the Passage but also conceptually as the Passage paralleled the mental vortex. In summary:

– Although animal images are depicted in both caves, a narrower range of species is depicted on the walls of Les Trois Frères than on the plaquettes of Enlène.

– Plaquettes are abundant in the depths of Enlène, and all but absent from Les Trois Frères.

– Anthropomorphic figures constitute a frequent theme on the Enlène plaquettes; they are rare on the walls of Les Trois Frères.

– On the Enlène plaquettes there is usually only one animal image per face; in the Les Trois Frères Sanctuary animal images are concentrated in a particular area and entangled one with another.

– The Enlène plaquette images are 'sketched', while those on the walls of the Sanctuary are done with continuous, sure lines.

– The Enlène anthropomorphs were different kinds of beings from the multi-species 'sorcerer'; they were attendant spirits or the ritual specialists who made the plaquettes.

– The visual and conceptual climax of the two caves is in the deepest part of Les Trois Frères – the dominating 'sorcerer'.

When they finally reached the Sanctuary, people found their attention focused on a single image – the elevated 'sorcerer' – and then, secondarily, on the less prominent tangled images below 'him'. How did they apprehend all those images? How did the images fit into an overall mental framework? Is even 'image', the most neutral word of which I can think, nevertheless tendentious?

Images or embodiments?

The answers to these questions bear directly on our understanding of subterranean religious experiences and practices. Clearly, the densely engraved panel in the Sanctuary is not a 'scene' in any Western art sense.[61] Indeed, the Sanctuary engravings display many of the characteristics of Upper Palaeolithic images in general. They are sometimes superimposed one on another. They are juxtaposed without attention to relative size. Many are fragmentary, the head being the most frequently depicted part of animals. Some images face down the rock surface, others upwards, while still others are upside down. No ground surface is depicted. Hoofs and, in the case of the bears, paws are not always distinctly drawn, but, when they are shown, they sometimes hang loosely rather than stand on an imaginary ground surface. Images are presented devoid of contexts, with no trees, grass or other surroundings.

This combination of features corresponds to characteristics of hallucinations.[62] Hallucinations of animals, people and monsters tend to be self-contained and, as we have seen, to be projected onto surfaces, such as walls or ceilings.[63] The psychiatrist Ronald Siegel[64] writes of 'a motion picture or a slide show,' while Heinrich Klüver writes, 'Sometimes the phenomena are definitely localized on the walls, on the floor or wherever the subject happens to look.'[65] Under these conditions, the images appear to float on the surfaces and are devoid of any context other than the surfaces themselves and images that may already be present on them – there are no ground lines, no trees, no background. Thus suspended, they rotate, move, fragment and combine.[66] Floating projected mental imagery, repeatedly recorded, would lead to the apparent confusion we see in the Sanctuary engravings. But for the image-makers there was no confusion, only a concentration of many revelations. Religious experiences were materialized by the practice of image-making.

Another frequent characteristic of Upper Palaeolithic imagery is the use of features of rock surfaces to provide the dorsal line, an eye, legs or other parts of an image; sometimes it is a shadow thrown by a light held in a specific position that provides the trigger for an image.[67] The image-makers merely added missing parts to create the full image. This characteristic is also found in the densely engraved panel below the 'sorcerer' as well as in other parts of the cave.[68] Images are thus integrated with the rock face, and the surface on which they are projected becomes their entire context: the rock face was not a meaningless tabula rasa.

To explain this feature, I suggest that people were searching for images of animals in the walls themselves. With sight, touch, light and, sometimes, projected mental imagery they scrutinized the rock for tell-tale signs of the presence of an animal image. Then, with painted or engraved lines, they 'fixed' their mental images, regardless of any relationship they may have seemed to have had with other painted or engraved images – at least any relationship that most Westerners would discern as meaningful. Images were thus materially embedded in particular places in a multi-component underworld. They were in some ways comparable to icons tucked away in specially dedicated side chapels in a cathedral.

An image-maker may sometimes have accomplished the fixing of projected mental images while in a lightly altered state; at other times a painter or engraver in a contemplative, introverted state may have made an image in order to recall a vision, perhaps locating a nodule or crack in the rock wall to locate and aid the reconstruction. We must also remember that, after a subject has returned to alert consciousness, the phenomenon of projection may continue intermittently.[69] The anthropologist Geraldo Reichel-Dolmatoff, for instance, found that South American Tukano shamans not only experienced their yajé-induced images projected onto plane surfaces; as 'after images', they recurred spontaneously in this way for several months after the altered state in which they were first seen.[70] People thus 'carried them around in their heads.' Such after-images may also have been fixed in Upper Palaeolithic caves. There is a powerful emotional bond between people and their hallucinations.

Materialized personal visions were, however, but one kind of image. As I have pointed out, elsewhere in Upper Palaeolithic caves (e.g., in parts of

Lascaux) images seem to have been communally produced and subsequently altered; their making entailed the labour of a number of people, the construction of scaffolds, and a concatenation of ritual preparation.[71] Rather than being single-experience images, these may have been summations of personal and much discussed and revisited revelations. Communal images related more to *shared* religious beliefs than to individual religious experiences: the making of both types was a religious practice.

A key conclusion to be drawn from this evidence is that, in Upper Palaeolithic religion, images were not 'pictures' of other things, such as animals living outside the caves. Rather, many were 'things-in-themselves', floating, but now fixed, visions embedded in special places (though some were communally produced). The fashioning of many Upper Palaeolithic images was a complex ritual performance during which image-makers engaged, controlled and embodied their visions. Still today, novice shamans learn to control the content of their visions by 'actively engaging and manipulating the visionary phenomena'.[72] Once visions were fixed, Upper Palaeolithic people responded to them in ways that went beyond looking and contemplation. They interacted in more physical ways with underground spaces and with what the images were in themselves: they touched them, marked and 'cut' them, altered them, and placed other images over and around them. In doing so, they interacted with them in profound ways because they believed them to be real things that could inhabit their own minds.

Once they had fixed their visions on the rock, as complete images or as fragments, Upper Palaeolithic image-makers probably had a special relationship with their particular spirit animals and the power that they bestowed: image-makers established personal and social bonds between this world and the beings and animals of a subterranean tier of the cosmos. Over the years, some panels grew in complexity as people resorted to them for power visions, no doubt often triggered by existing images that were themselves revivified by trance vision. Through this feedback process, the power, or theophany, of an activity area increased.

Upper Palaeolithic religion

The Volp Caves show that Franco-Cantabrian Upper Palaeolithic concepts of the underworld could be highly complex and multi-component, indeed as complex as the mythical underworld of Dante's *Inferno*. Caves became models of, and locales for, transcosmological journeys. People interacted with the immutable forms of the caves. We can see evidence for their religious beliefs and practices. We can also infer the kind of mental experiences that led to and accompanied those beliefs and practices. Importantly, the experiences, beliefs and practices that we can detect fit together to form a logical, coherent whole – a religion.

I argue that Upper Palaeolithic people felt that not only whole caves but also narrow passages within them were in some way parallel to, or even identical with, the neurologically engendered experience of a constricting vortex that separates mental experiences of markedly different kinds. Multiple cosmological tiers entail multiple vortices to connect them. For example, the Chukchee of north-eastern Asia believe in a cosmos that has between five and nine tiers that are said to be joined to one another by 'holes situated under the Pole Star'.[73] Similarly, a North American Navajo myth describes cosmological tiers being linked by reeds that supernatural beings can climb,[74] and some Siberian groups say that the multiple layers of the cosmos can be 'reached through a small hole'.[75] Constricting passages like the one between Enlène and Les Trois Frères and the one in the Les Trois Frères Sanctuary probably linked levels or sections of a highly complex cosmos. Upper Palaeolithic myths probably told of their protagonists moving from one cosmological space to another rather as Dante's *Inferno* leads the reader from one region of suffering to the next.

The therianthropy of Upper Palaeolithic images suggests an intense kind of transformation, an interaction of both spiritual and material animality with humanity. Such interaction points to mediation, first between human beings and spirit animals, and, secondly, between human beings thus endowed and another realm of existence at the end of the narrow vortex where the integration and the fragmentation of mental images is neurologically generated. This is a common mythological theme. In a sense, the Les Trois Frères 'sorcerer' sums up the concepts of enveloping, transforming animality and mediation to which the various practices conducted in Enlène refer in different, perhaps preliminary, ways. It projects those concepts in an especially striking visual way.

Everyone had the potential to verify the teachings of ritual specialists, not only by seeing the underground labyrinths for themselves (if they were permitted to do so) but also by their own parallel mental experiences 'wired into' the human brain: all people have some knowledge of other so-called realities, even if only through dreams. There was therefore a common neurological resource that people could have (probably always have) manipulated and exploited in processes of social differentiation that are founded on the existence of a supernatural realm and privileged access to it. Penetration into the depths of both caves and minds legitimated the interests of the few and distanced the rest of the community, as religion always does.

A related point is that the embellished caves resulted from interaction between their highly diverse topographies and human conceptions of, to put it broadly, the cosmos and humankind's place in it. Upper Palaeolithic people must have had a notion of what the cosmos was like, even as the Maya and other archaeologically known societies did. Once Upper Palaeolithic people had decided to exploit subterranean labyrinths, they had to adapt the immutable topographies – open chambers, narrow passages and small niches – to their conceptions of how the cosmos was ordered. At the same time, they had to adjust their myth-enshrined concepts of the cosmos and, especially, the manner of their contact with specific parts of it. For example, if there was no large chamber that would suit their ritual needs near the entrance to a cave, they had to modify their use of the cave accordingly. In this interactive process, people varied not only the ways in which they made subterranean images (engraved, painted, moulded in clay, etc.), but also other underground practices that left traces – and many that did not, such as dancing, chanting, and meditation.

Social implications

Distinct activity areas show that people did not wander freely through the caves performing rituals wherever they wished. Rather, it seems probable that the caves were controlled by a spiritual elite who at times guided newcomers from one area to another, explained the significance of each (probably by means of myths) and stipulated appropriate ritual responses. In this way, embellished caves provided not only conceptual and material frameworks for images but also instruments of social discrimination.[76]

At a broad level, there was probably a distinction between those who enjoyed access to various activity areas in the subterranean realm and the rest of the community who remained outside the cave.[77] Those who entered the two Volp caves faced a series of choices. It would, for instance, be possible to spend time in Enlène and then exit the cave or to proceed at once to Les Trois Frères. Then, having crawled through the connecting Passage, visitors would either enter or bypass the Chapelle de la Lionne on their route to the Sanctuary. Their choices would have been influenced by the social implications of the activities conducted in the various parts of the caves, not by the topography alone. I therefore suggest that different social groups were associated with distinct activity areas and that those social distinctions would have been manifested and reinforced by movement through the caves: movement was conceptually and socially meaningful.

In the practice of complex religious rituals, the Volp people thus determined who had access to what parts of the caves, and how that access was achieved by different social groups.[78] Bégouën and Clottes sum up this sort of difference and complementarity:

the portable art on stone in the Volp caves had different authors and probably different uses ... [I]t was accessible to individuals who had little or no access to the decorated cave walls while at the same time sharing a set of conventions and concepts with the artists who created the works of wall art.[79]

The exact nature of these implied social distinctions is hard to determine. At this point we move beyond the limitations of ethnographic precedents: Upper Palaeolithic religion and ritual has no exact present-day or historically recorded analogues. Nevertheless, some tentative suggestions can be made.

The absence of parietal images from Enlène and their presence in Les Trois Frères, together with the abundance of plaquette images in Enlène and their virtual absence from Les Trois Frères, is surely of major significance. If the walls, ceilings and floors of caves were, as I argue, a 'membrane' between subterranean activity areas and chthonic spirit world behind the rock, the placing of animal images directly on the mediatory surface would have implied a relationship between the maker of an image and the spirit realm. But the Enlène group did not go so far as to embellish the 'membrane' itself; they kept their images away from the walls and on plaquettes. They had a different, less direct, perhaps preparatory relationship with the spirit realm from those who made parietal images in Les Trois Frères. In marginally altered consciousness, people stand back from and 'see' their hallucinations; it is only in deeper stages that they participate in, and engage with, them.

Although some people may well have performed activities in both caves, it seems likely that those who frequented Enlène may have been novices seeking power or lesser ranking ritual specialists or, perhaps, a dedicated 'chapter' within the religious hierarchy of the community that had its own responsibilities. A small group of supportive people may have congregated in Enlène amid the odour of burning animal bone while a few ventured through the Passage and then on to the lower depths of Les Trois Frères, where they experienced more intense mental states. If complex post-vortex hallucinations were experienced in Enlène, they did not result in the kind of activities that characterized Les Trois Frères, as the dense Sanctuary engravings suggest. For visitors to Enlène, deeper experiences and hallucinations were imperfectly engaged, controlled and understood: they were probably fleeting or even suppressed. Direct visionary contact with the 'membrane' was therefore not considered appropriate.

In Les Trois Frères, on the other hand, profound hallucinations may have been associated with direct contact with the 'membrane' and the fixed visions of other ritual specialists and thus with the spirit world beyond it. Transition to this kind of deeper spiritual experience with its advantageous social concomitants may have been suggested by crawling through the low Passage that joins Enlène to Les Trois Frères, by the fairly steep slope down to the deeper Sanctuary, and, finally, by the narrow tunnel with its parietal images that leads up to the climactic 'sorcerer'. Rarely, especially privileged people carefully wormed their way up to the ledge and engaged directly with him.

It would be wrong to stop short at this point. It seems likely that the caves were not only templates for the reproduction of social distinctions. There were probably times when individuals and groups contested the conceptual and physical barriers and, by implication, social order. Instruments for maintaining social distinctions can also be used to challenge them[80] – the story of the Christian Church through the ages.

Before we leave the Volp Caves a few caveats are in order. I do not argue that people never experienced deep, post-vortex states in Enlène. Nor that the full progression from alert consciousness to deeply altered states was never experienced in Les Trois Frères. Nor that everyone who entered the caves experienced altered states. Nor that all Upper Palaeolithic images were made by people while they were experiencing altered consciousness. Nor that the same means of inducing altered states were used in the two caves. Nor that people experienced altered states only underground. Nor that religion remained unchanged throughout the long Upper Palaeolithic period, even though some components seem to have endured.

Conflict and tensions

Even at this very early time, there was dissent in Eden. It seems that the exclusive superiority of the 'knowledgeable' Upper Palaeolithic elite was sometimes challenged by people who tried to emphasize aspects of religious experiences that the elite rather downplayed. Why do we suspect this sort of social conflict in a society that existed so long ago?

Overwhelmingly, the images of Upper Palaeolithic cave art are of animals; there are, overall, very few anthropomorphic depictions. It was within this category of apparently human figures that some minority image-makers fashioned what we may call 'images of dissent'. The small group of distinctive images to which I am now referring shows people apparently pierced by spears; they are found in the caves of Pech Merle and Cougnac in the Lot region of France.[81] These images seem to manifest one of the experiences of somatic hallucinations – bodily sensations of stabbing and pricking. This sort of experience was not depicted by the general run of image-makers. How successful were these few who chose to identify themselves by making stabbing sensations an important criterion of spiritual advancement? The little evidence we have suggests that those who emphasized this aspect of spiritual experience were not successful in their challenge, at least not in the long term. Their failure is suggested by the short period and limited geographical area to which depictions of stabbed bodies were restricted. People soon returned to the usual depictions of animals.

The important point here is that Upper Palaeolithic people harnessed spiritual knowledge, 'pictures' and subterranean spaces not only to reflect and create social distinctions but also to contest – to protest against – those distinctions. The sections of the caves where they placed their 'stabbing' images were thereby set apart from other areas in which there were no such images. If certain spaces are closely associated with special people, the distinction between the elite and the common people is visually dramatized and thus reinforced every time members of the community take up their allotted positions, as we saw in the case of a cruciform church. But those very social distinctions can be challenged by dissenting people who dare to cross over into forbidden spaces. A dramatic example from recent history would be the common people forcing their way into palaces that were reserved for royalty and their lackeys.

Past and present religions

What, then, can we say about Upper Palaeolithic religion? With some degree of confidence, we can hazard hypotheses about what was happening in the Franco-Cantabrian caves. Those hypotheses uncover patterns of belief and behaviour, not confused muddles. Stone Age people knew what they were doing, and they did it methodically. Religious experience, belief and practices, as I have described them, account for what we find in the caves.

We can now gather together the principal points that our study of the Volp Caves has made so that we can begin to compare them with present-day religions. At a few places in this chapter I have referred to parallels between Upper Palaeolithic religion and Christianity. Readers will probably have spotted many more. How many of the following six features of Upper Palaeolithic religion have parallels in present-day religions?

– Upper Palaeolithic people believed in a tiered cosmos.

– They believed that spirit beings and animals lived in the nether realm but nonetheless influenced their lives.

– They divided caves into areas in which they performed rituals of different kinds.

– These rituals were designed to contact and to manifest spiritual entities in a variety of ways.

– Some rituals included the acquisition of revealed knowledge in the form of visions.

– The whole religious enterprise was implicated in social discriminations.

It is defeatist and unrealistic to claim that we shall never know anything about Upper Palaeolithic religion and to believe that modern religious experiences, beliefs and practices are altogether different and more advanced. If I am correct in arguing that the wiring of the human brain accounts for some (not all) components of religious experience and belief, can those fundamental neurological functions be detected in more recent religions as well? To explore this issue further in the next chapter I consider two vastly different religious contexts and show that, despite all their obvious differences, certain identifiable neurological events occur in both.

CHAPTER 9

Hildegard on the African Veld

Religious experience is generated by the human nervous system. This explains why religious experiences reported by devotees of diverse religions exhibit parallels, no matter what the specific cultural input may be. Culture does not swamp neurology.

To illustrate this conclusion I select an indisputably neurologically generated motif. It is both sufficiently complex and, in the two cultural circumstances I discuss, precisely contextualized to dispel worries that it may have been accidentally stumbled upon in different cultures – unlike, say, a simple circle or cross. I certainly do not argue that the motif had the same *significance* in different cultural contexts (or necessarily for all individuals in a given society): the form is neurologically generated, while its meaning is contingent. This, in short, is the history of religion.

The two cultures to which I refer are vastly different and widely separated in space and time. They are the medieval Christian milieu of western Europe and the Later Stone Age San communities of southern Africa (Bushmen). That people in both these contexts should believe in supernatural beings and forces comes as no surprise. By contrast, the repetition of the motif I discuss is intriguing. It exemplifies a key point that I made in earlier chapters: much (not all) of religion is concerned with explaining not merely the natural world, as we are repeatedly told, but rather with coming to terms with mental experiences. In studying the depiction of this motif in medieval Europe and Stone Age southern Africa, we encounter an illuminating amalgam of biology and culture: although for the purposes of discussion we distinguish between biology and culture, you cannot have one without the other.[1]

Hildegard of Bingen

The word 'illuminating' brings us in two senses to Hildegard of Bingen (1098–1179), the medieval Christian abbess and mystic who is today well known as a composer of 'spiritual' music and is a focus of feminist interest.[2] She claimed to be 'illuminated' by a bright, visionary light that clearly recalls

Paul's, Constantine's and other religious devotees' experiences. For insight into her experiences, we turn to the illuminated manuscripts that not only record but also depict Hildegard's visions. It is unlikely that she herself made the illuminations. Some of these beautifully and brilliantly coloured pictures have a subsidiary 'box' showing Hildegard receiving inspiration from on high and telling scribes what to write down. Just as she dictated the texts to scribes, it is virtually certain that she supervised the making of the pictures, not as mere decorations but to clarify her sometimes startling visions: readers could see (more or less) what Hildegard had seen.[3]

The social circumstances of Hildegard's life throw light on how she achieved fame in her own time throughout western Europe.[4] Her renown cannot be separated from her influential origins: she was no God-chosen waif. She came from a background used to power, the tenth child of a wealthy landowning family living near Bermersheim in what is today western Germany. Early in her life, her parents resolved to place her as a 'tithe' in a convent. This decision may be read as an act of devotion or as a way of dealing with a growing and expensive brood: at least, as a nun, Hildegard would not have to be provided with a dowry.

Later in her life, Hildegard criticized the custom of placing children in nunneries and monasteries, though she did not distance herself from her own parents' action. Two of Hildegard's brothers and one of her sisters willingly devoted their lives to the Church, so, even though influence within the Church was (and sometimes still is) desired by powerful families, there can be no doubt about the devout and socially significant background from which she came. She had a good start in life.

The process of her dedication to the Church began when she was eight years old. At this time, her parents put her in the personal care of Jutta of Spanheim, a pious noblewoman. Then, in 1112, taking the child with her, Jutta became a recluse at an anchorage that later developed into a convent associated with the flourishing monastery of Disidodenberg. There, Hildegard was immersed in the offices of the Benedictine order. Daily she was surrounded by the architecture, liturgies, chants, music, discipline, teachings and intense communal life of the order. For anyone brought up in so claustrophobic a situation it must be hard to view the world and its doings in any other than a religious way: for such people, the hand of God is discernible in everything.

When Jutta died in 1136 the nuns elected Hildegard as her successor; she must have grown into a forceful woman with what are today called 'leadership qualities'. We see this aspect of her character in her bold decision to move away from Disidodenberg, despite opposition from Abbot Kuno, the supreme ecclesiastic at Disidodenberg. Contrary to Kuno's wishes, she was determined to set up her own foundation at nearby Rupertsberg and announced that God had told her to do this. Notwithstanding her frequent self-description as a poor, weak maid, power and authority were attractive to her. In the event, all did not go well. First, the move was strongly resisted not just by Kuno but also by the rest of the monks. Her fame had by that time attracted wealthy visitors, political influence and valuable donations. The monks did not wish to forfeit these advantages. For her own part, she did not enjoy seeing her fame benefiting too many 'undeserving' peripheral people and an institution that she did not control. Secondly, after the move, some of Hildegard's followers defected, much to her dismay. It can be difficult to live with forceful personalities, especially if they are pious.

Soon, Rupertsberg fell on hard financial times, but, ever resourceful, Hildegard called in favours from the much put-upon Kuno. Wealthy families like Hildegard's gave land to monasteries and thus secured influence in Church affairs. It so happened that some of Hildegard's nuns' families had donated land to Disidodenberg, and Kuno felt obliged to avoid estranging them. He therefore gave the requested money to Hildegard, thus saving her nunnery. According to charters drawn up in 1158, Kuno's successor at Disidodenberg remained responsible for priestly duties at Rupertsberg, the nuns themselves being unable to administer the sacraments. This financial matter was not the only problem that Hildegard encountered. Later, she was involved in a fierce dispute over the burial of a young man in sacred ground. Medieval monastic life had its share of internal strife, and controversy was an integral part of Hildegard's long life. She died aged 82 on 17 September 1179.

Hildegard's life was less cloistered than this brief biography may suggest. Her location in the Rhineland provided her with easy access to many places of importance: the Rhine was the highway of western Germany. She was also strategically situated between two major spheres of political power: Rome to the south, and the Holy Roman Empire to the east. Hildegard took advantage of this geographical position and maintained correspondence with the leaders of both centres of power. They included St Bernard of Clairvaux, four successive popes and the Holy Roman Emperors Conrad and Barbarossa.

Even more remarkably for a woman, Hildegard went on preaching tours of western Germany. Some of her sermons were intended to drum up support for the bloody twenty-year-long crusade against the Cathar Christian sect in south-western France: on one occasion alone the Catholic Church burnt 200

Cathars alive. Hildegard also wrote a tract denouncing the Cathars: she believed that the orthodox clergy were not opposing the sect with sufficient enthusiasm. In mitigation, some writers claim that Hildegard herself was opposed to the sort of Christian mayhem that was visited upon the Cathars,[5] but she does not appear to have taken any significant stand against it. This is another of the disjunctions that we find repeatedly throughout the history of the Christian Church. The most saintly of believers find it easy to condone and indeed perpetrate the vilest tortures on those whom they consider heretics. Politics and power, especially when combined with supernaturally revealed knowledge, can so easily mute generosity of spirit.

Hildegard's prominence was due not only to her aristocratic origins and strategic location. She claimed that her first vision came to her before the age of five. This event may have been instrumental, among other inducements, in her parents' decision to dedicate her to the Church. There was a superstitious predisposition during the Middle Ages for people to accept accounts of visions at face value, especially if they seemed to support one's own interests.

At first Hildegard did not speak of her visions. She claimed to be too embarrassed and afraid. Later, when she did bring herself to tell others about them, she either wrote accounts of them herself or, as we have seen, had a scribe do so. Her own command of Latin was rudimentary. Her principal visionary books are known as the *Scivias* (written between 1141 and 1150) and the *Liber Divinorum Operum Simplicis Hominis* (written between 1163 and 1170). Eventually, news of her revelations reached the Church hierarchy and even the Pope himself; people began to seek her advice on various matters.

Interestingly, Hildegard did not distinguish between revelations about religion and those that concerned the physical structure of the cosmos: both came to her in the same supernatural way. From her point of view, all knowledge came from God, whether it be religious or what we today call scientific. As we saw in Chapter 2, it was this view of the indivisibility of knowledge that made it possible for the Church to persecute those who challenged belief in a heliocentric solar system.

There is an intriguing aspect of Hildegard's life that is not often explored by her biographers: they are rather baffled by what today seems to be a childish project. She constructed a secret script and language, a *lingua ignota*.[6] It was a kind of code to be understood only by her and, apparently, some of her nuns, though it did not survive her death. She invented 23 rather squiggly letters, *litterae ignotae*, and just over 1,000 words. Her grammar seems to be derived from Latin. Though Hildegard said that her language came directly from God, just like her devotional music, a great deal of cerebral ingenuity went into its construction. Why did she bother? A great advantage of secret knowledge is that it cannot be challenged and seen to be trivial. The surviving *lingua ignota* texts contain little of interest, as is often the case with secret knowledge. One canticle simply contains flowery divine praise: 'O boundless Ecclesia, girded with the arms of God, and arrayed in hyacinth, you are the fragrance of the wounds of nations…'.[7] For all its convoluted symbolism, this sort of thing is superstitious mumbo-jumbo, and many Christians would have denounced it as such had they come across it in some 'primitive' religion.

The visionary nature of Hildegard's acquisition of religious knowledge placed her at odds with the growing influence of the schools (eventually universities), especially the one in Paris, where Thomas Aquinas would later build his reputation. Monastic wisdom bestowed by the Holy Spirit came to be contrasted with the scholastic tradition with its emphasis on logic and laborious learning.[8] Volmar, a Disidodenberg monk who was Hildegard's teacher and amanuensis, wrote to her scathingly about scholastic intellectuals who indulge in 'litigious declamations of disputes…[and] sweat over profundity or rather conundrums of opinions…for private advancement'.[9] Then as now there was a split between intellectual theologians and charismatic preachers who believed that God communicated directly with them.

For Hildegard, a vision was not a rhetorical device of the kind to which I referred in an earlier chapter and that was common during Old Testament times and indeed the Middle Ages: it was a profound physical and psychological experience. But what sort of experience exactly? Her descriptions of her illnesses and visions, together with the pictures of them that appear in her books, have led researchers to conclude that Hildegard suffered from migraine.[10] She experienced bouts of ill-health from an early age, but nonetheless lived into her eighties. Her longevity was more consistent with migraine than with a congenital wasting disease. Today this verdict is widely accepted: Hildegard interpreted migraine attacks as God speaking to her and showing her 'diagrams' or 'models' of the cosmos and of Christian doctrines.

Migraine

To understand the illuminations in Hildegard's visionary books, we need to examine the experiential components of migraine.[11] First, we must recognize that migraine is not a synonym for a headache. It is possible to experience a migraine attack without having a headache – and vice versa. Moreover, a headache is never the sole symptom of a migraine. Instead of focusing on headaches, we need to attend to what Oliver Sacks, a well-known professor of clinical neurology and writer, calls 'a whole world – the cosmography of oneself'.[12] Referring to the aura that is part of migraine, he writes, '[T]here will have occurred, perhaps, in the space of twenty minutes, such a revelation of bewildering (and perhaps beautiful) complexity as the mind may never be able to forget.'[13]

Sacks's final phrase captures the overwhelming significance that such experiences can have for some people. Indeed, a migraine aura may be 'so strange as to transcend the powers of language'.[14] He quotes a nineteenth-century physician's report of a migraine aura; it contains elements that will be crucial to our understanding of Hildegard's visions. A patient said:

I have frequently experienced a sudden failure of sight. The general sight did not appear affected; but when I looked at any particular object, it seemed as if something brown, and more or less opaque, was interposed between my eyes and it, so that I saw it indistinctly, or sometimes not at all…. After it had continued for a few moments, the upper or lower edge appeared bounded by an edging of light of a zigzag shape, and coruscating nearly at right angles to its length…. The cloud and the coruscation…would remain from twenty minutes, sometimes to half an hour.[15]

Notice four components of this account:

- the area of blindness that blotted out perceptions of surroundings,
- the zigzag form,
- its scintillating nature, and
- the duration of the attack.

Migraine visual perturbations of this kind are not a modern ailment. The Cappadocian physician Aretaeus described the visual aberrations of migraine nearly 2,000 years ago, comparing the colourful zigzags to a rainbow.[16] He also realized that the migraine aura is comparable to hallucinations experienced by epilepsy sufferers, another condition that he studied. Aretaeus realized that various conditions can trigger the same interferences with normal sight. Migraine has been with humankind for thousands of years.

Analysing the visual effects experienced in migraine attacks, Sacks identifies two main components: both are directly relevant to our understanding of Hildegard's visions.

First, there is 'a dance of brilliant stars, sparks, flashes or simple geometric forms across the visual field'.[17] Sacks calls these 'stars' phosphenes and adds that they are usually white but may be brightly coloured. Although they tend to swarm across the visual field, one 'star' may detach itself and become brighter than the others. At other times there 'may be only a single, rather elaborate phosphene', sometimes comprising a number of 'pointed leaflike projections, alternately red and blue', which, after moving to and fro, may disappear 'leaving a trail of dazzlement or blindness in its wake'.[18] Patients may elaborate phosphenes into recognizable images: for instance, one patient 'described small white skunks with erect tails, moving in procession'.[19] In addition to 'stars', migraine phosphenes may take other forms: these include latticed, faceted and tessellated motifs that recall mosaics, spider webs, honeycombs or Turkish carpets.[20] The vortex that we discussed in Chapter 6 is also present in migraine experiences: 'Circles may spin, rotate into spirals, a spiral may deepen into a vortex, a large vortex may break up into little scrolls or eddies.'[21]

The second major component that Sacks identifies is the migraine scotoma, also known as the fortification illusion (or spectrum). 'Scotoma' means darkness or shadow. Within a scotoma we can distinguish two elements. First, it may start as a brilliant point of light and then expand across the visual field into a huge crescent before it disappears into the periphery (Fig. 35). The scotoma arc has a jagged, scintillating, zigzag outer edge that is brilliantly coloured. It may also take the form of a central 'rainbow' that seems to hover above objects in veridical sight (Fig. 36). The second component of a scotoma lies within the arc: it is an area of invisibility that, in a rather uncanny fashion, blots out veridical sight – the 'darkness' implied by the word scotoma.[22] For instance, the sufferer may see a person without a head. Not only within, but also before or after the scotoma, patients may experience an area of blindness. The perceptual alterations experienced in a migraine attack also include micropsia (objects become smaller) and macropsia (objects grow in size). Together, these perceptions may create a zoom effect.

As with phosphenes, objectification may play a role when people experience a scotoma. Sacks, in association with Ralph Siegel of the Center for

Molecular and Behavioural Neuroscience, Rutgers University, quotes a report that describes an objectification of a migraine scotoma:

When it was at its height it seemed like a fortified town with bastions all around it, these bastions being coloured most gorgeously.... All the interior of the fortification, so to speak, was boiling and rolling around in a most wonderful manner as if it was some thick liquid all alive.[23]

For many migraine suffers the matter ends with visual perturbations, but some also experience sounds, such as hissing, growling and rumbling. Others experience olfactory hallucinations of revolting odours.[24] Somatic sensations may also precede, accompany or follow migraine attacks; these include corporeal vibrations that seem to be of the same frequency as the scintillating scotomata. One patient described a sensation of 'vibrating wires' in the pit of his stomach. Others speak of vertigo, staggering and nausea.[25]

In sum, we may say that some sufferers experience only the aura and the scotoma passing across their vision; many experience attendant hallucinations in the other senses as well as considerable pain and nausea. As a generalized sequence of a migraine attack Sacks suggests the following:

- Simple visual effects of the aura (dots, lines, stars, etc.)
- A scintillating scotoma
- Bizarre alterations of perception
- Elaborate illusory images or dreamlike states[26]

As with the stages of altered consciousness that I discussed in Chapter 5, these migraine stages are not ineluctable: for example, some sufferers move directly into the scotoma stage; others do not experience dreamlike states.

All in all, migraine auras and scotomata and their sequence recall the entoptic phenomena described in Chapter 5. The scintillating geometric

35, 36 OPPOSITE *A scotoma develops from a small break-up of vision to a scintillating zigzag with an inner area of invisibility.* LEFT *The scotoma with its area of invisibility sometimes hovers over objects perceived in normal sight. Vision and reality combine.*

forms of the aura and the scotoma dovetail with the six categories of entoptic types that I listed in that chapter (bright dots, sets of parallel lines, grids, zigzags, nested catenary curves and glowing filigrees). Moreover, and importantly from the perspective of our present discussion, there is a tendency for people in altered states and for migraine sufferers to construe their visual motifs as things in the material world or things in which they deeply believe. The neural structure of the brain is clearly the decisive, generating factor in both altered states and migraine attacks.

Hildegard's visions

As I have mentioned, Hildegard was subject to illness and visions from an early age. She suffered attacks, sometimes prolonged, but she was nevertheless able to continue with her duties:

In this affliction I lay thirty days while my body burned with a fever.... [W]hile I did not die, yet did I not altogether live. And throughout those days I watched a procession of angels innumerable who fought with Michael against the Dragon and won the victory.... And then the whole troop cried out with a mighty voice... 'Arise maiden, arise!' Instantly, my body and my senses came back into the world.... Thus was my body seethed as in a pot.... But although I was thus tortured, yet did I, in supernal vision, oft repeat, cry aloud, and write those things which the Holy Spirit willed to put before me.[27]

This description is clearly consonant with severe migraine attacks, though Hildegard's mention of 30 days seems excessive. Perhaps she meant that she experienced a series of migraine attacks, each of which left her debilitated yet able to conduct daily affairs.

She went on to describe more specifically how she saw her visions:

These visions which I saw I beheld neither in sleep, nor in dream, nor in madness, nor with my carnal eyes, nor with the ears of the flesh, nor in hidden places; but wakeful, alert, with the eyes of the spirit and with inward ears, I perceived them in open view and according to the will of God.[28]

She was neither asleep nor in the kind of trance that would cause her to lose contact with her surroundings. She thus distanced herself from the contemporary visionary Elizabeth of Schönau, who experienced her visions in a trance. Unlike Hildegard, Elizabeth was an ascetic whose hunger and isolation may have triggered her visions. In a letter, Hildegard warned her fellow nun against excessive punishing of her body, which is, she says, 'a fragile vessel'.[29]

Hildegard's visions, 'perceived in open view', fall into three interconnected categories:

- Points of light that shimmer and move across the field of vision.
- One large, brilliant light set in concentric circles.
- Well-delineated fortification figures, also brightly coloured, that radiate from a coloured area.

What did she make of these visions? She wrote of a vision of the first kind:

I saw a great star most splendid and beautiful, and with it an exceeding multitude of falling stars which with the star followed southwards.... And suddenly they were all annihilated, being turned to black coals…and cast into the abyss so that I could see them no more.[30]

The illumination depicting this vision appears in the third book of the *Scivias*. As Charles Singer, a medical doctor who was interested in the history of science, and Sacks both argue, Hildegard saw a 'shower of phosphenes' followed by a negative scotoma that obliterated the vision[31] (Fig. 37). But for Hildegard, steeped in medieval Christianity and ever on the lookout for divine intimations, the moving stars were not merely phosphenes: they represented 'The Fall of the Angels'.

'Swarms of stars' become an identifiable element that Hildegard incorporated in other illuminations. For instance, in the remarkable one that pictures Adam, the creation of Eve and the temptation in Eden, stars appear in two places (Fig. 38). In the upper register, the stars represent angels and saints in Heaven. Below, a wing-shaped emission from the recumbent Adam's left side is filled with stars. Genesis 2:21–22 was Hildegard's source: God 'took one of his ribs…[and] he made a woman'. The stars in Hildegard's depiction of this event represent the children to be born of Eve: they will be as radiant as the angels above and represent a potential Heaven on earth. In medieval belief, the angels that fell from Heaven will be replaced by humankind: Eve's children will replace the stars that have fallen.[32]

The bright, blinding light that may follow the falling stars also features in many of Hildegard's visions. It is often represented by brilliant concentric circles (Fig. 39). These, she insists, she sees while her 'outward eyes remain open and the other corporeal senses retain their activity'. She names this light the 'cloud of the living light', the memory of which 'remains long with me'. In a comparable vision, she sees God seated on a throne floating above the concentric circles of light.[33]

37, 38 LEFT *Hildegard saw her brilliant migraine phosphenes as 'an exceeding multitude of falling stars' and, among them, 'a great star most splendid and beautiful'. Human neurology provided raw material for her visions.* RIGHT *In Hildegard's* Scivias *a recumbent Adam listens to the voice of God, while Eve issues as a cloud of stars from his body, and the serpent's tongue infects her. The stars (phosphenes) above are angels; those in the cloud representing Eve are 'the whole multitude of mankind'.*

The overpowering brilliance of the light led Hildegard to take it to be God, on whose face human beings can hardly look. Paul's Damascus road 'light from heaven' (ACTS 9:3) was no doubt in her mind. It was God who, in the beginning, said, 'Let there be light' (GENESIS 1:3) and who 'commanded light to shine out of darkness' (2 CORINTHIANS 4:6). As the Psalmist wrote, the face of God is radiant: 'Make thy face to shine on thy servant' (PSALM 31:16). When Jesus was transfigured on 'an high mountain', 'his face did shine as the sun,

39 'A most shining light and within it the appearance of a human form of sapphire colour which glittered with a gentle but sparkling glow.' In many of Hildegard's illuminations, blinding circles were the sphere of the Trinity.

and his raiment was white as the light' (MATTHEW 17:2). Hildegard took up this idea and wrote of 'a most shining light and within it the appearance of a human form of a sapphire colour which glittered with a gentle but sparkling glow'.[34] Figure 39 (from the *Scivias*) shows Hildegard taking the concentric circles with their human figure to represent the Trinity. Her construal of the light as a human figure is thus in accordance with the way in which migraine sufferers may sometimes interpret what they see: Hildegard's highly imaginative intellect was nurtured by her cloistered life.

The fortification scotoma also appears in a number of Hildegard's visions (Figs 40, 41 and 42). The simplest (Fig. 40) shows typically migrainous fortifications radiating from a central point. At the apex of the 'battlements' is a winged head. Of this vision, Hildegard wrote:

I looked and behold, a head of marvellous form…of the colour of flame and red as fire, and it had a terrible human face gazing northward in great wrath. From the neck downward I could see no further form, for the body was altogether concealed…but the head itself I saw, like the bare form of a human head. Nor was it hairy like a man, nor indeed after the manner of a woman, and very awful to look upon.

It had three wings of marvellous length and breadth, white as a dazzling cloud. They were not raised erect but spread apart one from the other, and the head rose slightly above them…and at times they would beat terribly and again be still. No word uttered the head, but remained altogether still, yet now and again beating with its extended wings.[35]

This is what Hildegard calls *Zelus Dei*, the Jealous God.

A similar vision shows a human figure sitting on the apex (Fig. 41). The figure holds a book and on his lap is a model of the City of God. The vision is known as *Sedens Lucidus* and comes from the *Scivias* II:1. What appears to be a conflation of this and other visions is shown in Fig. 42. Here the *Zelus Dei* vision is combined with *Sedens Lucidus* to construct a complex image of the City of God. Hildegard is using her separate visions as building blocks in the same way that a theologian uses selected biblical texts to construct an argument.

A particular state of mind that often accompanies variously generated hallucinations is clearly important: migrainous visions may be accompanied by feelings of exaltation. Sacks writes that Hildegard's visions 'provide a unique example of the manner in which a physiological event, banal, hateful, or meaningless to the vast majority of people, can become, in a privileged consciousness, the substrate of supreme ecstatic inspiration'.[36] Aptly, he

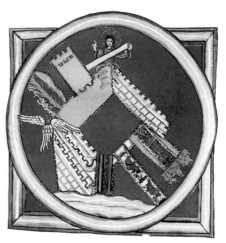

40, 41, 42 ABOVE In one of Hildegard's visions, fortification figures radiate from a central point, at which she sees the 'terrible human face' of the triple-winged Jealous God. ABOVE RIGHT In a variation on the basic fortification migraine scotoma, Christ, the cornerstone of the Holy City, is at the apex.
RIGHT Sometimes, Hildegard combined a number of visions in a complex structure, as a theologian combines selected biblical texts to construct a framework for belief. Visions are always raw material awaiting interpretation by the visionary.

quotes a passage in which the Russian novelist Dostoevsky describes his own migraine attacks:

There are moments, and it is only a matter of five or six seconds, when you feel the presence of eternal harmony…a terrible thing is the frightful clearness with which it manifests itself and the rapture with which it fills you. If this state were to last more than five seconds, the soul could not endure it and would have to disappear. During these five seconds I live a whole human existence, and for that I would give my whole life and not think that I was paying too dearly.[37]

These words could have easily been written by Hildegard. They show just how intense revealed knowledge can be and how utterly persuaded the recipients can be that God has indisputably spoken to them – and that others should listen dutifully. It was this feeling that prompted Hildegard to speak out even though she was a woman. A heavenly voice said to her:

Though as a woman you are uneducated in any doctrine of fleshly teachers in order to read writings with the understanding of the philosophers, nevertheless you are touched by my light, which touches your inner being with fire like the burning sun. Shout and tell! … Therefore, paltry soul, instructed as you are in your inner being by mystical inspiration, and although you are trampled by the male form because of Eve's transgression, speak nevertheless of the fiery work of salvation which this most certain vision reveals to you![38]

Here Hildegard cites her supernatural visions to exonerate herself from any obligation to follow the philosophical reasoning of the growing scholastic tradition and to strike at the male hierarchy of the Church. In her view, her Heaven-sent light was supreme, and she used it to cow the male clergy.

In addition to light, blood seems to be widely associated with such intense feelings. It featured prominently in medieval Christianity, and Christian art of the period frequently dwells on blood. For instance, one of Hildegard's illuminations shows a crowned woman, Ecclesia (the Church) holding a bowl in which she is collecting the crucified Christ's blood as it spurts from the wound in his side (Fig. 43). In the lower register of this illumination the power of the blood pours down to a chalice on an altar. Here, Ecclesia (though a woman) is a priest celebrating Mass.

Medieval theology seems to have been almost perpetually caught between reality and symbolic interpretations of ritual and biblical events: Christian theologians wanted to have it both ways.[39] The Church's fascination with blood (literal or symbolic) certainly recalls the Maya rituals I described in the Prolegomena. In those rituals, priests fed the blood of butchered captives to the gods. Both the Maya and the Christian owners of phials of Christ's blood saw the efficacy of blood literally.

43 *In another of Hildegard's illuminations, Ecclesia, the female personification of the Church, catches the blood of Christ in a chalice. In the lower register, the blood flows to another chalice on an altar, where Ecclesia celebrates Mass.*

San visions

The social and cognitive world of the San could hardly be more different from that of Hildegard of Bingen. Yet there is an important parallel. Both the San and Hildegard lived in encompassing intellectual worlds in which belief in the supernatural and its contacts with human beings was taken for granted. People in both milieux had supernatural interventions built into the ways in which they understood life. They were constantly looking out for manifestations of the supernatural and their bodies were conduits of divine revelations.

Originally, the San lived over the entire southern African subcontinent and spoke many mutually unintelligible languages. Today, viable San communities exist only in the Kalahari Desert of Namibia and Botswana where they are fast being absorbed into a Western-oriented lifestyle (Fig. 44).[40] Formerly, they were exclusively hunters and gatherers who lived in small bands of approximately 25 or 30 people. These mobile bands were closely interrelated rather than exclusive. At certain times of the year a band nexus amalgamated, and the people enjoyed seeing relatives, brokered marriages and performed rituals, the chief of which was the great dance, the healing dance or the trance dance, as it is today variously known.

This dance was – and in some parts of the Kalahari Desert still is – the one ritual that all people, men, women, children and visitors, attend. It has been comprehensively described in many publications, so only a brief outline is needed here.[41] The women sit in a close circle around a central fire; they sing and clap the rhythms of special songs believed to contain a supernatural potency that has been likened to electricity. Depending on the San language

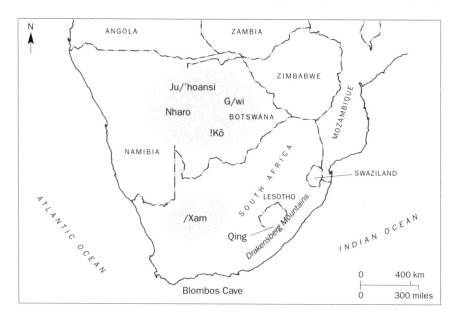

in question, it is variously known as *n/om*, *!gi:*, *//ke:n* and other words. The men dance in a circle around the seated women, now clockwise, now anti-clockwise. The clapping, singing and intensely rhythmic dancing causes ritual specialists to enter an altered state of consciousness. These specialists are usually men but may also be women. In the Kalahari Ju'/hoan language they are known as *n/om k"xausi* (owners of potency; singular *n/om k"au*). To the south, the nineteenth-century /Xam San called them *!gi:ten* (people who are full of potency; singular *!gi:xa*). I shall use the second of these two terms; the now-extinct /Xam language was widely spoken in a range of dialects in the parts of southern Africa with which I deal.

In the most intense cases, *!gi:ten* fall to the ground insensible, even cataleptic. They believe that in this state their spirits leave their bodies to visit God, to fight off malign spirits of the dead, to heal the sick, to guide animals into the hunters' ambush, and so forth. Others, who have learned to control the level of their altered state, remain conscious; their hallucinations are projected onto their surroundings. Although the dance affords the San's principal access to visions, they also glimpse the spirit world in dreams, special curings in which only one or two *!gi:ten* participate, and what we may call waking dreams that, they say, may come as they are walking through the veld. The spectrum of consciousness is clearly involved.

San 'illuminations'

In the form of paintings (pictographs) or engravings (petroglyphs), San rock art is found all over southern Africa, except for the sandy Kalahari Desert where there are virtually no rocks on which to make images. It is with paintings rather than engravings that we are here concerned. San rock paintings have become world-renowned for their remarkable delicacy, accuracy and variety, as well as for the close parallels between them and the voluminous San folklore that was recorded in the nineteenth century.[42]

At present it is impossible to date large numbers of the paintings by direct means. But we know that the most recent images in southern Africa were made towards the end of the nineteenth century or in the first decade of the twentieth century. The art extended back millennia from this time, but the open nature of the rock shelters in which images were painted suggests that most cannot be more than a few thousand years old.

44 Southern Africa. San linguistic groups formerly lived throughout the entire subcontinent. Today their languages and way of life, already much modified, survive only in the Kalahari Desert of Namibia and Botswana.

San rock art frequently depicts the great dance, either in its entirety or by means of single dancers in recognizable dancing postures. It also depicts animals with which the dancers engage and from which they derive potency. Especially interestingly for our present discussion are paintings of visions and transformations that !gi:ten experience.[43]

The class of visions with which we are now concerned is depicted in numerous paintings. These images are construals of what in other publications I have called the 'navicular entoptic phenomenon'; indeed, some of the paintings had been mistaken for depictions of boats.[44] To avoid potential confusion, I here use terms I have employed in this book: 'scotoma' and 'fortification illusion'. Some complex construals of the scotoma depict naturally formed honeycombs complete with bees individually painted. Other images suggest that those who painted them were concerned primarily with the area of invisibility within the arc of the scotoma: in some instances, the legs and bodies of animals protrude from the 'black hole' of the scotoma. The painters, who were in many cases !gi:ten, combined their experienced area of invisibility with their belief that the spirit realm lay behind the rock face. Painters used their art, inequalities in the rock surface and the area of invisibility in a scotoma to represent 'portals' to this realm. The rock was a penetrable 'veil' between daily life and the seething supernatural world.

I now draw attention to two panels of rock paintings that depict the scotoma in a precise way. I do not argue that these images necessarily resulted from migraine attacks, though that may well have been the case; but rather that the painters construed the fortification illusion, however induced, in a way that was specific to the San belief system and that therefore contrasted with Hildegard's Christian construals. Neurological 'raw materials' are processed differently, yet still recognizably, in different social contexts.

One panel was published at the beginning of the twentieth century in an incomplete tracing.[45] The other has not been previously published. One of the reasons why I have selected these two panels is that they were painted some 800 km (500 miles) apart: the San artists could not have been in direct contact with one another and probably spoke different dialects. Nonetheless, ethnographic studies have shown that their overall belief system and many rituals were virtually pan-San.[46] Both panels merit close examination: they are filled with allusions to San religious experiences, beliefs and practices. Yet each was not created by a single painter. Rather, the varied images in them probably accumulated over a long period: they were the contributions of individuals to communal knowledge of the spirit realm.

The first and more densely painted of the two panels is dominated by a prominent form that is remarkably similar to a scotoma: it is a long, thick, slightly undulating black line with bold castellations (part of it has weathered away) (Fig. 45). Elsewhere in the same rock shelter there are two more castellated lines, but without the addition of other images. There are also upward of 30 human figures and numerous images of antelope, some only partially preserved. To avoid the loss of clarity inevitably caused by severe reduction only part of the panel is shown here.[47]

The antelope images are of eland and some small buck; one white image (right of centre and accompanied by a white human figure) may be of an ox and would therefore have been painted more recently than two or three hundred years ago when Bantu-speaking farmers brought cattle into the region. The eland images are especially significant because we know that the San considered this animal, the largest and fattest of all African antelope, to be exceptionally imbued with the potency they activate in the great dance.[48]

Here we again encounter the theme of blood. The San sometimes mixed the blood of a recently killed eland with their paint so that the animal's potency would, via its blood, enter the depiction itself. When dancing in a rock shelter, San dancers turned to these images to draw from them more potency so that they could enter trance and see visions. The parallel with Hildegard's image of the crucified Christ's blood being caught in a vessel and then its potency, or grace, draining down to the chalice on the altar below where the faithful will drink it is striking (Fig. 43). Like holy relics and Orthodox icons, San paintings of eland were reservoirs of potency. Then too we can recall the way in which the Maya caught the blood of their human sacrifices in bowls as an offering to the gods. In all three contexts, Christian, San and Maya, blood was a vehicle for power.

In a nearby part of the panel not shown here there is a large seated figure with one knee drawn up in a commonly painted posture. A stream of small flecks seems to come from its nose or mouth, another frequently painted feature. When San !gi:ten entered trance, they suffered a nasal haemorrhage. This human blood, like eland blood, was believed to contain potency. !Gi:ten rubbed it on people in the belief that its potency would keep sickness at bay.

In sum, we can say that the powerful paintings of eland that form part of the congeries of images shown here provided the foundation for the transforming experiences and visions that are depicted along with them.

The human figures carry the idea of transformation further and make it more explicit. Three human figures, two above and one below, appear to

emerge from the castellated line; all three have outstretched arms. (The small squatting figure to the right was painted earlier than, and is superimposed by, the line.) They recall the human figures Hildegard associated with her depictions of the fortification illusion: like them, the San images seem to emerge from the castellations (Figs 45, 46). Numerous features associate the other human figures with the San's central means of access to the spirit realm: the great dance. At least seven of them are in a clapping posture, their fingers individually drawn; three of these stand upright, while three are seated. They are activating potency by clapping the rhythm of potency-filled songs – a circumstance that is often painted.

Two human figures bend forward at the waist (one is clearly clapping). San dancers say that they adopt this posture when their stomach muscles contract painfully as their potency begins to boil and to rise up their spines. Just to the left of the clapping figures, a person has fallen into a kneeling position: he or she holds a fly whisk, an artefact that the San use only in the great dance and that is commonly painted in dance scenes.

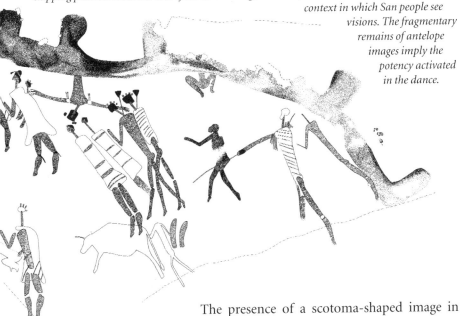

45 *A San rock painting of a fortification hallucination. Two human figures with outstretched arms rise from it, while another emerges from below it. Numerous people are in dancing and clapping postures characteristic of the San healing, or trance, dance. This is the principal context in which San people see visions. The fragmentary remains of antelope images imply the potency activated in the dance.*

The presence of a scotoma-shaped image in the midst of a dance which we know induced altered states of consciousness is compelling. Taking the panel as a whole, I argue that the castellated image represents a scotoma that a *!gi:xa* construed as an entrance into the spirit realm behind the rock face. I suspect that, as with Hildegard, San people in trance sometimes experienced their mental imagery with their eyes open and projected onto their surroundings: in this instance, onto the rock face. People felt themselves drawn to the rock face, for them the 'veil' between this world and the spirit realm, so that they could be swallowed up by the area of invisibility in their own projected scotoma. Moreover, the three figures with outstretched arms that emerge from (or sink into) the scotoma are probably *!gi:ten* enabled by the dancing and clapping people to enter the spirit world. The short lines from their exposed armpits probably represent sweat, another bodily substance associated with the strenuous dance; it too was believed to contain potency.

The second panel that I consider takes this interpretation further (Fig. 46). It is painted, not in an open rock shelter, but in a narrow cleft in a rocky protrusion from a low ridge. The whole panel is painted in black. The prominent thick line has the same sort of castellations as the one I have just described. There are, however, no naturalistically painted animals: the two that are painted are chunky, non-realistic quadrupeds. The human figures here are also less explicitly diagnostic of the great dance than those I have described. One of them may be clapping and one is prostrate with heels in the air. San people in deep trance fall to the ground, usually face down.

Here we have some nuances that are not present in the first panel that I discussed. But these also place the castellated image in the overall context of the great dance and visions of spiritual transformation. Nine bags seem to be hanging from the scotoma. Some of them are not well preserved, but eight clearly have tassels hanging down from them, just as San animal-skin bags still do today. Why did the San frequently paint bags? In their myth and art, bags were and still are symbols of potency and transformation. The interior of a bag is dark and smells strongly of the animal from which it was made. Because the San believe smell to be a vehicle for potency, there is, for them, an overwhelming concentration of potency in a bag. We see evidence for this belief in a nineteenth-century southern San myth. /Kaggen, the trickster deity, gets into a bag and is transformed into a flying creature.[49] Significantly, both transformation and flight are depicted in the striking winged figure left of centre. San *!gi:ten* commonly transformed into birds.

Why do the bags appear to hang from the scotoma? It was common practice for San to hang their bags from wooden pegs hammered into cracks in rock shelter walls. In some shelters the remains of ancient pegs can still be found. I suggest that, in this instance, the painter was thinking of the entire shelter as being an entrance into the spirit realm: he or she equated the arch of the

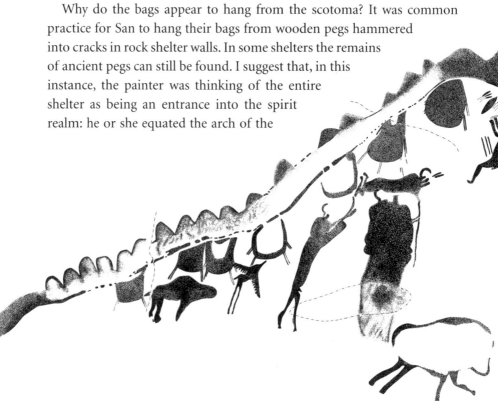

rock shelter with the curve of the scotoma; the interior of the shelter, where the great dance and transformation into a flying creature could take place, was equivalent to the obliterating part of the scotoma, the 'black hole'. San *!gi:ten* commonly speak of entering holes in the ground in order to visit the spirit realm: a rock shelter was such a hole.[50]

Before leaving these San paintings, I note that comparable castellated arches painted in rock shelters of the Lower Pecos River in Texas have also been interpreted as entrances into the spirit world.[51] Indeed, caves in many parts of the world are regarded as entrances into a supernatural realm. As we have seen, the Maya thought of the built caves under mountain-pyramids as entrances of this kind: they were not alone.[52]

Religious experience, belief and practice in medieval Europe and Stone Age southern Africa

The Hildegard illuminations and two San rock painting panels that I selected for discussion show how a neurologically wired mental image can be interpreted in different ways in different cultures. In all such cases, the religious concept represented by the motif is an amalgam of physiological and cultural factors. A religious visual experience (the scotoma) is interpreted in terms of a belief system and manifested in religious practice (illuminating a manuscript and making a rock painting).

The social roles of visions in the two societies are also comparable. As I have pointed out, there is little doubt that Hildegard's fame and political influence were

46 *Typical San leather bags hang from a fortification hallucination. The winged figure suggests the experience of flight of which San shamans speak. The fortification with its 'black hole' of invisibility may represent an entrance into the spirit realm behind the rock face.*

founded in large measure on her visions: she had direct access to divinely revealed knowledge. Small wonder then that many people wrote to her seeking her advice and even the Pope listened to her.

The San present a more complex situation. Unlike hierarchical medieval Europe, San communities are widely recognized as essentially egalitarian. They have no chiefs, and decisions (such as when to move camp) are taken collectively. On the surface, it appears that San *!gi:ten* do not enjoy the sort of status that priests in Christian churches have, or that Hildegard enjoyed. They perform all the usual daily tasks: if they are men, they track and hunt animals; if they are women, they gather plant foods. They do not wear distinguishing clothes; nor do they live in separate places. In a camp, visitors can discern no special deference accorded to *!gi:ten*.

Yet they are respected. Indeed, some become famous and may be summoned from afar to treat a sick person. The status of *!gi:ten* in general is best seen after a dance when people sit and listen attentively to their reports of activities and beings in the spirit realm. They tell of what they saw, the evil spirits (sometimes in the form of lions) that they fought off, what is happening at distant camps, what will happen in the future and how they and others are transformed into animals.[53]

In addition, we must consider the social impact of the images that the San painted in rock shelters. Because shelters also served as living spaces, the accumulated paintings formed a backdrop to daily life. They constantly reminded people not only of the closeness of the spirit world behind the 'veil', but also that their *!gi:ten* had access to that realm. The images spoke of the power of *!gi:ten* to move between realms, to traverse the tiered cosmos. Similarly Hidegard's illuminations brought her texts to life. Readers could see what she saw and be awed by the insight, majesty and terror of the illuminations.

The secrecy that so often accompanies divine revelations of this kind requires further comment in both the medieval and the San contexts. Secrecy and power go hand in hand.

Hildegard spoke of her *secreta Dei*, secrets of God:

You, human creature! In the way of humans, you desire to know more about this exalted plan, but a seal of secrecy will be imposed on you; for you are not permitted to investigate the secrets of God more than the divine majesty wishes to reveal, because of his love for believers.[54]

Caught in the trammels of medieval ignorance and power struggles, she says that the loving God wishes people to be ignorant of many things.

I have referred to the secret language that Hildegard invented. I suggest that it was a manifestation of her dilemma: she had to broadcast her revelations because God required that of her and because her position of influence depended on it. But if she gave everything away she would lose her uniqueness. So she concocted a language that only she and some of her nuns understood. Here is an important point: it did not really matter *what* was written in the secret language. It was the very existence of the language that advertised her unique relationship with God. Far from their being a personal and esoteric pastime, Hildegard's efforts at cryptology should be seen in the wide social context of divine revelation and the secrecy, total or partial, that special revelation implies. Knowledge, as is widely accepted, is power – but not if you give it all away.

Among the San the problem of secrecy that is created by supposed divine revelation played itself out in a slightly different way. To understand that way we need to go back to the nineteenth century. In 1873, Joseph Millerd Orpen, a British magistrate, interviewed a San man named Qing about rock paintings in the Drakensberg Mountains. Qing explained much to Orpen, but when Orpen asked him where Coti, a mythological figure, came from, he replied that 'only the initiated men of that dance know these things'.[55] He implied that he was not one of 'the initiated men'. The dance to which he was referring was the healing dance that I have described.

How do we square Qing's disclaimer with what we know about the openness with which Kalahari *!gi:ten* today speak about their visionary experiences? We must allow that the Drakensberg *!gi:ten* may have been more secretive than those in the Kalahari, but I suspect that there is more to it than that. I think that Orpen was asking about something that nobody, not even *!gi:ten*, knew about or, for that matter, had even wondered about. In all religions, there are questions that outsiders ask but that devotees consider to be of no interest. Faith obliterates curiosity. As in *all* religions, we have here an example of select people – 'the initiated men of that dance' – supposedly knowing more than other people and being respected for their knowledge. In all religions, ordinary people mistakenly think that ritual specialists know more than they in fact do know.

We must also remember that the ways in which visionaries apprehend religious knowledge differ from the ways in which ordinary people learn about these things. For visionaries, sacred information is not second-hand, learned from others. Visionaries actually *see* things first-hand, face to face. And part of their 'secret' is the mode of acquiring their visions. Among the

San, most young men strive to become *!gi:ten* by dancing with an experienced *!gi:xa*, but many abandon their attempt because, so they say, the transition to the spirit realm is too painful and terrifying. The ultimate 'secret' thus remains with the brave few.

With 'seeing' comes another feature of intense religious experience. Hildegard says she 'saw, heard, and understood in an instant'.[56] She experienced a sudden, engulfing understanding of her visions. This feeling of overwhelming certainty recalls Wordsworth's secular interpretation of his state of Absolute Unitary Being (Chapter 5). He felt that he was seeing 'into the life of things' and that 'the heavy and weary weight / Of all this unintelligible world is lightened.' Exalted certitude is also part of the migraine syndrome. Sacks writes of the migraine sufferer's 'feelings of sudden familiarity and certitude' and in some cases 'states of profound awe or rapture' and of 'oneself being newly minted'.[57] He also points out that migraine visions come with an intense sense of actuality.[58] He writes of a physician who had much experience of migraine attacks and the illusory nature of migraine auras but who would nevertheless involuntarily search for the cause of the odour that was part of his aura. All this is true of hallucinations triggered by other factors as well: from the Old Testament prophets to the Maya to modern-day televangelists, those who experience revealed knowledge preach with terrifying fervour. Similar feelings of certitude were probably common to both Hildegard and the San. Migraine, one of a range of similarly affective altered states of consciousness that are built into the human brain, teeters on the brink of religion.

The differences between visions in Hildegard's medieval Germany and in the San's Stone Age rock shelters were, I argue, negligible. In both contexts, the scotoma, whether triggered by migraine or rhythmically induced trance, was taken to be evidence for supernatural revelation. But what exactly could it mean? People in both communities construed their neurologically generated hallucinations in accordance with the beliefs of their communities: religious experience, belief and practice interacted in similar ways. In that, they were not much different from Palaeolithic cave painters, the Maya, biblical prophets, and present-day charismatic Christians.

CHAPTER 10

God's Empire Strikes Back

The question of whether there is a God, and what that God might be like, has not – despite the predictions of over-confident Darwinians – gone away since Darwin, and remains of major intellectual and personal importance.[1]

Many people are dismayed by the current clash between religion and science. They see so much of what they and their forebears valued in their religious tradition slipping away. Some who do not themselves believe see their religious friends assailed from all sides and feel the need to spring to their defence. It is not only believers who desire some sort of rapprochement between science and religion. Indeed, some non-believing scientists lean over backwards to accommodate believers. They imply that the awe that many of us experience when we contemplate the vastness and the intricacy of the universe points to the reality of religion and God. At the same time, many people who are not especially given to religion worry that science has become a run-away human endeavour. They argue that it needs to be reined in and guided by specialists in morality and religion (note the link between morality and religion).

Long ago, Thomas Aquinas thought it necessary to provide a series of arguments to prove the existence of God (Chapter 2). His strategy shows that, even in the thirteenth century, there were those who doubted the existence of God. Indeed, atheism has a long history. As long ago as half a millennium BC, the Psalmist wrote, 'The fool hath said in his heart, There is no God' (PSALM 14:1; also PSALM 53:1). But atheism has never been as strong as it is today in the West, and non-believers have turned up the intensity of their attack since the end of the twentieth century with the publication of bestsellers like Richard Dawkins's *The God Delusion*.[2]

When all is said and done, we face an undeniable tectonic shift. The continental plates have, unquestionably, moved. Why is it so hard to believe in God in (many *milieux* of) the modern West, while in 1500 it was virtually impossible not to do so?[3] The initial chapters of this book provide an answer to the more substantive implications of this question: quite simply, the growth of science has changed people's mindset. The change came about not because sharp-minded atheists stunned the public with their logic and philosophical arguments. Rather, the difficulty of believing in supernatural things has grown up, slowly and with setbacks, over a period that lasted from Thales to the present: the gradual advance of science has brought with it a way of looking at the world and conducting one's life that does not require belief in supernatural entities, indeed that renders their existence highly unlikely (Fig. 47).

It was not long before the Church struck back in the current phase of the debate – though no religious book has sold as well as *The God Delusion*. Religious responses have taken various forms and adopted divergent tones.[4] Some believers accuse writers like Dawkins of being sneering and arrogant. The debate easily degenerates into name-calling, 'fundamentalist' (with its modern connotations of suicide bombers) now being just about the worst thing even an atheist can be. God's warriors can be just as virulent as they allege Dawkins to be. It is they who preach forbearance and love, yet, as history has shown, they can abandon these virtues at the drop of a hat and organize an *auto-da-fé* (literal or figurative).

47 In 2009 the science versus religion controversy spilled over onto London public transport with an advertising campaign promoting atheism. Soon, believers responded with their own slogans on buses. The debate became very public.

Indeed, the Church has for 2,000 years denounced non-believers as heretics and sinners, excommunicating them, persecuting them and burning them. Nowadays it stages vast evangelical rallies (in addition to routine Sunday services) at which non-believers are branded as 'sinners' desperately needing the 'salvation' that only the Church offers. Yet, if the 'sinners' indignantly respond to these insults, the Church tells them that they are ignorant, hubristic, overweening, offensive, insulting, disrespectful, and, yes, 'fundamentalist'. Reading these writers of faith has, I confess, been a depressing experience. Nevertheless, the Church's various responses merit consideration. I do not wish to be denounced as a 'fundamentalist'.

Denial

Denialists are perhaps the believers who are most difficult to understand. They claim that there is no problem: for them, Church membership is not falling, and atheism is not gaining ground in the West.

John Gray, Professor of European Thought at the London School of Economics, is an articulate denialist. For him, it is not difficult to believe in supernatural entities in the modern world. He wrote a long article for the *Guardian* (15 March 2008) in which he combined a number of strategies to denigrate writers like Dawkins, Christopher Hitchens and Daniel Dennett, all of whom he clearly and passionately dislikes. He ended his article with a breathtaking denial:

The attempt to eradicate religion, however, only leads to it reappearing in grotesque and degraded forms. A credulous belief in world revolution, universal democracy or the occult powers of mobile phones is more offensive to reason than the mysteries of religion, and less likely to survive in years to come. The Victorian poet Matthew Arnold wrote of believers being left bereft as the tide of faith ebbs away. Today secular faith is ebbing, and it is the apostles of unbelief who are left stranded on the beach.

Gray ignores the west European Church's own alarm at falling membership, diminishing authority and lessening influence on politics. His dislike of punchy, bellicose atheistic writers seems to have prevented him from seeing the reality of a waning Church. Gray is also wrong to argue that an absence of religion 'only' leads to its reappearance 'in grotesque and degraded forms'. The old religion (acceptance of supernatural agency in the material world) has not been replaced by a new religion that 'worships' reason, 'the occult powers of mobile phones', and other abstractions. Rationalists do not 'worship' anything, let alone anything supernatural. Blurring the meanings of words is a common debating trick.

The United States of America is another matter altogether. Why does the American population overwhelmingly still claim to believe in God and a spirit realm? Over half the population rejects evolution; many want Intelligent Design, Reverend William Paley's grandchild, to be taught in schools in place of, or at least alongside, evolution. They say, 'Teach the debate!' But, despite creationist claims, there is no purely *scientific* debate about evolution as the mechanism which led to life on earth as we see it today and as we find it in the fossil record, though details of when certain changes took place are still open to research. Arguments like the one that fossils on hilltops are a result of Noah's Flood are so absurd that most scientists will not waste their time refuting them.[5]

High American political office is, even after the Republicans' defeat in the 2008 election, effectively closed to any who openly proclaim atheism. The phrase 'under God' was inserted into the Pledge of Allegiance as a result of a sermon that Reverend George Docherty preached before President Dwight Eisenhower. Docherty argued that the phrase was acceptable to Christians, Jews and Muslims. It also excluded atheists, who, Docherty believed, 'fall short of the American ideal of life'.[6] How has this state of affairs, so different from what one finds in western Europe, come about?

Alister McGrath, an Oxford scientist who has become a leading Christian theologian and denialist, entitled one of his books *The Twilight of Atheism: The Rise and Fall of Disbelief in the Modern World.*[7] He argues that atheism is linked to a particular time and place, that is, western Europe, where for centuries the Church closely aligned itself with oppressive secular regimes and so gave cause for revolt.[8] In America, by contrast, constitutional separation between state and Church meant that the conditions in which European atheism thrived were absent. Enlightenment ideas had reached America by the time that its constitution was formulated. So America has escaped atheism. I am not persuaded. I find it hard to accept that disbelief in a supernatural realm is merely a product of a special point in European political history. McGrath's view can be entertained only if we disregard the long history of science and the total absence of any scientific, testable evidence for supernaturalism.

How then did religion come to dominate the United States? Tentatively, I offer one possible explanation (although it is more likely to have been a conjunction of factors[9]). Religionists who feel attacked often raise the bar and claim to believe more and more impossible things as a demonstration of their steadfast faith. Here is a bizarre example. In the second half of the nineteenth century some exceedingly unwise people who felt themselves under attack from science and Darwin argued for a flat earth.[10] So, too, present-day American religious conservatives, likewise feeling under attack by science and perhaps feeling rejected by third world people, turn in upon themselves and emphasize their supposedly God-given mission to enlighten the rest of the world. As they do so, they raise the bar and move towards fundamentalism. Many now claim divine approbation of their lifestyle and embrace hardline, anti-science fundamentalism.

There are those, of course, who are critical of their nation's foreign policy, militarism and consumerism. They tend not to belong to the religious right. They are not blinded by the belief that, born again, they are heirs to supernatural blessings and missions. Perhaps now, in the aftermath of the 2008 election, fundamentalist American religion will decline. Waxing and waning religious revivalism has long been part of American history.

The American anomaly illustrates the intricately interacting nature of religious experience, belief and practice in a particular political and social context. Often, politics and economics play such a prominent role in the amalgam that it is hard to know whether we should use the word 'religion' at all. Nevertheless, the power of faith in otherwise unbelievable things should not be underestimated.

Whatever may be the case in America, the Church itself knows that religion is waning in western Europe. Empting churches are plain to see. *Pace* Gray and McGrath, it is not the secular point of view that is ebbing like Matthew Arnold's tide on Dover beach. It is belief in supernaturally revealed knowledge that is evaporating. Even in devout Ireland, the Church confronts diminishing support. And the benefits are patent. The weakening of both the Catholic and the Protestant Churches' hold on the Irish people probably contributed to the viability of a political solution to that riven land's problems. Certainly, after so many centuries of Irish strife, the Church cannot claim to be a blessed peacemaker. Then, too, the Italian and Spanish governments are no longer intimidated by the Vatican. In the United Kingdom, the established Anglican Church, the most accommodating of all churches, is disintegrating: congregations are diminishing and as many as a third of the bishops failed to turn up for the 2008 international Lambeth Conference, a gathering that was supposed to manifest the worldwide unity of the Anglican Communion. Anglicans cannot agree on whether there should be female bishops. Nor can they reach agreement on the status of gay members and clergy. Urgent matters that cry out for moral and ethical clarity pass unresolved: the supernatural revelation that the Church claims to have counts for nothing when believers come up against the really hard issues of life. If, in this age of multiculturalism, the various branches of the Christian Church (of all institutions) cannot agree on important issues, we are entitled to question the very notion of supernaturally provided knowledge, insight and guidelines for behaviour.

Walking away

A Teflon approach to the past is not new. Church history written by church-men (and also by writers of less faith who nevertheless accept the in-house propaganda) has more often than not presented the development of the institution in two ways. First, as triumphantly spreading a benign message and, secondly, as a matter of gradually getting theology right and thereby achieving an increasingly wise and accurate understanding of God and 'true' religion. Is there any virtue in this two-part view of Church history?

Understandably, believers point to the good that the Church has achieved. Often, they cite the abolition of slavery and the preservation of knowledge during the Middle Ages as shining examples. But the position is not that straightforward.

In biblical times, slavery was taken for granted: neither the Old nor the New Testament condemns slavery. On the contrary, Paul and the writer of the Epistle of Peter both urged that servants (for which read 'slaves') should obey their masters 'with all fear' and not answer them back. Slaves should accept cruel treatment: they should be submissive to owners who were not only 'good and gentle, but also to the froward [harsh]' (1 PETER 2:18).[11] Rephrased, this reads: 'Accept the lash meekly. Do not revolt!' For many centuries thereafter the Church approved of slavery, as indeed did Islam. In the Middle Ages, feudalism was a thinly disguised form of slavery in which the Church acquiesced. In the United States of America, religion stood on both sides of the debate (and war) about slavery, as in South Africa the Church was split between those who denounced apartheid and those (the Afrikaans-speaking Dutch Reformed Churches) who openly supported the policy.

The Church's inertia in the face of such blatant evil was remarkable and stands as a monument to its moral ineffectiveness. In the end, it took conscience-stricken individuals, not the united institution, to get the antislavery movement on the road. Even those churchmen who in the nineteenth century successfully opposed slavery can be said to have waited until profound economic changes led to the growth of capitalism and the need for a large wage-earning and purchasing public. Along with those changes went the great ground-swell of the Enlightenment and the French Revolution: slavery could have no place in the ethos of *Liberté, Egalité* and *Fraternité*. The abolition of slavery is more persuasively explained by economics and secular philosophy than a sudden awakening of piety after centuries of dormancy.

Comparable explanations cover the matter of literacy and the medieval preservation of knowledge. Let us remember that the medieval Church kept literacy very much to itself: secular authorities often relied on the Church to keep their records and accounts. When lords and princes sealed pacts (often to secure the crushing of any sign of nationalism in a province), the Church was there to record it all. If the Church were to teach the peasantry to read and write, these poor people could be employed by secular authorities, perhaps more economically. The Church also realized that if the peasantry learned to read, they would be able to form their own understanding of the scriptures. So, when literacy did become more widespread, various measures were enacted to prevent the laity from reading the Bible. When the great Reformers urged ordinary people to read the Bible, they were, ironically, encouraging a social class that would work for profound economic change and a weakening of not only the Catholic Church's power but, moreover, their own Protestant influence as well. The medieval Church was not the benign institution of popular imagination.

The issues of slavery and literacy are but two of a number of tainted arguments for the good that the Church is supposed to have wrought. Despite the Church's appalling record of its 1,800 years' acceptance of slavery, obscurantism, dogmatism, torture, burnings, persecutions, full-scale wars, rejection of science, and so on, many believers still stoutly maintain an in-house attitude to this sorry tale: disingenuously, they say that all that horror was 'then'; now is now. For the most part, the Church simply walks away from its dismal past: present-day Christians protest that they are not responsible for the Church's past mistakes. Perhaps. But is that the point?

What, then, of the Church's claim to have refined its beliefs over the centuries? As we saw in Chapter 1, the Council of Nicea, called by Constantine in 325, is frequently taken to have been a fundamental event in the defining of 'true' Church doctrine. Writers omit to mention that Nicea, together with so much else of Church history, was deeply embedded in politics and all-too-human personalities. The historian Charles Freeman has tried to correct the balance in his books *The Closing of the Western Mind and AD 381: Heretics, Pagans and the Christian State.*[12] He points out that the Council was not fully representative of the Church, the Roman bishops being absent, and that Constantine presented himself as the image of God on earth. Freeman's endeavour has earned him considerable ecclesiastical opprobrium. What actually did happen at Nicea and subsequently? The Church achieved a semblance of unity not through amicable agreement but by denouncing variations in belief as 'heretical'. The unity of the Church was and still is an illusion.

So bleak a past cannot be simply swept under the carpet: *it is telling us something.* But what? An institution that claims divine revelation and preaches love must expect to be assessed by its own standards. The historical evidence shows that charity – love – was, and still is, practised as much (arguably more) outside the Church as in it. The notion of the Church as a possessor of divinely revealed knowledge must, on grounds of its own report card, be discounted. That some secular organizations (Stalinist communism, for example) are just as bad is not the point. Rather, we have the right, based on the Church's own claims to revealed knowledge and divine empowerment, to expect something noticeably better than the record of any non-religious organization we may care to name. The counter argument that revealed knowledge and the 'grace' (divine assistance) to implement it should be seen in a context of human frailty and be forgiven – as it is by the Church itself – is unconvincing. As Jesus himself put it, 'Ye shall know them by their fruits' (MATTHEW 7:16). Not, be it noted, by just some of their fruits.

Of course, Church history does not in itself disprove the existence of God and the supernatural. Nor should we ignore the good that religion has achieved: especially since the nineteenth century the church has pursued a programme of social services. In doing so, it has probably mapped out its future. But there are two points we must bear in mind. First, the horrors of Church history necessarily diminish our confidence in the Church's claims to contact with a benign deity. Secondly, the good that the Church, or groups within the Church, has achieved does not point to the existence of a supernatural realm. There have been so few 'poor Parsons', perhaps more outside the Church than within it. I repeat: the generally dismal history of the Church is telling us something.

Experience

In responding to atheistic and agnostic disbelief in supernatural things, devout writers have tended to ignore religious experience and to concentrate on logical and philosophical issues. At first glance, this appears strange. If, as I argue, certain mental experiences are the foundation of religion – though not something to which it can be reduced – why do defenders of the faith not use experiences more frequently to persuade non-believers of the existence of God? One reason is that most religious writers who pick up the gauntlet are intellectuals. Prudently, they leave overpowering religious experiences in evangelists' tents and TV shows.

An author who does not follow this trend is David Robertson, a pastor of the Free Church of Scotland. Taking up what seems to have become an obligatory starting point, Robertson distances himself from an understanding of religious experience as the hearing of voices and the seeing of visions, though he does not discount the validity of at least some of these experiences. Instead, he points to:

- answered prayer,
- a sense of God ('truly God is among you'),
- experience of the miraculous,
- experience of the truths and truthfulness of the Bible and
- the experience of being filled with the spirit.[13]

Robertson says that this is the sort of thing that ordinary people experience and which leads them to belief in God and the supernatural.

We are clearly dealing here with sensitive issues. I certainly do not wish to ridicule the devout for whom the experiences that Robertson lists mean a lot. Still, I must point out that 'a sense of God' and 'being filled with the spirit' are of the same category as the voices and visions of which he is suspicious: all these experiences are produced by neurological events in the brain that have been identified and studied. There is no need to invoke the supernatural to explain them, though I know that some would say that God works through neurology to reach us. If we take that fall-back position, we create further problems. Personal mental experiences are internal; they are not persuasive to onlookers, some of whom are aware of exactly the same thing in secular contexts (think of the voices of schizophrenia and Wordsworth's sense of the 'sublime'). Then, if we allow that someone's mental experiences point to the existence of God, we have to ask about the implications that I have noted elsewhere in this book. If a supernatural being or realm intervenes in the material world, those interventions must be clearly and unequivocally observable by everyone, not just locked up in the heads of believers. If they were generally observable, we would all be believers. Inner experience, though the foundation of religion, is therefore not a persuasive argument for the existence of a supernatural realm. It results from one way of interpreting the electrochemical functioning of the brain.

The notion of prayer brings its own poignant difficulties. Why, we may ask, is it necessary to beseech God to heal a sick child? Why do we encounter the belief that the more earnestly and repeatedly we pray and the more people who pray the more likely that God is going to answer our entreaties and heal the child? Why is God waiting for us to pray? If we do not pray, will the child go on suffering? This is not merely an age-old enigma that we can pass over as an inexplicable 'mystery'. The questions I pose powerfully undermine the very notion of a supernatural, omnipotent, loving God.

Some positive responses

Making the best of a bad job is a position adopted by some within the Church. They know all too well that the tide is ebbing 'down the vast edges drear / And naked shingles of the world'. We can consider a few of their strategies.

Roy Strong, a former director of the Victoria and Albert Museum and evidently a practical man, wonders what will happen to English parish churches, those ancient and beautiful buildings for which the young Darwin yearned on his *Beagle* voyage. They cannot be allowed to fall into decay.[14] They were once the centre of a kind of village life that no longer exists and, I may add, has never existed in many parts of the world where the Anglican Church established itself. Who will pay for the upkeep of the many medieval churches that are no longer required by believers is, of course, a troubling matter. Strong recommends that Christian denominations should bury their differences, that ugly Methodist and Baptist chapels should be closed and converted into residential apartments (something that is already happening), and that the remaining believers should take turns at using the ancient parish church buildings and accept responsibility for their upkeep. Something like this may well happen, but the point is that Strong, unlike McGrath and Gray, is aware that faith in unbelievable things is ebbing. Sharing buildings will not reverse the trend. It will merely ease the decline.

A more sanguine outlook comes from Stephen Platten, the Bishop of Wakefield.[15] He observes that English cathedrals appear to be doing better than humble depopulated parish churches. He also notes that the Orthodox Churches in eastern Europe are doing well; emigrants from those countries are in fact shoring up the falling membership of highly liturgical Catholic churches in their host countries in western Europe. Platten's main point is that the eastern churches are steeped in complex, elaborate rituals. He concludes that the Anglican Church should revitalize itself by going down the High Church road of impressive, awe-inspiring rituals (the famous smells and bells) that will, he believes, seize people's imaginations. Believers like Platten fail to understand disbelief. He thinks that, if only the Church can get it right (whatever 'it' and 'right' may be), people will return to the fold. More realistically, people who cannot believe in a supernatural realm and divinely revealed knowledge will not find renewal of their faith in a theatrically spiced-up Church, even though some may enjoy its music and ceremony. Furthermore, removed from the transient political situation in eastern Europe, Catholic immigrants will not indefinitely support an institution that is founded on falsehoods and ecclesiastical power structures.

Another response is diametrically opposed to the one that the Bishop of Wakefield advocates. This difference of opinion further illustrates the theological and strategic disarray in which the Church finds itself. Richard Turnbull, principal of an Anglican theological college in Oxford, insists that the Church must return to Evangelical values.[16] For him, the trappings of the High Church with its elaborate rituals and emphasis on Church tradition are destroying the authentic institution. Arguing that the 'Low' Evangelical wing of the Anglican Church is the true Church, he emphasizes the Bible as the only foundation for 'true' belief and practice. This foundation includes Jesus' sacrifice on the cross as a satisfaction of God's wrath, personal religious experience (conversion and an emotional relationship with Christ), and an implacable desire to spread this view of Christianity.

Is what the bishop preaches a viable strategy? It seems unlikely that belief in a divinely inspired Bible, with all its inconsistencies, contradictions and downright unpleasantness (Lot included) is ever going to attract an unbelieving public. Despite the way in which televangelists brandish it, the Bible is a dead duck in the educated Western world, not as an influential piece of literature, but as a unique, inerrant, divine revelation. For instance, Hector Avalos, professor of religious studies at Iowa State University, is calling for an end to biblical studies as we know them.[17] He is not an ignorant atheist: he has a doctorate in biblical studies from Harvard. He argues that, except for cherry-picking acceptable bits, the Bible is irrelevant to modern society.

48 *A fourteenth-century illuminated manuscript from Bologna presents the Church, in the lower register, paralleling God's judgment in the upper register. The Pope, God's viceroy on earth (at least in the Roman Catholic Church), has the power to torture and burn heretics and sinners.*

He goes on to point out what should be obvious: study of the Bible is kept alive by religiously motivated agendas emanating from institutions that depend on its continued veneration.

The Bible-centred Evangelical wing of the Church of England is indeed fighting back. In an interview published in *New Statesman*,[18] Tom Wright, the conservative Bishop of Durham, came up with a take on life after death that seems to be of the 'raise-the-bar' variety. The Last Days, death and what happens to the body when, at the end of time, Judgment Day dawns, have long been topics of Christian speculation, ones which painters and poets have graphically and sometimes violently explored (Fig. 48). John Donne wrote of 'numberless infinities of souls' flying to their 'scattered bodies' at the end of time, joyously for those whose pardon had, as he put it, been sealed with Christ's blood. Of the others, the less said the better. Their hellish

agonies are, however, vividly depicted in the Sistine Chapel and in those medieval wall paintings and sculpted tympanums that escaped destruction.

For the most part modern Christians prefer faith and hope to lurid supposition concerning eschatology (the study of death and the end times), but unity of doctrine on this point has proved elusive. As is routine these days, the Bishop of Durham pooh-poohs Sunday school and nursery ideas of angels sitting on clouds. Claiming to take the Bible as his source, the bishop says that, after a literal Second Coming of Christ and Day of Judgment, there will be a *physical* resurrection of human bodies and a remaking of the material world with room for everyone. The Resurrection of Jesus and the resurrection of human beings are both to be taken literally. Wright, like John Donne, has no time for mealy-mouthed metaphors. Physical resurrection 'is actually the key thing,' he says. 'We are talking about a good physical world which is to be remade, not a bad physical world which is going to be trashed in favour of a purely spiritual sphere.'[19] Will this higher leap of faith attract non-believers? Will it be palatable even to moderate believers? (I think I prefer angels sitting on clouds.) Retreat into a literalist medieval mind-set is not a viable option for the Church. Today brazen irrationality (read: faith) is no longer a safe bolthole. Neither the Bishop of Wakefield's ritualism nor the Bishop of Durham's and Richard Turnbull's fundamentalism is a cure for the current decline in religious belief in western Europe.

A more innovative response to dwindling interest in the supernatural is to link faith to a dramatic, apparently secular, concern that is, for the most part, uncontroversial. These writers adopt the rhetorical strategy that Paul used in his address to the Athenians: they start from what they think will be agreed ground and segue to their real message. Nick Spenser, a former market researcher who now works for the Jubilee Centre (Christian discussion arena), and Robert White, Professor of Geophysics at Cambridge University, argue that something must be done about climate change and deteriorating global environments.[20] They claim that more than science is needed to combat global disaster. No question about it. But they then go on to take concepts from the Old Testament about the world belonging to God and people's responsibilities to keep it in good order. Readers are caught in the middle. They may agree with Spenser and White that 'urgency' should be our watchword, but will they buy into Christianity simply because they think doing so will assist in global management? Given its history of tardy responses to its own internal inhumanities, is the Church likely to have any real impact on efforts to combat climate change? I think not. Diminishing belief in what the Old Testament teaches is something that Spenser and White simply ignore: it undermines their otherwise laudable aims.

One of the most artful responses comes from Damian Thompson, who holds a London School of Economics doctorate in the sociology of religion. His book is entitled *Counterknowledge: How we Surrender to Conspiracy Theories, Quack Medicine, Bogus Science and Fake History.* He defines counterknowledge as 'misinformation packaged to look like fact'.[21] He wittily exposes the lunacy of Creationism, pseudohistory, alternative medicine and so forth. In doing so, he asks why people believe such palpable nonsense. The reader is taken in by all this scepticism: surely we are dealing with a no-nonsense, down-to-earth author who cannot be bamboozled, one whom we can trust. Then we find that Thompson is not only himself a religious believer but an influential Catholic: he is editor-in-chief of the *Catholic Herald.* Presumably, he believes in the Virgin Birth of Jesus, the Immaculate Conception of the Blessed Virgin Mary, the infallibility of the Pope pronouncing *ex cathedra*, the role of saints in the Catholic firmament, and so on. Seemingly, he does not realize that all this is as much counterknowledge as are Creationism, the belief that doses of vitamins can cure AIDS, and homeopathy, all of which he castigates. His denunciation of quackery and foolishness is merely a cloak for Catholic counterknowledge.

Complexity

A less devious approach is what I call the Prufrock strategy. In T. S. Eliot's poem a bored Prufrock says,

> Would it have been worth while
> If one, settling a pillow or throwing off a shawl,
> And turning toward the window, should say:
>   'That is not it at all,
>   That is not what I meant, at all.'

T. S. ELIOT, *The Love Song of J. Alfred Prufrock*, 106–110

Taking this line, theologians respond to writers like Richard Dawkins, Christopher Hitchens, Lewis Wolpert and Daniel Dennett[22] by claiming that these ignorant atheists present a caricature of Christianity: 'What you say we believe is far too naïve. It is not what we believe at all.' I am, of course, aware that believers will respond to this book by saying that I trivialize Christian theology and that I do not distinguish between fundamentalist ravers and subtle theologians.

Given the present-day flowering of atheism and the wide publicity that it garners, it is understandable that theologians are now side-stepping what I argue is the most fundamental – indeed, simple – issue of all: the very existence of a supernatural realm. Instead of tackling the matter head-on, they talk of metaphors and follow convoluted routes to definitions of God and 'transcendence' that, in their view, make criticisms of supernaturalism appear simplistic and ignorant. This is merely another strategy. Theologians go for complexity, metaphors and slippery definitions: some claim that anything as complex as the notion of God must be inexplicable in simple language. They thus return to the ancient notion of religious 'mysteries' – factitious enigmas that are created by religious belief and that become raw material for the theological industry. The existence of a parallel supernatural

realm and the attendant notion that people can appeal to this realm for interventions in the material realm are simple and fundamental issues that
uneducated people all over the world and throughout history have been able
to understand. Prufrock notwithstanding, either there really is such a realm
or humanity has by some means or other created the illusion of its existence.

The intellectual believer's response that God is an extremely complex 'idea'
that may be encountered in 'social practice', rather than an aged anthropomorphic being in the sky, begins with a claim that recalls the stance taken
(laughingly) by some string theorist physicists: what we study, they say, is too
complex for ordinary people to understand or, to take the joke further, is
intrinsically too complex for anyone to understand. As Prufrock put it:

> It is impossible to say just what I mean!
>
> T. S. ELIOT, *The Love Song of J. Alfred Prufrock*, 4

Some of the counter-attacks by religious people are abstruse in the extreme.
For instance, Sir Michael Dummett,[23] a respected British philosopher, argues
from a mathematical and philosophical perspective that is situated in the long-
standing realist *versus* anti-realist dispute. Taking a well-known philosophical
conundrum, he asks: are numbers real? We talk about one, two and three,
but do these numbers exist outside of people's heads or are they 'constructs' –
mere ideas that have no independent reality? If we take an anti-realist position,
as Dummett does, and conclude that the world exists only if, and as, it is
apprehended by people, we have to conclude that there could not have been a
cosmos without someone to think about it and thus to call it into existence.
If thinking makes things exist and they do not exist independently of thinking,
and, further, if there are (potentially) as many different ways of apprehending
the world as there are people and creatures, there must be 'something' that
apprehends the whole cosmos but not in any particular way. And that 'something' must be God. *Ergo*, God exists. Few of Dummett's 'ordinary' readers
will wish to follow all this. Certainly – and this is an important point – they will
have trouble moving from the 'something' God to God the Father who hears
prayers and intervenes in human lives.[24] Similarly, scientific speculation about
possible parallel universes has no bearing on belief in supernatural interventions by a beneficent god in the affairs of the universe we know. Any parallel
universes, if they do indeed exist, are probably as material as our own.

One of the complex 'mysteries' with which Christian theologians wrestle is
the Trinity. Since the Council of Nicea, it has remained a cornerstone of

orthodoxy, the touchstone of who is for us and who is against us. How do we get around its complexity and contradictions? Does its very complexity imply its truth? G. K. Chesterton is said to have claimed that all great truths are paradoxes. Perhaps. But the converse is certainly not so. In one of the many interviews I had in preparation for this book I asked an Anglican priest if I could join his Church if I did not believe in the Trinity. He replied that belief in the Trinity is fundamental. But, in true Anglican fashion, he hastened to add that that need not bother me too much. He argued that in today's world we encounter God not in miracles or personal epiphanies, but rather in human relationships (presumably only certain relationships). God is in how we react to other people, and vice versa. The Trinity, he then explained, is all about relationships – between God the Father, God the Son and God the Holy Spirit. It is a model by which we can understand how we encounter God in our daily human relationships. This is ingenious – but also typical of theological ducking and diving. Can people pray to a model? If God is encountered only in human relationships, Christianity has been deluded for 2,000 years. One wonders how many Christians believe this sort of thing and how many remain true to their traditional beliefs, whether they understand them or not.

Complexity of thought is certainly to be valued, but it can become obfuscation. In the end, we are tempted to say, well, if so-and-so, who is obviously very intelligent, perhaps Augustine, Aquinas or our local priest, understands it all, we can go ahead and simply accept it. *Really* believing (Chapter 6) need not, indeed cannot, enter into it. I am reminded of an Anglican reading group, a meeting of which I attended many years ago. The leader of the group explained that he had read something very profound about the Trinity and, for a fleeting moment, he could understand the mystery of the Three-Personed God. Apparently, this was some sort of theophany. Later, he was unable to recall any of the thoughts he had read and that had led to his acceptance of the Trinity. But that did not worry him because, he said, someone had thought it through and that was good enough for him. He suggested that we do the same, accept his report of his own fleeting understanding and go ahead and accept belief in the Trinity. The group responded by nodding slowly and sagely.

It is true that there are some things that we are willing to accept without understanding them. They often feature in specious analogical arguments for the naturalness of faith. I accept that my computer works without knowing why it does. Therefore I should be able to accept that God exists – and the theology and religious practices that flow from that starting point. But working a computer in ignorance of its electronics is quite different from accepting that God exists or subscribing to the doctrine of the Trinity. There is abundant evidence for the effectiveness of my computer, all of it out in the open. That evidence is based not on personal epiphanies but on overall scientific theories in which there is no place for supernatural interventions. When Einstein set out his famous equation $E=MC^2$, he did not add a rider: 'save for the intervention of the deity'. Microsoft does not say that pressing 'delete' will eliminate a word, save for the intervention of the deity. The evidence for the working of my computer can be, and has been repeatedly, tested. Is there comparable evidence for the existence of God? Of course not. Testing the existence of God through subjective religious experiences is not testing at all. If God exists and sometimes invades this world, we should be able to see evidence for his invasions, whether we are believers or not. So whether we understand the essence of God or not is not the point: where is the evidence for a supernatural realm and its interventions in the material world?

Making things seem so complex that only the specially trained, or the especially devout, can understand them is something that we encountered in our overview of the development of science within religion's cocoon. In medieval times it was the cloistered clergy who guarded and understood revealed knowledge; the illiterate, ragtag faithful were simply urged to have faith. The position that pleads complexity thus achieves two things: it establishes a power base for clergy and theologians, and it invites faith even as it devalues questioning. Theologians, writhing within their self-constructed cage, try to appear complex (as they say befits discussions of what they see as ineffable) and simultaneously simple (as befits 'poor parsons'). They forget Paul's distrust of philosophy and, ironically, clothe his insistence on faith in richly embroidered intellectual garments. It is not a skein of tangled cerebral paradoxes that we need, but clarity.[25]

While the Church's responses to waning belief in supernatural things deserve to be considered, presented together, as they often are, they are clearly an evasion tactic designed to avoid responding directly to central issues that are unacceptable and unbelievable in the twenty-first century. Theology is a continuously tended and trimmed hedge that screens off fundamental and really quite straightforward matters.

Viewing the disputatious history of theology, we may again recall Prufrock:

> In a minute there is time
> For decisions and revisions which a minute will reverse.

T. S. ELIOT, *The Love Song of J. Alfred Prufrock*, 47–48

Spirituality

Sidestepping the intricacies of theology, the Pope urges youth to reject the 'spiritual desert' of contemporary society. It is an injunction that still goes down well in a post-Woodstock era. But what does the Pontiff mean by 'spiritual'? Few people would deny that there is much to deplore in our world today, but we must ask whether, in the Pope's view, 'spiritual' includes his own infallibility on certain issues, the Assumption and Immaculate Conception of the Blessed Virgin Mary, and the banning of condoms in the face of millions of unwanted pregnancies and the AIDS epidemic? 'Spiritual' is an attractive word, embellished with alluring connotations, but we need to push aside those accretions and ask for more precision. With good reason, spirituality has been called 'religion's poor cousin'.[26]

A common notion abroad today is that humankind has an intrinsic spiritual component that craves to be satisfied. The popular cliché 'mind, body and spirit' encapsulates this idea and implies that humankind's supposed 'spirituality' is as real and obvious as its physical bodies. Descartes lurks in the background. This is, of course, a downgraded version of the 'eternal soul' of religious teaching. Anticipating Descartes, Jesus told his disciples to fear Satan, who 'is able to destroy both body and soul in hell' (MATTHEW 10:28). Paul wrote to the Thessalonians: '[Y]our whole spirit and soul and body be preserved unto the coming of our Lord Jesus Christ' (1 THESSALONIANS 5:23). Defenders of religion against what they present as the aridity of science are much enamoured of the idea that people have not only bodies but also souls and therefore a kind of intrinsic spirituality. It is not only religious people, in the traditional sense of the word, who repeat this platitude. 'Spirituality' is a broadly popular notion. The problem lies with how we understand the word. It has two principal meanings.

First, it refers to spirits, that is, to supernatural, person-like beings – hence Wallace's spiritualism. *The Shorter Oxford Dictionary* puts it like this: 'The fact or condition of being spirit or of consisting of an incorporeal essence.' 'Incorporeal' is the key word. This is where religion finds its niche: it tells us about spirits and helps us to engage with them, be they God, Jesus, the Holy Spirit, saints, the Devil, or, as the writer of the Epistle to the Hebrews puts it, 'the cloud of witnesses' with which 'we also are compassed about' (HEBREWS 12:1).

Hildegard may have had no problem with this kind of 'spirituality', but today more and more people in the West are wary of the notion. Nevertheless, they want to retain something of the 'other worldliness' (to use a somewhat dismissive phrase) suggested by 'spirituality'. They therefore embrace the second meaning of the word, one that is difficult to define without sounding 'spiritual' oneself. In this definition of 'spiritual' we find phrases such as the 'inner nature of man' and 'having the higher qualities of the mind' (*Concise Oxford Dictionary*). Both 'inner' and 'higher' are elusive metaphors that recall theological formulations about the 'ground of all our being' (Chapter 6). What secular 'spirituality' boils down to in practice is an interest in music, art, literature, philosophy, the beauty of nature, intense personal relationships, and that sort of thing. There is nothing wrong with taking 'spirituality' to mean this – as long as we remain aware that it does not imply beliefs in an incorporeal soul and spirit beings. The worrying thing here is that many today seem to believe that wonder, awe and reverence are inescapably indicative of a spirit realm and God.

'Spirituality' in the 'higher-qualities-of-the-mind' sense seems to me to be culturally situated rather than an inescapable product of evolution. Those in the Western tradition are more or less sympathetic to this meaning of the word, and they seize on it when they encounter it in Eastern religions and philosophies. Those in other traditions, say the Maya, have found it hard to get past the notion of spirit beings; for them 'the higher qualities of mind' and 'the inner nature of man', if they exist at all, look quite different from what we in the modern West value.

One way out of the dilemmas posed by the notion of spirituality is to say that the first meaning of the word (belief in incorporeal beings) is a metaphor, as are (for at least some *avant-garde* theologians) the Resurrection of Jesus or the Virgin Birth. Just how confusing current appeals to metaphor can be is illustrated by Mary Midgley, a moral philosopher. It has long been accepted that metaphors play a key role in the formation of our worldview and how we approach problems; even in science, researchers are aware that they employ metaphors to understand difficult concepts.[27] Midgley writes: 'Myths are not lies. Nor are they detached stories. They are imaginative patterns, networks of powerful symbols that suggest particular ways of interpreting the world. They shape its meaning.'[28] So far so good: this is a common understanding among anthropologists and folklorists. But she glosses over an important point. For the most part this is *not* how people respond to their own religious myths: on the contrary, they believe that their myths give them a divine right to legislate on how *all* people should behave. Myths even accord them the right to kill to achieve their divinely ordained ends. People do not easily abandon the supposed historical foundations and the divine status of their myths. If, in fifteenth-century England, you claimed that the Resurrection was a myth or a metaphor, you would have been burnt alive by Christians who believed that it was a fact. Today, if you claim in some countries that Muhammad was not actually taken up into Heaven, you may suffer a comparable fate. Midgley may regard myths, such as biblical narratives, as 'networks of powerful symbols'; but most of the people who align themselves with those myths regard them as divine revelations vouchsafed by a supernatural realm. A university seminar is a very different place from a souk in Tehran or a televangelist's packed auditorium in Texas.

Midgley gives an example of what she is talking about: 'For instance, machine imagery, which began to pervade our thoughts in the seventeenth century, is still potent today.'[29] All this leads her on to question what science can do for us. It cannot, she says, provide objective statements on morality. That, presumably, is the task of moral philosophers like Midgley. Indeed, science does not in any way claim to be a system of moral guidance, though many scientists would say that the debates about morality would benefit from the type of thinking that characterizes the daily course of science. And, more importantly, metaphors like 'machine imagery' are not the same thing as religious myths that talk about spirits and their doings.

Somewhere in between the two meanings of 'spirituality' lies that seemingly ever-present but highly tendentious question about the 'meaning of life'. If we ask, 'What is the meaning of life?', we must realize that the question presupposes the existence of a particular kind of 'spiritual' answer. The question implies the category of thought and experience in which the answer is believed to lurk: we have only to be specific within that category. We should therefore note that the question itself comes out of our own religious tradition and expects an answer within that tradition. When brave religious souls do enunciate what they believe to be the 'meaning of life', it turns out to be both trivial and unbelievable. By contrast, evolution, founded on meaningless, random mutations (witness the long-gone dinosaurs), has no need of 'meaning'. We cannot say that the 'meaning of life' is the replication and mutation of genes. It would be better to think about what constitutes altruism and simple kindness in specific situations. We should think of what leads to a fulfilled life.[30] The question of the meaning of life is meaningless.

The Anglican theologian Keith Ward recognizes the value of simple kindness. In his sensitive book *Is Religion Dangerous?*, he allows that religion can be dangerous in certain circumstances. But he contends that, ultimately, religion is not about profound experiences or sticking to doctrines; rather, it is about 'the search for supreme goodness, a life lived for the sake of goodness alone'.[31] Would it were so. Like many present-day theologians on the defensive, Ward winnows religion out of existence. Surely, anyone who has been to a church, a synagogue or a mosque, will see that religion is about much more than goodness. For one thing, in all three edifices there is the expounding of sacred scriptures as God's unique and inviolable message to a select group of humanity. For another, the congregation prays for multiple supernatural interventions in the affairs of the material world. 'The search for supreme goodness'? Thinking only of goodness, one could go to any religious gathering vainly 'hoping it might be so'.

If religion is only a search for goodness, how does it differ from any number of secular, humanist philosophies? Anyone who has studied the history of Christianity (for it is about Christianity that Ward is primarily speaking) will warm to his statement about living for the sake of goodness but at the same time will see little connection between it and what has happened over the last 2,000 years. Religion becomes dangerous when people *really* believe in a supernatural realm and revealed knowledge. It seems to me that the difference today between Christianity and Islam is that many Christians no longer *really* believe while most Muslims do.

Respected faith and nasty religion

'Spiritual' also carries connotations of 'respect'. It almost goes without saying that to disrespect something spiritual is shocking and sacrilegious. In medieval times, a lack of 'respect' for religious things could lead to a public burning. Today, the notion of 'respect' is a bulwark against questioning other people's religions. Thus, in both the science *versus* religion debate and the clash between religions themselves we hear repeated calls for 'respect'. If people say that they believe the world to be flat, we can laugh and simply walk away; we need not 'respect' such nonsense. But if these people add that their *religion* tells them that the world is flat (or that the whole creation process lasted only six days, or that the Virgin Mary floated up into the sky, or that God will punish non-believers in an afterlife), we are expected to respect their views and not laugh or even argue.

Why should the concepts of 'spirituality' and 'respect' stifle debate? This is a question that Richard Dawkins asked in *The God Delusion*.[32] He was castigated for doing so. 'Spirituality' and 'respect' have jointly become an impregnable defence of irrationality. If we 'respect' that defence and grant that religious beliefs, even those that clearly trespass on the realm of science and the material world, must be 'respected', we open the floodgates to irrationality and bigotry.

When Abraham and his wife were escaping from Egypt, when Jesus was preaching in Palestine, and when Muhammad was living in Arabia, there were vast numbers of human communities all over the world that passionately espoused a religion of one sort or another (a thought that struck Darwin during his *Beagle* voyage). Why should we keep quiet (exercise respect) when people tell us that the God who created the vast universe and our spinning galaxy selected only one small group of people to be recipients of divinely revealed knowledge about his nature and motives? And, moreover, that that blessed group is the one into which they were born and in which they grew up (or to which they converted). *Many* widely differing groups of people and individuals believe themselves to be special conduits of God's revelation to the whole human race. Apart from being palpable nonsense, this situation calls into question the viability of the notion of 'respect'. If we 'respect' other 'faiths', we have to allow that everything, from the creation of our peripheral earth to multiple supernatural revelations to war in the Middle East, *may* be part of a divine plan. 'Implausible' is too weak a word. There is no reason why we should respect the religion in which we grew up, or, for that matter, other religions, any more than we should respect beliefs about a flat earth.

'Faith' deserves further comment. Coleridge wrote about 'that willing suspension of disbelief for the moment, which constitutes poetic faith'.[33] Today the word has become a euphemism for religion. Fundamentally, it means believing something without being given good reasons for doing so. If there were good reasons for believing something, there would be no need for faith. Let us bear that in mind when we speak of the Christian or the Muslim faith: in Coleridge's words, faith requires the 'willing suspension of disbelief'. It is therefore strange that the British Government (along with some other governments) encourages the foundation of 'faith-based schools'. The idea of 'faith-based education' is an oxymoron. 'Faith-based' means 'Do not question'. Education in today's world should surely mean no-holds-barred criticism that accepts *nothing* at face value or on the authority of the Church or any other organization.

But that is not how many people understand the word 'faith'. They think of the more pleasant connotations that are attached to 'faithful' and allow them to colour the notion of religious 'faith'. People may well be faithful to their religion by sticking to it through thick and thin. But, ultimately, their faith consists in obdurately believing things for which there is no evidence.

The tit-for-tat argument that belief in evolution is a comparable matter of faith is, of course, nonsense. Science does not request faith. The evidence for evolution is before the public, and it is overwhelming.[34]

In 2008, Prince Charles, the heir to the British throne, took up the warm and woolly connotations of 'faith', as he had done some years earlier. Instead of being crowned in due course 'Defender of the Faith' (meaning the Church of England), he said he would prefer to be 'Defender of Faith'. He had not, it seems, considered the matter very deeply. We may ask, which faith? The Maya? Fundamentalist Islam? Will he, as King of Great Britain, decide which faiths, or which parts of faiths, he will defend?

I end this section with six questions on 'respect' and 'faith'. If you respect people of other faiths, does that mean you won't try to convert them? If you try to convert them, are you not denigrating their faith? If you respect them, does that mean that you respect *all* manifestations of their faith, including the oppression of women, the banning of contraceptives and suicide bombers? If you respect only the milder manifestations of their faith, are you basing your discrimination on your own selected criteria? If your respect means that you will leave them alone to practise their religion without the seeds of discord that your evangelism may sow, does that mean that Christianity was wrong for 2,000 years in its endeavours to convert them? If you respect them and leave them alone, but in your heart you believe them to be wrong to the extent that they are not 'saved', what sort of 'respect' is that?

Morality

So far in this book I have not said much about morality except by implication. It has, however, become the last lifeline to which believers cling as the floods of rationality sweep over them. Where, they ask, would we be without religion to supply a moral code and terrifying sanctions to go with it? Without religion there would be anarchy. In the same way that Wallace (and today his Intelligent Design descendants) thought that human consciousness could not have evolved without divine intervention, so, too, some argue that morality must have come from outside the line of natural evolution – from God. The question here is: does morality have to be derived from supernatural revelations or can people, including scientists, work out for themselves what constitutes humane behaviour?

Dawkins entitled his first book *The Selfish Gene*.[35] In his Introduction to its 30th anniversary edition, he considered the misleading impression that the title gave.[36] Dismayed by the misinformed criticism that it attracted, he wondered if some readers had got past the title. In response, he pointed out that the book is more about altruism than selfishness. His title does not imply that human beings are unavoidably rather horrible, selfish creatures. Contrary to this misunderstanding, there is no need to think that morality (let us say, consideration for others) could not have been part of natural selection and human evolution.

We did not evolve from incorrigibly selfish antecedents into a self-centred, innately aggressive species that needed an injection of religion to keep it in line and to stop it from destroying itself. On the contrary, the animal kingdom exhibits instances of what we may call altruism, whether or not the creatures themselves think of their actions as altruistic. As I briefly mentioned earlier in this book, the self – the notion of ourselves as people *vis à vis* others – evolved in social settings. The self, from whence comes the notion of selfishness, is necessarily embedded in a social environment. What we have come to call kindness was therefore an essential part of human evolution. Darwin himself pointed out in *The Descent of Man* that collaborative and self-sacrificing behaviour was present in numerous species.[37]

Indeed, it has since been argued that altruism was part of the evolving wiring of the human nervous system.[38] Frans De Waal, Professor of Primate Behaviour at Emory University, cites research on the neural basis for moral judgments. Neuroimaging has shown that moral judgment involves a number of brain areas, some very ancient in the evolutionary line.[39] From this finding, De Waal concludes that 'neuroscience seems to be lending support to human morality as evolutionarily anchored in mammalian sociality':[40] 'the building blocks of morality clearly predate humanity.'[41] That is a key statement. Human goodness is not a mystery, a glimpse of God in whose image we are made, though from which we have fallen. Survival of the fittest was, by all accounts, a pitiless business: there was no room for the weak (as assessed by their adaptive potential). But that is only part of the story: there was also cooperation, empathy and reciprocity. Early human communities going back to Blombos times must also have embraced '"survival of the kindest" aimed at family and potential reciprocators'.[42] As De Waal points out, the product of evolution (us) is not necessarily a cruel beast lurking beneath a thin veneer of culture and religion.[43] By teaching the moral of the Good Samaritan parable, Christianity (together with some other religions) is not calling for the injection of something new into humankind. Rather, it is emphasizing something that is already there, and was there even before people were people. The topic of altruism is more a matter of biology than theology. Yet the Christian Church and other religions, unlike science, present themselves as creators of morality.

Nowadays, the Church has no clear moral statement on the issues of the day: contraception, stem cell research, the AIDS pandemic, capital punishment, and, within its own bailiwick, gay and female clergy. When it comes down to specifics, the Church's attitude to the great moral issues of the present is as equivocal and reprehensible as it has ever been. Despite its own claims, the history of the Church up to and including its present-day dilemmas shows that religion does *not* provide useful moral guidelines. On the contrary, religious morality *creates* moral dilemmas.

Richard Harries, a former Bishop of Oxford, tries to get around the problem of Christianity's poor showing when it comes down to moral specifics by claiming that we should look rather at the overarching command to love, as preached by Jesus.[44] But the covering-law of love is as much humanism as it is Christianity; it does not require what Harries calls the 're-enchantment of morality'. This is really nothing more than the appropriation of morality by supernaturalism. The overarching command, so attractive in its do-unto-others phraseology and the beatitudes, fails when it comes down to the sort of specifics that I have noted. What exactly *is* the moral way of dealing with, for example, homosexuality in the Church (and outside it as well, if some clerics had their way)? Religious morality does not contain any answers. Harries's own Church, the Church of England, is rent asunder when it comes down to moral specifics.

For many, morality consists in choosing those parts of the Bible that are congenial and ignoring those that are plainly disgusting. Few Christians and Jews nowadays believe that adulterers 'shall surely be put to death' (e.g., LEVITICUS 20:10). But on what criteria do we make such discriminations? Clearly, they must come from outside the Bible, that is, from simple humanism. Philip Pullman, now famous for the trilogy *His Dark Materials*, puts the matter pithily: '[M]oral codes are a very interesting example of how religion, I think, gets it wrong, and how I wish religious people would be a bit more modest.'[45]

Today we cannot conclude that the Church was, and still is, the foundation of Western morality. The future will probably regard the present Church leadership, membership and institution to have been as desperate for power as they were in any period in the past. It seems highly likely that future generations will look back at the invasion of Iraq as a time when the Christian

Church *as a whole* abandoned its cloak of morality and acquiesced in violence and greed: the Western armies were accompanied by chaplains. The 2008 Republican candidate for the vice-presidency of the USA, Sarah Palin, preached (I choose the word carefully) that American soldiers in Iraq are God's army doing God's will – as it has been revealed to her and her fellow religionists. The invasion of Iraq, and more especially its aftermath, was a time when world leaders Tony Blair and George W. Bush, though elected by people, claimed that they were taking their orders not from their constituents but from God. When they now speak of having no regrets about dodgy dossiers, they mean that they are still confident that they did God's will. The future will probably see this time as a repetition of the Crusades (as a slip of the tongue by Bush put it), even if many individual Christians may have had reservations about the invasions of Afghanistan and Iraq. Small wonder, then, that Tony Blair seeks sanctuary from irrationality in even deeper irrationality – by raising the bar of belief. His much publicized conversion to Roman Catholicism means that he has signed up for belief in the Immaculate Conception of the Virgin Mary, the transubstantiation of bread and wine into flesh and blood, the Infallibility of the Pope, Augustine's notion of no salvation outside the Church, and all the other Catholic doctrines.

How can apparently intelligent national leaders subscribe to such non-sense, such counter-knowledge? The head-turning effect of great power has been the undoing of many leaders. Once they have set foot on the sacred road, many seem unable to withdraw into more sober assessments of affairs. This is what David Owen calls the 'Hubris Syndrome'. According to him, the mental health of George W. Bush and Tony Blair was undermined by the intoxicating belief that they had a direct line to God.[46]

Briefly put, science itself is amoral. That does not mean that it is wicked. Simply that it has nothing to say about what is acceptable behaviour. But the grossly illogical inference that *scientists* are amoral is absurd. At least some people have been formulating and living by viable, non-supernatural morality for centuries. People of good will must challenge the belief that a supernatural realm has revealed itself to a select few on earth so that they may impose their 'moral revelation' on everyone. Secular kind-heartedness is more humane and adaptable to specific circumstances than supernatural sanctions. True, there will always be situations in which it will be hard to know what actions will, in retrospect, turn out to have been genuinely kind. Deciding what is 'right' behaviour in many tricky situations will remain problematic and open to debate. But it helps to have the fundamental notion of morality freed from supposedly supernatural origins. We can discuss morality without feeling obliged to find support for our views in some ancient, ambiguous text that claims to reveal the mind of God.

A gift of God?

In the turmoil of the current debate sincere Christians are often tormented by the fear that they may not, after all, believe. Some preachers counter this fear by insisting that 'honest doubt' inevitably goes hand in hand with belief throughout our lives. They are probably right, though I cannot speak for charismatic televangelists whose tirades probably have nothing to do with their own belief or disbelief but everything to do with money and personal aggrandizement. More thoughtful and honest preachers (the 'poor Parsons' of Chaucer's pilgrimage) know the reality of doubt and the dark moods (as they see them) that can creep up on them. Why, I ask myself, do they take doubt to be a lapse, a failing to be combated? Scientists welcome doubt because it leads to further discoveries. Why do religionists not take belief in a supernatural realm to be the real trouble, a craven attempt to escape reality? Disbelief in supernatural forces and beings is not despair: it is liberation, a position from which one can tackle the realities of real life.

Paul recognized the agony of personal belief and disbelief, though what he said about it hardly solves the problem: 'For by grace are ye saved through faith; and that not of yourselves: it is the gift of God' (EPHESIANS 2:8). This is the sort of statement that leads to belief in predestination: God has inscrutably chosen a select few for salvation; the rest, those on whom he has not bestowed the gift of faith, are heading for the bonfire. If faith in the supernatural efforts of God to secure salvation for human beings is itself a 'gift of God', we cannot manufacture it by reading Augustine or anyone else. Nor can we be blamed for not having it. Evaporating faith cannot be blamed on a perverse unwillingness to believe or on the sinfulness of the individual, as many a 'poor Parson' genuinely seeking faith will tell you. Paul seems to have realized that some people believe easily while others simply and honestly cannot believe that there is a parallel, supernatural realm in which there is a Being who wants to save them. An explanation for this particular state of affairs stumped him; he managed only to come up with the notion that faith is a gift of God, an explanation that creates more theological problems than it solves. He, of course, did not know anything about labile, shifting consciousness and the hugely complex social circumstances in which people try to come to terms with it. For him, every human dilemma must have a supernatural explanation.

Is reconciliation possible (or desirable)?

This book has been about the clash between science and religion. Now, near the end, I face a dilemma. For many well-meaning people it has become *de rigueur* to end such discussions on a positive or at least open-ended note. Wealthy organizations like the Templeton Foundation have spent vast sums to encourage academics to reach pleasing conclusions that will satisfy both religionists and scientists. Blessed are the peacemakers. Not to be a peacemaker seems churlish, insensitive and downright dangerous. We must now come to the final question: is rapprochement between science and religion possible?

In fact, two kinds of reconciliation are involved. One conflict is between different religions; the other is between religion and science. Each conflict mirrors the other.

The first tells us something about the second. Reconciliation between religions can be achieved only at the price of covering up real, fundamental problems. Doing so has become an industry. Words, ambiguous and mellifluous, simply buy a little time: the underlying contradictions will not go away. In order to effect rapprochement between religions many people of goodwill emphasize similarities. They tell us that many roads lead to God, that – fundamentally – all religions are dedicated to peace and that the disunity we see so dramatically played out in the Middle East and elsewhere results from misunderstandings of religion. If only we can learn to be tolerant of what people of other religions believe and do, all will be well. This 'respect' point of view is understandable enough, given the literally explosive times in which we live. But it is a case of peace at all costs. If reconciliation between religions is based on silences that filter out awkward but actually fundamental beliefs, the desired unity of humankind will remain elusive. Goodwill by itself, untrammelled by critical thought, can only lead us from one blindness to another. Until the rabbis and mullahs in the Middle East sit down together to talk things over amicably and reach some sort of reconciliation, they cannot ask us to respect them as men of God.

It is unrealistic to think that people who have been taught to believe that they are a God-chosen people, that their holy book is the final Word of God, that they are the sole bearers of the full Truth, and that there is therefore no salvation outside their own religious community will live amicably side by side with non-believers without attempting to persuade their neighbours to abandon their beliefs and families for the 'true' religion. Worse, when persuasion fails, history shows that the divinely inspired devout consider (logically enough) force to be a legitimate option – or, at the very least, they will try to make life difficult for non-believers. The features I have listed here – God-chosen people, a holy book and so forth – are the fundamentals of most of the religions that are today competing for power and influence. Ultimately, differing religions can never reconcile as long as they believe in their own supernaturally provided revelation – that is to say, their very foundation. Indeed, supernaturalism and multiculturalism are incompatible. Supernaturalism is the death knell of multiculturalism.

The second kind of rapprochement is between religion and science. Here too we find papering over of cracks. As far as I can make out, most scientists do not bother much with religion. They simply get on with their work and shake their heads sadly at arguments against evolution or attempts to try to demarcate the areas of research that they may address. Most scientists are not concerned with trying to accommodate believers. In any event, the principles of science do not allow for adjustments to research findings in the light of other people's supposed divine revelations. As we saw in Chapters 1 and 2, whenever there has been conflict it has always been the religionists who, in the end, have had to adjust their beliefs. Given this history, many scientists see no reason why they should bother about believers and supernatural realms. If believers are troubled by science, that is their problem, not the scientists'.

To defuse the conflict, the influential palaeontologist Stephen Jay Gould has written of *magisteria* and popularized the view that science and religion occupy two separate dimensions of knowledge. He claims that these dimensions are complementary, not contradictory.[47] To make the idea memorable, Gould followed a modern trend and devised an acronym: NOMA. It stands for non-overlapping *magisteria*. The *magisterium* of religion, he says, covers issues of ultimate meaning and moral values. Science deals with the material world. On the face of it, this sounds a fair conclusion that will please everybody and obviate conflict. You can believe in both *magisteria*.

Difficulties start to arise when we look more closely at Gould's use of 'non-overlapping'. As we have seen throughout this book, religion repeatedly impinges on the *magisterium* of science: the idea of a parallel supernatural realm is, in effect, scientific; virgin births and resurrections are scientific statements. Answered prayer implies supernatural intervention in the material world. That religious morality should, as some nowadays forcefully argue, govern the direction of scientific research is another overlap. It is not simply a case of religion telling us that God created the world and science telling us how he did it. Religionists are never satisfied with that naïve view. For them, God must be seen to be active within the material world. Indeed, religion could not exist without overlaps.

How do we deal with these difficulties? For some religious warriors, 'humility' is a battle-cry. Who wants to be thought arrogant? We (i.e., scientists) should be humble and admit that we do not know everything and therefore God and a whole supernatural realm may well exist. Let us examine this argument from ignorance.

The argument sees what we know and what we do not know as two piles of poker chips. The pile that represents what we know is considerably smaller than the one that represents what we do not know. There is a vast number of things that scientists would like to know but do not. That being so, we must admit that God and the supernatural may be one of the chips in the higher pile. When the argument is presented along these lines another factor manifests itself. Contemplation of the soaring pile of ignorance makes us feel humble. To claim that something does not exist seems an act of unconscionable *hubris*. We feel obliged to present ourselves as humble, modest and generous – and largely ignorant.

The fallacy of this argument is exposed when we question the notion of two piles of facts in the form of poker chips, one pile of known facts, the other of unknown facts. A moment's thought shows that this is not how knowledge works. Our knowledge is not a pile of disparate facts. All our known facts are interlinked. Instead of imagining two piles of poker chips, we should think of a closely woven fabric of interrelated facts. The point is that, because facts are connected and patterned, they preclude a whole range of possibilities. Once people accepted the reality of a heliocentric solar system, they stopped talking about crystal spheres and the beautiful music that they emitted. We do not have to be humble and keep an open mind about any of this. What we know rules out a whole range of what were at one time considered possibilities.

Today we can see that what we know – the fabric of interrelated facts – leaves no room for belief in a supernatural realm that intervenes in the affairs of this world, either by causing people to turn into pillars of salt or by causing virgins to become pregnant or by planting thoughts (revealed knowledge) in people's brains. Never has the ever-growing fabric of knowledge had a gap (it would have to be a gaping rent) that suggests that supernatural agents are at work. Religion necessarily postulates supernatural interventions and consequent disjunctions in the material world, be they the creation of separate species or answered prayer. None have been discovered.

As long as religion holds that there is a supernatural realm and that beings in that realm intervene in the material world no reconciliation between science and religion is possible. Nor is it desirable. All the initiatives that are today pursued suppose that dialogue between science and religion is necessary and beneficial. It is not. The notions that science and religion are compatible or complementary or non-overlapping or that religion takes over where science leaves off or that religion answers questions that science cannot are all groundless. They are parts of a façade that keeps belief in a supposed supernatural realm alive and apparently respectable.

I end this chapter with an anecdote that illustrates some of the points I have made. Joseph LeDoux, Professor of Science at the Center for Neural Science at New York University, attended a conference sponsored by the Vatican Observatory and the Centre for Theology and Natural Sciences.[48] The topic of the conference was 'Neuroscience and divine action'. Participants were invited to address the problem of how God communicates with people. Some speakers argued that, if the soul is immaterial, there is no problem, at least initially, because God is also immaterial. But if the soul, along with the daily thoughts and emotions that we believe to be part of the soul, part of our 'self', is created by neural activity in the brain, then we must allow that God in some ways suspends the laws of physics and tampers with our synapses, neuro-transmitters and the electrical impulses that pass along neural pathways. But scientists have found no evidence for such tampering. This is, of course, the essence of the conundrum in which belief in supernatural things lands us and that has underlain much of this book.

One notion put forward at the Vatican conference was that God *interacts* but does not *intervene*. In the beginning, God set up the material universe in accordance with laws of physics that he devised. Now he just lets it run: he does not *intervene* in what is, in effect, the realm of science. He does, however, *interact* (communicate) with human beings. How he manages to do this without tinkering with our material brains, which, we know, produce religious experiences and thoughts, is unexplained. Theologians speak as if practical problems can be solved by semantics. The distinction between *intervene* and *interact* is absurd. Here we have 'advanced' theologians at work; as LeDoux points out, their congregations back at home would be bemused by their distinctions. LeDoux was not surprised to find that no consensus could be reached. He was witnessing one of the many unsolvable problems that religion creates and then worries about.

Darwin was right: arguments do not accomplish much. As he said in the words I quoted as one of the epigraphs to the Preface of this book, the best thing to do is to get on with research and let people see the whole sweep of science and its open-ended history. Repeatedly, science has modified religion, never vice versa. If adjustments still have to be made, it is religion that will have to make them: this is not science's problem. Moreover, religion must cease making claims about reality. 'Reconciliation' and 'rapprochement', ideas that imply concessions from both sides, are not realistic. Those who spend vast sums of money on conferences and publications aimed at 'reconciliation' are in fact creating an illusion of a two-way interaction that simply does not exist; they are achieving nothing for science. As two scientists, Matthew Cobb and Jerry Coyne, neatly put it in a letter to *Nature*, 'In reality, the only contribution that science can make to the ideas of religion is atheism.'[49]

Epilegomena
Of Babies and Bathwater

Many people gradually come to realize that there are no such things as supernatural realms and mystical beings who intervene in worldly affairs. They then begin to worry about the baby in the bathwater. If we drain off the bathwater of supernaturalism, what of all the wonderful music, literature, architecture, art and all the family traditions that have been such an integral part of our heritage? Must all that go too? Fearful, we sometimes feel drawn to our religious past, almost 'hoping it might be so' – at least in part.

Unquestionably, the baby need not be lost. We can wonder at the universe glimpsed in the night sky without being religious about it. Awe does not entail belief in supernatural things, despite what some well-meaning scientists nowadays come close to implying. We can be stirred by a Bach cantata without buying into Reformation Protestantism. Verdi, atheist and anti-cleric, was able to compose a resplendent *Requiem*. Offering solace for the living, Brahms's more humane *Requiem* has no prayers for the dead. Moved by the death of his father, Fauré, an agnostic church organist, wrote an orthodox *Requiem*, but one that that plays down the terror of the *Dies Irae* and tenderly emphasizes eternal rest. We can marvel at the Sistine Chapel ceiling and its adjacent *Last Judgment* without believing that all the events depicted actually happened or will happen. We can revel in the King James translation of the Psalms without the triumphalism of believing that God will smite our enemies and humiliate them by preparing a table before us in their presence. We can gratefully say, 'There but for the grace of God go I', without believing that a supernatural God has intervened in our lives. When we see bickering bishops, we can even exclaim, 'Thank God I am an atheist!' The baby can indeed be rescued from its bathwater.

The realization that religion was transforming from supernaturalism into something richer and more inclusive was underlined in 1882. The Dean of Westminster, who was in France at the time, received a telegram from the President of the Royal Society: '[I]t would be acceptable to a very large number of our fellow-countrymen of all classes and opinions that our illustrious countryman, Mr Darwin, should be buried in Westminster Abbey.' Later, the Dean recalled that he did not hesitate. He telegraphed back that his 'assent would be cheerfully given'[1] (Fig. 49). The irony suggested by the Dean's word 'cheerfully' was probably unintended, but it hints at the possibility that some righteous criticism would be directed at him.

Since then the Anglican Church has become a good deal more inclusive than it was when Henry VIII established it in persecution, blood and destruction. While I cannot see grounds for any unity, or complementarity, between science and supernaturalism, it seems to me that other churches, even Rome and the fundamentalists, will eventually have to concede that they have lost their long battle with science. Notwithstanding claims to the contrary, their religious beliefs do extend into the field of science and must therefore be open to scrutiny. They cannot demand privileged 'respect'. They will have to follow the Abbey's increasingly inclusive example: eventually, they will have to let belief in supernatural things be a private, optional extra to religious practice.

Westminster Abbey is now a symbol of the way that religion is heading. In addition to Darwin, numerous people who entertained no religious beliefs have been interred in the Abbey or have had monuments erected to them. In an astonishing about-face, the Abbey now includes a memorial to Oscar Wilde. Less controversially, the grave of the Unknown Soldier, whose religion, if any, is of course irrelevant, symbolizes profound sentiments that are not rooted in supernatural realms but are nevertheless appropriate in the Abbey.

49 Charles Darwin's memorial plaque in Westminster Abbey shows how far the Anglican Church has come in diminishing its demand for religious belief. Should acceptance of supernatural realms and beings be an 'optional extra' in Christian belief?

In 1928, Thomas Hardy's ashes were laid to rest in the Abbey, his heart having been buried at Stinsford in Dorset. During the Abbey service, the Dean sprinkled a handful of the writer's beloved Wessex earth on the casket. Born in 1840, Hardy had seen the publication of Darwin's *Origin of Species* and had lived through the tumult that followed it. He became a confessing atheist and sharp critic of religious hypocrisy. But he was also acutely aware of the old religious tradition. As a young architect, he had been involved in the restoration of ancient churches; history and culture meant a lot to him. In his poem 'God's Funeral', he watches distraught mourners bearing the God of Christianity to his grave. But some 'in the background' refuse to believe that God is dead. They cry out, 'Still he lives to us!' Hardy does not mock them. Instead, he writes:

> I could not buoy their faith: and yet
> Many I had known: with all I sympathized.

Thinking of Darwin and the great evolution debate, Hardy had God ask a key question in 'A Plaint to Man':

> When you slowly emerged from the den of Time,
> And gained percipience as you grew,
> And fleshed you fair out of shapeless slime,
>
> Wherefore, O Man, did there come to you
> The unhappy need of creating me –
> A form like your own – for praying to?

Today we have a better idea of why humankind created God. Belief in supernatural beings to whom we can pray is no longer viable. Does this mean that humanity is lost? Also in 'A Plaint to Man', Hardy points not to despair but to an obvious truth: history has shown that religion does not have a monopoly on goodness. It is possible to be a thoroughly decent person without God breathing down your neck:

> On the human heart's resource alone,
> In brotherhood bonded close and graced
>
> With loving-kindness fully blown,
> And visioned help unsought, unknown.

Still, people do sometimes feel the tug of tradition. Perhaps – I do not know – believers can set aside supposedly revealed knowledge about super-natural realms, together with all the zealousness and exclusivity that it inevitably implies. Darwin's 'freedom of thought' awaits them. Perhaps we can, in some sense, all go 'in the gloom', even if we know that, despite our hoping, the 'meek mild' things of our 'childhood' will not be so. The human heart's resource of loving kindness will prove more than adequate.

> Christmas Eve, and twelve of the clock.
> 'Now they are all on their knees,'
> An elder said as we sat in a flock
> By the embers in hearthside ease.
>
> We pictured the meek mild creatures where
> They dwelt in their strawy pen,
> Nor did it occur to one of us there
> To doubt they were kneeling then.
>
> So fair a fancy few would weave
> In these years! Yet, I feel,
> If someone said on Christmas Eve,
> 'Come; see the oxen kneel,
>
> 'In the lonely barton by yonder coomb
> Our childhood used to know,'
> I should go with him in the gloom,
> Hoping it might be so.
>
> THOMAS HARDY, 'The Oxen', 1915

Notes

PREFACE (PAGES 6–10)
1 Quoted in Gould (1999b).
2 Dobzhansky (1945, 75).
3 Whitehead (1926, 229).

PROLEGOMENA (PAGES 11–22)
1 Recent discoveries at another site, Pinnacle Point, show that a fully modern way of life was being practised on the southern Cape coast as long ago as 164,000 years before the present. See Marean *et al.* (2007).
2 D'Errico *et al.* (2005); Mellars (2006); Henshilwood *et al.* (2002); Jacobs *et al.* (2006); Henshilwood (2007).
3 E.g., Hovers *et al.* (2003).
4 Since the Blombos discovery slightly earlier finds of shell beads have been made at Oued Djebbana, Algeria and Skhul, Israel: see Vanhaeren *et al.* (2006).
5 McBrearty & Brooks (2000); McBrearty (2007); Barham (2007); Mellars *et al.* (2007).
6 E.g., Knight (1991).
7 Rappaport (1999, 16).
8 Coe (1992); see also Stuart & Stuart (2008); Miller & Martin (2004).
9 Freidel *et al.* (1993, 202); Schele & Freidel (1990); Schele & Miller (1986).
10 Leach (1969, 1970, 1976); Lévi-Strauss (1967).
11 On the history and mythology of Israel see: Finkelstein & Silberman (2002), Hillel (2005), Hoffmeier (2005), Leach (1969), Liverani (2003).
12 It is apparently to this incident that the writer of Hebrews (13:2) refers: '[S]ome have entertained angels unawares.' Dawkins (2006a, 242) draws attention to a comparable incident in Genesis 20:2–5 and wonders about textual reliability. I am inclined to suspect that the multiple versions of myth is the principle at work.

1 A NEW WAY OF THINKING (PAGES 23–48)
1 For example, Freeman (2003, 9–10); Kirk & Raven (1957); Wolpert (2006, 207–209); Yonge (2006). For overviews of the history of science see, among many others, Gaukroger (2008); Bowler & Morus (2005); Gribbin (2002); Watson (2005); Freeman (2003).
2 Watson (2005, 129); Schrödinger (1996, 55–58).
3 In recent years there has been much debate on realism and relativism (or constructivism) in science; though leaning decidedly towards realism, I do not debate the issues here. Whatever role the social context of science may play – and we should not underestimate it – I

believe that there was a past and that our research can bring us closer to a more accurate account of it. On logic and scientific thought see Copi (1982); Kuhn (1970); Hempel (1966); Chalmers (1999, 1990); Hacking (1983); Nagel (1997); Laudan (1996); Boghossian (2006).
4 Aristotle, *History of Animals*, 588b, 4–14. In Barnes (1984, 1, 922).
5 Plato, meaning 'broad', as in 'broad-shouldered', was a nickname. His real name was less euphonious: Aristolcic.
6 This is discussed in Plato's *Phaedo*.
7 Plato, *The Republic* VII:530; see Plato (1935, 225); parenthesis added.
8 Plato's theory of forms is discussed in *The Republic*.
9 Pagels (1980); Krosney (2006); Kasser *et al.* (2006).
10 The King James translation is less accurate in today's language: 'ye are too superstitious'.
11 Haeckel (1883, 2:33).
12 Williams (1985).
13 Marcus Aurelius, a Stoic who became Emperor of Rome in 161, wrote of the virtues of facing death in a noble way: 'But this readiness must come from a specific decision: not in mere revolt, like the Christians, but thoughtful, dignified, and – if others are to believe it – undramatic.' *Marcus Aurelius* (2006).
14 Williams (1985, 153); quoted by Freeman (2003, 78).
15 The early Church Father Tertullian (*c.* 160– *c.* 225); see his *Apologeticus*, Chapter 50, Section 13.
16 Quoted in Freeman (2008, 1).
17 Lewis (2003, 384).
18 Cantor (1963, 50).
19 Burkert (1987).

2 FROM SKIES TO SPECIES (PAGES 49–86)
1 E.g., Cantor (1963); Davis (1988); Jordan (2001b).
2 Lewis (2003, 413)
3 Augustine of Hippo (1991, 145).
4 This appears in Augustine's *De Baptismo contra Donatistas*, Book 4, Chapter 17.
5 Lewis (2003, 425).
6 Brown (1999).
7 Augustine of Hippo (1993, 368).
8 Augustine of Hippo (1991, 210–12).
9 Herrin (2007).
10 Cantor (1963, 508).
11 Porter (2005).
12 Dawkins (2006a, 77–79).

13 Aquinas (1911).
14 For an excellent discussion of the problems with the First Cause argument see Everitt (2004, 59–84).
15 Barnes (1984, 1:483).
16 Kuhn (1966); Lindberg & Westman (1990); Henry (2008).
17 Biagioli (1993).
18 McMullin (2006).
19 Rowland (2008).
20 Finocchiaro (2006).
21 Drake (1957, 183).
22 Drake (1957, 186).
23 Shea & Artigas (2003).
24 Redondi (1987).
25 Bowler & Morus (2005).
26 The Brahe family fortune derived from taxing ships entering or leaving the Baltic Sea; Thoren (1990).
27 Losee (1980, 46–47).
28 Bowler & Morus (2005, 46).
29 Westfall (1980).
30 Fauvel *et al.* (1988).
31 Gribbin (2002, 190).
32 Here we encounter such names as Immanuel Kant (1724–1804), Baruch Spinoza (1632–1677), Voltaire (whose real name was François-Marie Arouet; 1694–1778), Charles-Louis Montesquieu (1689–1755), Jean-Jacques Rousseau (1712–1778), David Hume (1711–1776), Adam Smith (1723–1790) and a much less well-known person, Charles Darwin's grandfather, Erasmus Darwin (1731–1802).
33 Spinoza (1883).
34 Hume (2004).
35 Wordsworth (1888, 1:500).
36 Foucault (1984).
37 Ruse (2001, 9).
38 Darwin (1818, 1:507).
39 Darwin (1800, 115, 103).
40 Darwin (2002b, 44–45).
41 Ruse (2001, 149).
42 Millhauser (1959, 6).
43 Secord (2000).
44 In Clark & Hughes (1890, 2:83, 84); first parenthesis added.
45 Emphasis added; Darwin (1968, 459).
46 On the complexity of Victorian Christianity's reception of Darwin see Brooke (2003).
47 Kingsley, quoted in Endersby (2007, 3).
48 Van Wyhe (2006, 6).
49 Avise (2001).
50 James (1907, 198).
51 On the infamous 'monkey trial', see Larson (1998) and Settle (1972).
52 Drummond (1894, 439).

3 A Tale of Two Scientists (pages 87–114)
1 Darwin (2002a, 49–50).
2 Wallace (1869), in George (1964, 234)
3 For biographies of Darwin and summaries of his thought see George (1982), Howard (1982), Browne (1998), Eldredge (2005), Desmond *et al.* (2007), Gribbin (2002, 339–58), Boulter (2008).
4 Darwin (2002a, 7).
5 Darwin (2002a, 8).
6 Darwin (2002a, 10).
7 Darwin (2002a, 85).
8 Darwin (2002a, 10).
9 Darwin (2002a, 10).
10 Darwin (2002a, 22).
11 Darwin (2002a, 23).
12 Darwin (2002a, 29).
13 Paley (1802).
14 E.g., Ayala (2007); Ings (2007).
15 Dawkins (1986).
16 Everitt (2004, 85–111); Coyne (2009).
17 Darwin (2002a, 34).
18 Darwin (2002a, 39).
19 Darwin (2002a, 39).
20 Darwin (2002a, 39–40).
21 For the complete poem, see http://dwardmac.pitzer.edu/Anarchist_archives/shelley/maskofanarchy.html
22 Darwin (2002a, 42).
23 Darwin (2002a, 44).
24 From *Meteorology*. In Barnes (1984, 1:573).
25 Darwin (1934, x).
26 Moorehead (1969, 83).
27 Darwin (1934, 159–60).
28 Sulloway (1982).
29 Darwin (2002a, 44).
30 Darwin (2002a, 49).
31 Darwin (2002a, 45–50).
32 Darwin (2002b, 40).
33 Burkhardt & Porter (1997).
34 Van Wyhe (2006, 6, 12).
35 Darwin (1982, 1).
36 Darwin (2002a, 73).
37 E.g., Wright (1996, 301–10).
38 Darwin (2002a, 74).
39 Wallace in an interview with W. B. Northrop (1913, 622).
40 Browne (1998).
41 Darwin (2002a, 50).
42 Raby (2001, 230).
43 George (1964); Raby (2001).
44 Wallace (1905).
45 Raby (2001, 202).
46 Raby (2001, 15).
47 George (1964, 7).
48 George (1964, 10).
49 George (1964, 10).

50 Raby (2001, 81).
51 George (1964, 53).
52 George (1964, 55).
53 George (1964, 56).
54 George (1964, 57).
55 George (1964, 59).
56 Wallace in George (1964, 59).
57 George (1964, 62).
58 George (1964, 236).
59 George (1964, 241).
60 George (1964, 247).
61 Wallace in Raby (2001, 202–03).
62 George (1964, 246); Raby (2001, 190).
63 Darwin in Raby (2001, 203).
64 Raby (2001, 190).
65 Darwin in George (1964, 246).
66 Raby (2001, 191).
67 Raby (2001, 192).
68 Raby (2001, 193).
69 Milner (1996); Gould (1999c).
70 Dawes (1979, 157).
71 Silverman (1996, 40–41).
72 Silverman (1996, 250).
73 Silverman (1996, 253).
74 Silverman (1996, 170).
75 Silverman (1996, 259).
76 Randi (1982, 1986).
77 Brown (2006).
78 Wiseman (1997).

4 EXPLAINING RELIGION (PAGES 115–138)
1 Oubré (1997); Insoll (2004).
2 Durkheim (1964, 1972, 219); Berger (1973).
3 Norenzayan & Shariff (2008).
4 Silverman (1996, 252).
5 Isaiah 42:3; Matthew 12:20.
6 Marx (1843–4): Introduction; Kingsley (n.d.)
7 Winkelman (2008, 43–44); my parenthesis rectifies a typographical error; see also Winkelman (2000).
8 On what has become known as the Goldilocks enigma, see Davies (2006).
9 Eogan (1986); O'Kelly (1982); Chippindale (2004); Lewis-Williams & Pearce (2005); Lawson (2007).
10 Hayden (2003, 400).
11 Schmidt (2001, 2006); Lewis-Williams & Pearce (2005).
12 Writers who do recognize the importance of religious belief include Dornan (2004) and Renfrew (2008).
13 Today social anthropologists are fast running out of 'pristine' small-scale societies to study; they therefore turn to the interaction between indigenous communities and world systems, and multi-cultural urban communities.

14 E.g., Malinowski (1922, 1954); Evans-Pritchard (1937, 1956); Lévi-Strauss (1963); Beattie (1964); Fortes (1966); Turner (1967); Rappaport (1999).
15 Frazer (1961).
16 Later, another Cambridge anthropologist, Edmund Leach (1961), wrote scathingly of Frazer's second-hand knowledge in his article 'Golden Bough or gilded twig?'.
17 Robertson (2007).
18 The New English Bible confirms the translation as 'hate', as does the translation known as Good News for Modern Man. J. B. Phillips's 1956 translation, entitled St Luke's Life of Christ, places 'hating' in inverted commas.
19 Hinde (1999, 10).

5 RELIGIOUS EXPERIENCE (PAGES 139–160)
1 James (1960, 482).
2 Fenwick (1996); Horrobin (2001).
3 For more on neurology and religion, see Atran (2002), Boyer (2001), Churchland (2002), d'Aquili & Newberg (1993, 1999), Fenwick (1996), Newberg, d'Aquili & Rause (2002), Hamer (2004), Shermer (2003), Persinger (1987), and Ramachandran & Blakeslee (1998).
4 On the structure of the brain and consciousness, see Gregory (1987), Pinker (1998), Dennett (1991), Calvin (1996), Greenfield (1997, 2000), Solso (1997), LeDoux (2002), Ramachandran (2004), Blackmore (2006), Linden (2007), Boden (2007).
5 For more on the consciousness spectrum see Lewis-Williams (2002); Lewis-Williams & Pearce (2005).
6 Ardener (1971, xxi).
7 Laughlin et al. (1992, 132).
8 Kroll & Bachrach (1982b, 711).
9 Kroll & Bachrach (1982a).
10 Lewis-Williams & Dowson (1988); Clottes & Lewis-Williams (1998); Clottes & Lewis-Williams (2001); Lewis-Williams (2001); Lewis-Williams (2002); Lewis-Williams (2004a); Lewis-Williams & Pearce (2005).
11 E.g., Klüver (1966).
12 Oster (1970); Bókkon (2008).
13 Dronfield (1995, 1996).
14 Lewis-Williams & Dowson (1988).
15 Reichel-Dolmatoff (1978, 32–34); see also Reichel-Dolmatoff (1988).
16 England (1968, 432); Biesele (1980, 56, 61); Katz et al. (1997, 108, 113); Keeney (1999, 62, 105, 109, 115).
17 Keeney (2003, 42, 77).
18 Rose, in Narby & Huxley (2001, 123).
19 Horowitz (1964, 514; 1975, 177, 178, 181).
20 Heinz (1986).
21 Horowitz (1975, 177); original parentheses.

22 Shanon (2002, 89).

23 Klüver (1966, 71–72).

24 James (1890); quoted in Siegel & Jarvik (1975, 104–05).

25 Krishna (1997, 11–12).

26 Krishna (1997, 12–13).

27 Bleek (1935, 22, 23); Lewis-Williams (1981, 78–79); for more on 'vertebral artery' see James (2001, 212).

28 Bleek (1956, 425).

29 Marshall (1999, xxxii–xxxv).

30 Lee (1967, 31).

31 Katz (1976, 286).

32 Katz (1982, 42); in this publication Lee omitted the click from *n/um*.

33 Katz (1982, 8, 44); Biesele (1993, 70 ff).

34 Katz (1982, 44).

35 Keeney (1999, 62).

36 Keeney (2003, 70).

37 Keeney (2003, 53).

38 Rahula (1974).

39 Batchelor (1997, 102).

40 Douglas-Smith (1971, 549).

41 James (1960, 374).

42 D'Aquili & Newberg (1993, 1999); see also Ramachandran & Blakeslee (1998).

43 Hinde (1999, 197–98).

44 Douglas-Smith (1971, 550).

45 E.g., La Barre (1970, 1972).

46 Linden (2007) discusses the formation of mind in the evolution of the brain.

47 Hobson (2002); Rock (2004); Siegel (1999); Stickgold *et al.* (2001). On Freud and Jung, see Crews (1998, 2006); but see Franklin & Zyphur (2005) on supposed evolutionary value of dreaming.

6 **RELIGIOUS BELIEF** (PAGES 161–183)

1 Singer (1958).

2 Siegel (1977); Drab (1981).

3 Freidel *et al.* (1993, 151).

4 E.g., Blackmore (1982); Fox (2003).

5 Moody, quoted in Fox (2003, 17).

6 Moody, quoted in Fox (2003, 21).

7 Moody, quoted in Fox (2003, 21).

8 Lucy Lloyd (L.2.6.669 rev.).

9 Keeney (2003, 80, 105–108, 127).

10 Keeney (2003, 109).

11 Biesele (1993, 72).

12 Vitebsky (1995, 70).

13 Eliade (1972, 234).

14 Eliade (1972, 240).

15 Halifax (1979, 38); parenthesis added.

16 Cohen (1964, 76–77).

17 Moody, quoted in Fox (2003, 21).

18 Halifax (1979, 38).

19 Biesele (1993, 71).

20 Sullivan (1988, 122).

21 Eliade (1972, 235).

22 Bressloff *et al.* (2000).

23 Sullivan (1988, 412).

24 Narby & Huxley (2001, 99).

25 Crocker (1985, 201).

26 Some researchers believe these may represent a hanging bear skin; Sutherland (2001, 138–140). The carvers may have intended the ambiguity; the supernatural power of bears enables people to fly.

27 Jordan (2001, 92).

28 Bleek (1935, 18).

29 Eliade (1972, 140).

30 Vitebsky (1995, 11).

31 Blanke *et al.* (2002).

32 James (1960, 78).

33 Vitebsky (1995, 15).

34 Halifax (1979, 1); emphasis added.

35 Huxley (1959); Smith (2003).

36 *Biographia Literaria* (1817, chapter 13); for more along this line see Jay (1999).

37 James (1960, 373).

38 Whitehouse (2000).

39 Albright (1961).

40 See Finkelstein & Silberman (2002, 27–47, 319–25) for a summary of the evidence. See also Liverani (2003). For a more conservative view see Hoffmeier (2005) and Hillel (2005).

41 Finegan (1954).

42 E.g., Leviticus 20:13.

43 Bettenson (1956, 89).

44 Robinson (1963).

45 Rolston (1999, 359).

46 Kelly (2006).

47 Kelly (2006).

48 See Augustine's *City of God*, Volume 2, 15:23.

49 Aquinas (1911).

50 Lewis (1957).

51 Batchelor (1997, 114).

52 Orru & Wang (1992).

53 On Eastern mysticism see Hitchens (2007, 195–204); Dawkins (2006, 394).

54 Taylor (2006).

7 **RELIGIOUS PRACTICE** (PAGES 184–206)

1 Whitehouse (2000).

2 Cantor (1963, 500–501).

3 Stoichita (1995, 48).

4 Freidel *et al.* (1993, 260, 267–70).

5 Miller & Martin (2004, 99–101, pls 49–51), Schele & Miller (1986, 186).

6 Freidel *et al.* (1993, 207–10).

7 Dobkin de Rios (1974, 1978); see also Dobkin de Rios (1990).

8 Lara (2008).

9 Gray (2008).
10 E.g., Harner (1972); Halifax (1979); Furst (1982); Wilbert (1987); Dobkin de Rios (1990); Goodman (1992); Rudgley (1993); Jakobsen (1999); Jay (1999); Pearson (2002).
11 Allegro (1958); Gosso & Camilla (2000).
12 Stoichita (1995, 162).
13 Quoted in Stoichita (1995, 186).
14 Quoted in Stoichita (1995, 186).
15 Stoichita (1995, 189–90).
16 Stoichita (1995, 192).
17 Stoichita (1995, 190).
18 For the paintings see Stoichita (1995, fig. 49).
19 Quoted in Stoichita (1995, 121–22).
20 Quoted in Stoichita (1995, 122–23).
21 Stoichita (1995, 161).
22 Lewis-Williams (2002, 216–20).
23 Schmidt (2001, 2006); for a summary and interpretations see Lewis-Williams & Pearce (2005).
24 Ball (2008).
25 'To distant saints, well known in various lands'. Chaucer, Prologue to *The Canterbury Tales*. In Skeat (1912, 419).
26 Eade (2000, 65).
27 Eade & Sallnow (2000).
28 Van Gennep (1960); Turner & Turner (1978).
29 Eade & Sallnow (2000).
30 Hitchens (2007, 49–50, 223–27).
31 Auslander (2008).
32 Ward (2006, 200).

8 STONE AGE RELIGION (PAGES 207–231)

1 My argument has been developed through a number of publications: Lewis-Williams (1990, 1991, 1994, 1995a, 1997a, 1997b, 2002, 2003b); Lewis-Williams & Clottes (1998); Lewis-Williams & Dowson (1988, 1993); Clottes & Lewis-Williams (1998, 2001); I am, of course, aware that this proposition has divided archaeological opinion. But, encouraged by recent neuropsychological research – Bressloff *et al.* (2000), Burke (2002), ffytche (2002), ffytche & Howard (1999), ffytche *et al.* (1998), Santhouse *et al.* (2000) – I believe that the more serious objections have been answered (Lewis-Williams, 2004a). Those who argued that the spectrum of human consciousness, as I described it, is invalid were, quite simply, wrong: human consciousness does grade through a series of stages. This point is now indisputable.
2 Wylie (1989, 1999, 2000, 2002).
3 Wylie (1999, 293).
4 Rappaport (1999, 219); cf. Hayden (2003).
5 cf. Mithen (1996).
6 Lewis-Williams & Dowson (1988); Lewis-Williams (1997b, 2002, 2004).
7 cf. Lewis-Williams & Pearce (2005, 141–60).
8 See Lewis-Williams (1997b, 2002) and Clottes & Lewis-Williams (1998, 2001) for the reasoning behind these conclusions.
9 Lewis-Williams (2002, 228–67).
10 Bégouën (1999).
11 Bégouën (1999, 141).
12 Bégouën & Clottes (1991).
13 Bégouën (1999, 142).
14 Bégouën & Clottes (1981b).
15 Clottes pers. comm.
16 Clottes pers. comm.
17 Bégouën *et al.* (1988–89, figs 2–6).
18 Bégouën & Clottes (1981a, figs 13, 22).
19 Bégouën & Clottes (1981a, fig. 1.1); Bégouën *et al.* (1993); Clottes (1995, 52–53); Clottes & Lewis-Williams (1998, figs 76, 77); Lewis-Williams (2002, pl. 24).
20 Clottes pers. comm.
21 Bégouën & Clottes (1981a).
22 Dobres (1995, 2000); Sinclair (1995); Jones & White (1988); Taçon (1991).
23 E.g., southern African San; Lewis-Williams & Biesele (1978); Lewis-Williams (1981); cf. the Huichol peyote hunt, Myerhoff (1974).
24 Needham (1967).
25 Reznikoff & Dauvois (1988); Scarre (1989); Waller (1993).
26 Bégouën *et al.* (1984); Bégouën & Clottes (1991).
27 Bégouën & Clottes (1991).
28 E.g., Leroi-Gourhan (1968); Vialou (1991, 171–94); Clottes & Lewis-Williams (1998); Lewis-Williams (2002); White (2003, 112–16); Bégouën & Clottes (1991, 73).
29 Lewis-Williams (1994, 1995a); Lewis-Williams & Pearce (2004).
30 E.g., Eliade (1964); Halifax (1979); Vitebsky (1995).
31 Smith (1992); Whitley (2000); on the San see Lewis-Williams (1986), Lewis-Williams & Dowson (1990); Lewis-Williams & Pearce (2004).
32 Noll (1985).
33 Shamanism is today a controversial category, e.g., Atkinson (1992); Hamayon (1998); Kehoe (2000); Lewis-Williams (2003a); various entries in Walter & Fridman (2004); Pearson (2002). Some researchers feel, perhaps rightly, that the many ethnographic instances that have been labelled 'shamanistic' are too diverse to be accommodated under one heading. For definitions of 'shamanism' see, among others: Shirokogoroff (1935); Eliade (1964); Lommel (1967); Hultkranz (1973, 1996); Furst (1982); Bourguignon (1985); Ripinsky-Naxon (1989); Winkelman (1990, 1992); Atkinson (1992); Vitebsky (1995); Townsend (1997); Hutton (2001); Jordan (2001); Price (2001); Hayden (2003).

Readers who dislike 'shamanism' may substitute a word of their own choosing. Despite valuable critical appraisal, 'shamanism' continues to be a useful category of belief and associated practices that has explanatory potential in the study of Upper Palaeolithic and other image-making, even if, like many useful categories, it frays at the edges. On shamanistic imagery in North America see, for example, Hedges (1983); Bean (1992); Whitley (1992, 1998a, 1998b); Francis & Loendorf (2002); Keyser *et al.* (2006); Loendorf *et al.* (2005); for southern African imagery see, for example, Lewis-Williams (1995b), Lewis-Williams & Dowson (1990), Blundell (1998, 2004), Lewis-Williams & Pearce (2004).

34 Clottes pers. comm.

35 Clottes pers. comm.

36 Bégouën (1999, 143)

37 Breuil (1979, fig. 124); Bégouën & Breuil (1999, pl. VII).

38 Bégouën & Clottes (1981a, fig. 4).

39 Bégouën & Clottes (1984, fig. 16).

40 Bégouën & Breuil (1999, fig. 1).

41 Bégouën & Clottes (1986–1987).

42 Bégouën & Clottes (1981a, fig. 1.2).

43 Bégouën & Clottes (1981a, fig. 6); Lewis-Williams (2002, pl. 25).

44 Lewis-Williams (1997b, 2002); Clottes & Lewis-Williams (1998).

45 E.g., Reichel-Dolmatoff (1975); Lewis-Williams (1981, 97).

46 Breuil (1979, fig. 123).

47 Fig. 4; Bégouën & Breuil (1999, pl. XVIIIa); Breuil (1979, 167–71).

48 Breuil (1979, 170).

49 Lewis-Williams (2002, 238–50, pl. 19).

50 Klüver (1966, 26, 31); Siegel & Jarvik (1975, 105); Lewis-Williams (2002).

51 E.g., La Barre (1980, 53–54).

52 E.g., Gabillou; Lewis-Williams (2002, 266–67).

53 E.g., the Apse in Lascaux; Leroi-Gourhan & Allain (1979, 191–299).

54 Bégouën & Clottes (1991, 70).

55 Bégouën & Clottes (1991, 74–75).

56 Bégouën & Clottes (1984, fig. 4).

57 Smith (1992).

58 Cartailhac & Breuil (1906); Kirchner (1952).

59 Bosinski (1986, pl. 5); Lewis-Williams & Clottes (1998, fig. 42); Müller-Beck & Albrecht (1987, fig. 37); Marshack (1991, fig. 231); Conard (2003).

60 Clottes & Lewis-Williams (1998, fig. 41); Chauvet *et al.* (1996, fig. 93); Clottes (2001, figs 137, 161, 162, 163, 164).

61 Conkey (1982, 1987, 1991, 1995).

62 Lewis-Williams & Dowson (1988); Clottes & Lewis-Williams (1998); Lewis-Williams (2002).

63 Klüver (1966, 14, 18, 30, 36); Knoll *et al.* (1963, 208); Siegel & Jarvik (1975, 109).

64 Siegel (1977, 134).

65 Klüver (1966, 27).

66 Lewis-Williams & Dowson (1988, 203).

67 Lewis-Williams (1997b, 1997c, 2002, pls 17 and 18); Clottes & Lewis-Williams (1998, figs 83, 84, 86, 87, 88, 89).

68 Bégouën & Breuil (1999, pl. XIIIa).

69 Klüver (1942, 179).

70 Reichel-Dolmatoff (1978, 8).

71 Lewis-Williams (1994, 1995a).

72 Noll (1985, 445–46).

73 Bogoras (1907, 330).

74 Zimmerman (1996, 118).

75 Vitebsky (1995, 10).

76 Bender (1989); Lewis-Williams (2002).

77 Lewis-Williams (1994, 1995a, 2002); Lewis-Williams & Dowson (1993).

78 Bender (1989).

79 Bégouën & Clottes (1991, 77).

80 Lewis-Williams (1997a).

81 Lewis-Williams (1997a, 2002).

9 HILDEGARD ON THE AFRICAN VELD (PAGES 232–256)

1 Turner & Bruner (1986).

2 Singer (1958); Newman (1987); Flanagan (1989, 1998); Beer (1992).

3 Flanagan (1989, 224, note 8).

4 Newman (1987); Flanagan (1989); Singer (1958); Beer (1992); Atherton (2001).

5 Singer (1958).

6 Higley (2007).

7 Newman (1987, 203).

8 Flanagan (1989, 212).

9 Flanagan (1989, 213).

10 Singer (1958); Selby (1983); Sacks (1993).

11 Hachinski *et al.* (1973); Richards (1971); Asaad (1990); Sacks (1993).

12 Sacks (1993, 51).

13 Sacks (1993, 280).

14 Sacks (1993, 52).

15 Sacks (1993, 53–54).

16 Sacks (1993, 263, 273).

17 Sacks (1993, 55).

18 Sacks (1993, 55).

19 Sacks (1993, 56).

20 Sacks (1993, 276, 278).

21 Sacks (1993, 278; pls 2B, 3B, 4B).

22 Sacks (1993, fig. 2C).

23 Sacks (1993, 274).

24 Wolber & Ziegler (1982).

25 Sacks (1993, 64–65).

26 Sacks (1993, 73, 275).

27 Singer (1958, 232–31).

28 Singer (1958, 231).

29 Atherton (2001, 79–81); Flanagan (1989, 197–98); on starvation and other triggering factors of medieval visions see Kroll & Bachrach (1982a).
30 Singer (1958, 232).
31 Singer (1958, 232); Sacks (1993, 301).
32 Newman (1987, 100–103).
33 Flanagan (1989, fig. 10).
34 Singer (1958, 232–33).
35 Singer (1958, 233).
36 Sacks (1993, 301).
37 Sacks (1993, 301).
38 *Scivias* II:1; Atherton (2001, 10–11).
39 Bynum (2007).
40 Wilmsen (1989); Barnard (1992); Lee (1993); Marshall Thomas (2006).
41 Katz (1982); Marshall (1999); Biesele (1993); Guenther (1999); Keeney (1999); Lewis-Williams & Pearce (2004).
42 Lewis-Williams & Pearce (2004).
43 Lewis-Williams & Pearce (2004); Lewis-Williams (1981, 1987).
44 Lewis-Williams (1995d).
45 Tongue (1909, pl. 20).
46 Lewis-Williams & Biesele (1978); Lewis-Williams (1981, 1992); Lewis-Williams & Pearce (2004).
47 For a complete copy visit: http://www.sarada.co.za and search for RSA GRE1.
48 Lewis-Williams (1981); Lewis-Williams & Pearce (2004).
49 Lewis-Williams (1996); Lewis-Williams & Pearce (2004, 120).
50 Biesele (1993, 72); Lewis-Williams & Pearce (2004, 54–55).
51 Boyd (2003).
52 Lewis-Williams (2002, 169–70).
53 Lewis-Williams & Pearce (2004).
54 *Scivias* II:12; Atherton (2001, 17).
55 Orpen (1874, 3).
56 Flanagan (1989, 211).
57 Sacks (1993, 97, 98).
58 Sacks (1993, 66).

10 GOD'S EMPIRE STRIKES BACK (PAGES 257–289)
1 McGrath (2005, 158–59).
2 Dawkins (2006). See also Boyer (2001), Brockman (2006), Linden (2007), Kurtz (2003), Dunbar (2004), Everitt (2004), Davies (2006), Dawkins (2003), Dennett (1995, 2006), Grayling (2007a), Hitchens (2007), Humphreys (2007), Newberg *et al.* (2002), Wolpert (2006).
3 Taylor (2007).
4 See, for example, McGrath (2004, 2005), McGrath & McGrath (2007), LeRon Shults (2006), van Huyssteen (2006), Ward (2006),

Robertson (2007), and certain chapters in Bakewell (2006).
5 Ruse (2001); Kurtz (2003).
6 *Time*, 15 December, 2008, p. 11.
7 McGrath (2004); see also Robertson (2007, 20–28), McGrath (2004, 162).
9 See, for example, Schama (2008); Bageant (2007).
10 Garwood (2008).
11 See also Ephesians 6:5, Colossians 3:22, Titus 2:9.
12 Freeman (2003, 2008).
13 Robertson (2007, 57), my lineation, his parenthesis; see also p. 140.
14 Strong (2007).
15 Platten (2007).
16 Turnbull (2008).
17 Avalos (2007).
18 *New Statesman*, 14 April, 2008, pp. 30–33.
19 Byrnes (2008, 30).
20 Spenser & White (2007).
21 Thompson (2008, 1).
22 Dawkins (2006a); Dennett (1995, 2006); Hitchens (2007); Wolpert (2006).
23 Dummett (2006).
24 For a comparable approach see Rolston (1999).
25 See Searle (2006).
26 Cobb & Coyne (2008).
27 Lakoff & Johnson (1980).
28 Midgley (2007, 28).
29 Midgley (2007, 28).
30 Flanagan (2008).
31 Ward (2006, 200).
32 Dawkins (2006a).
33 Coleridge (1817); in Enright & de Chickera (1962, 191).
34 See, for example, Dennett (1995), Ruse (2001), Dawkins (2003), Brockman (2006).
35 Dawkins (1976).
36 Dawkins (2006b, vii–xii).
37 Darwin (1982).
38 Decety & Chaminade (2003).
39 Greene & Haidt (2002).
40 De Waal (2006, 56).
41 De Waal (2005, 214).
42 De Waal (2005, 172).
43 De Waal (2006, 57–58); Wright (1996).
44 Harries (2008).
45 Pullman (2006).
46 Owen (2008).
47 Gould (1999a).
48 LeDoux (2002, 14–16).
49 Cobb & Coyne (2008, 1049).

EPILEGOMENA (PAGES 290–293)
1 http://www.westminster-abbey.org/our-history/people/charles-darwin

Bibliography and Guide to Further Reading

Albright, W. F. 1961. Abraham the Hebrew: a new archaeological interpretation. *Bulletin of the American Schools of Oriental Research* 163, 36–54.

Allegro, M. J. 1958. *I Rotoli del Mar Marto.* Florence: Sansoni.

Allegro, M. J. 1970. *The Sacred Mushroom and the Cross.* London: Hodder & Houghton.

Aquinas, T. 1911. *Summa Theologica.* Trans. Fathers of the English Dominican Province, London: R&T Washbourne.

Ardener, E. 1971. Introductory essay: social anthropology and language. In Ardener, E. (ed.) *Social Anthropology and Language,* pp. ix–cii. London: Tavistock.

Asaad, G. 1990. *Hallucinations in Clinical Psychiatry: A Guide for Mental Health Professionals.* New York: Brunner/Mazel.

Atherton, M. (ed.) 2001. *Hildegard of Bingen: Selected Writings.* London: Penguin.

Atkinson, J. 1992. Shamanisms today. *Annual Review of Anthropology* 21, 307–37.

Atran, S. 2002. *In Gods we Trust: The Evolutionary Landscape of Religion.* Oxford: Oxford University Press.

Augustine of Hippo. 1991. *Confessions.* Trans. and ed. H. Chadwick. Oxford: Oxford University Press.

Augustine of Hippo. 1993. *De Libero Arbitrio.* Trans. T. Williams. Indianapolis: Hackett.

Auslander, S. 2008. *Foreskin's Lament.* London: Picador.

Avalos, H. 2007. *The End of Biblical Studies.* Amherst: Prometheus Books.

Avise, J. 2001. *The Genetic Gods: Evolution and Belief in Human Affairs.* Cambridge, Mass.: Harvard University Press.

Ayala, F. J. 2007. Darwin's greatest discovery: design without a designer. *National Academy of Sciences and Engineering* 104 (suppl. 1), 8567–73.

Bageant, J. 2007. *Deer Hunting with Jesus: Dispatches from America's Class War.* London: Random House.

Bakewell, J. 2006. *Belief.* London: Duckworth Overlook.

Ball, P. 2008. *Universe of Stone: Chartres Cathedral and the Triumph of the Medieval Mind.* London: Bodley Head.

Barham, L. 2007. Modern is as modern does? Technological trends and thresholds in the south-central African record. In Mellars, P., Boyle, K., Bar-Yosef, O., & Stringer, C. (eds) 2007. *Rethinking the Human Revolution: New Behavioural and Biological Perspectives on the Origin and Dispersal of Modern Humans,* pp. 165–76. Cambridge: McDonald Institute.

Barnard, A. 1992. *Hunters and Herders of Southern Africa.* Cambridge: Cambridge University Press.

Barnes, J. (ed.) 1984. *The Complete Works of Aristotle.* Princeton: Bollingen Series LXX1.

Batchelor, S. 1997. *Buddhism without Beliefs: A Contemporary Guide to Awakening.* New York: Riverhead Books.

Bean, L. J. (ed.) 1992. *California Indian Shamanism.* Menlo Park: Ballena Press.

Beattie, J. 1964. *Other Cultures: Aims, Methods and Achievements in Social Anthropology.* London: Routledge.

Beer, F. 1992. *Women and Mystical Experience in the Middle Ages.* Woodbridge: Boydell Press.

Bégouën, H., & Breuil, H. 1999 [1958]. *Les Cavernes du Volp: Trois Frères – Tuc d'Audoubert.* Paris: Flammarion. (Republished by American Rock Art Research Association, Occasional Paper 4)

Bégouën, R. 1999. Postscript. In Bégouën, H., & Breuil, H. *Les Cavernes du Volp: Trois Frères – Tuc d'Audoubert,* pp. 140–44. Tucson: American Rock Art Research Association.

Bégouën, R., Brois, F., Clottes, J., & Servelle, C. 1984-5. Art mobilier sur support lithique d'Enlène (Montesquieu-Avantès, Ariège): Collection Bégouën du Musée de l'Homme. *Ars Praehistorica* 3–4, 25–80.

Bégouën, R., & Clottes, J. 1981a. Apports mobiliers dans les cavernes du Volp (Enlène, Les Trois Frères, Le Tuc d'Audoubert). *Altamira Symposium,* pp. 157–87.

Bégouën, R., & Clottes, J. 1981b. Nouvelles fouilles dans la Salle des Morts de la caverne d'Enlène à Montesquieu-Avantès (Ariège). *Congrès Préhistorique de France,* pp. 33–56.

Bégouën, R., & Clottes, J. 1984. Grotte des Trois-Frères. In *L'art des Cavernes: Atlas des Grottes Ornées Paléolithique Françaises.* Paris: Ministère de la Culture.

Bégouën, R., & Clottes, J. 1986-7. Le grand félin des Trois-Frères. *Antiquités Nationales* 18, 109–13.

Bégouën, R., & Clottes, J. 1991. Portable and wall art in the Volp Caves, Montesquieu-Avantès (Ariège). *Proceedings of the Prehistoric Society* 57(1), 65–79.

Bégouën, R., Clottes, J., Giraud, J.-P., & Rouzaud, F. 1984. Compléments à la grande plaquette gravée d' Enlène. *Bulletin de la Société Française* 81, 1–7.

Bégouën, R., Clottes, J., Giraud, J.-P., & Rouzaud, F. 1988–9. La rondelle au bison d'Enlène (Montesquieu-Avantès, Ariège). *Zephyrus* 41–2, 19–25.

Bégouën, R., Clottes, J., Giraud, J.-P., & Rouzaud, F. 1993. Os plantés et peintures rupestres dans la caverne d'Enlène. In Delport, H., & Clottes, J. (eds) *Congrès National des Sociétés Historiques et Scientifiques* 118, 283–306.

Bender, B. 1989. The roots of inequality. In Miller, D., & Tilley, C. (eds) *Domination and Resistance*, pp. 83–93. London: Unwin & Hyman.

Berger, P. L. 1973. *The Social Reality of Religion.* Harmondsworth: Penguin.

Bettenson, H. 1956. *The Early Christian Fathers: A Selection from the Writings of the Fathers from St Clement of Rome to St Athanasius.* Oxford: Oxford University Press.

Biagioli, M. 1993. *Galileo Courtier: The Practice of Science in the Culture of Absolutism.* Chicago: Chicago University Press.

Biesele, M. 1980. Old K"xau. In Halifax, J. (ed.) *Shamanic Voices: A Survey of Visionary Narratives*, pp. 54–62. Harmondsworth: Penguin.

Biesele, M. 1993. *Women Like Meat: The Folklore and Foraging Ideology of the Kalahari Ju/'hoansi.* Johannesburg: Witwatersrand University Press.

Blackmore, S. 1982. *Dying to Live: Science and the Near-Death Experience.* London: Grafton.

Blackmore, S. 2006. *Conversations on Consciousness: What the Best Minds Think About the Brain, Free Will, and What it Means to be Human.* Oxford: Oxford University Press.

Blanke, O., Ortigue, S., Landis, T. & Seeck, M. 2002. Stimulating illusory own-body perceptions. *Nature* 419:269.

Bleek, D. F. 1935. Beliefs and customs of the /Xam Bushmen. Part VII: Sorcerors (sic). *Bantu Studies* 9, 1–47.

Bleek, D. F. 1956. *A Bushman Dictionary.* American Oriental Series Vol. 41. New Haven: American Oriental Society.

Blundell, G. 1998. On neuropsychology in Southern African rock art research. *Anthropology of Consciousness* 9(1), 3–12.

Blundell, G. 2004. *Nqabayo's Nomansland: San Rock Art and the Somatic Past.* Uppsala: Uppsala University Press.

Boden, M. A. 2007. *Mind as Machine: A History of Cognitive Science.* Oxford: Clarendon Press.

Boghossian, P. 2006. *Fear of Knowledge: Against Relativism and Constructivism.* Oxford: Oxford University Press.

Bogoras, W. 1907. *The Chukchee, Part II: Religion.* New York: Memoirs of the American Museum of Natural History, Vol. 11.

Bókkon, I. 2008. Phosphene phenomenon: a new concept. *BioSystems* 92, 168–74.

Bosinski, G. 1986. *Die Grosse Zeit der Eiszeitjäger: Europa zwischen 40 000 und 10000 v. Chr.* Römisch-Germanisches Zentral Museum, Mainz.

Boulter, M. 2008. *Darwin's Garden: Down House and The Origin of Species.* London: Constable.

Bourguignon, E. 1985. Comment on Noll 1985. *Current Anthropology* 26:451–52.

Bowler, P. J., & Morus, I. R. 2005. *Making Modern Science: A Historical Survey.* Chicago: University of Chicago Press.

Boyd, C. E. 2003. *Rock Art of the Lower Pecos.* College Station: Texas A&M University Press.

Boyer, P. 2001. *Religion Explained: The Evolutionary Origins of Religious Thought.* New York: Basic Books.

Bressloff, P. C., Cowan, J. D., Golubitsky, M., Thomas, P. J., & Wiener, M. C. 2000. Geometric visual hallucinations, Euclidean symmetry and the functional architecture of the striate cortex. *Philosophical Transactions of the Royal Society, London*, Series B, 356, 299–330.

Breuil, H. 1979. *Four Hundred Centuries of Cave Art.* New York: Hacker Art Books.

Brockman, J. (ed.) 2006. *Intelligent Thought: Science Versus the Intelligent Design Movement.* New York: Vintage Books.

Brooke, J. H. 2003. Darwin and Victorian Christianity. In Hodge, J., & Radick, G. (eds) *The Cambridge Companion to Darwin*, pp. 192–213. Cambridge; Cambridge University Press.

Brown, A. 1999. *The Darwin Wars: How Stupid Genes became Selfish Gods.* London: Simon & Schuster.

Brown, D. 2006. *Tricks of the Mind.* London: Random House.

Browne, J. 1998. I could have retched all night: Charles Darwin and his body. In Lawrence, C., & Shapin, S. (eds) *Science Incarnate: Historical Embodiments of Natural Knowledge*, pp. 240–87. Chicago: University of Chicago Press.

Browne, J. 2007. *Darwin's 'Origin of Species':
A Biography*. New York: Atlantic Monthly Press.

Burke, W. 2002. The neural basis of Charles
Bonnet hallucinations: a hypothesis. *Journal of
Neurology and Neurosurgical Psychiatry* 73,
535–41.

Burkert, W. 1987. *Ancient Mystery Cults*.
Cambridge, Mass.: Harvard University Press.

Burkhardt, F., & Porter, D. 1997. *The
Correspondence of Charles Darwin*, Vols 14–15.
Cambridge: Cambridge University Press.

Bynum, C. W. 2007. *Wonderful Blood: Theology
and Practice in Late Medieval Northern
Germany and Beyond*. Philadelphia: University
of Pennsylvania Press.

Byrnes, S. 2008. 'Jesus will appear again as judge
of the world and the dead will be raised.' *New
Statesman* April, 30–34

Calvin, W. H. 1996. *How Brains Think: Evolving
Intelligence, Then and Now*. London: Phoenix.

Cantor, N. F. 1963. *Medieval History: The Life and
Death of a Civilization*. London: Macmillan.

Cartailhac, E., & Breuil, H. 1906. *La Caverne
d'Altamira en Santillana del Mar*. Monaco:
Imprimerie de Monaco.

Cauvin, J. 2000. *The Birth of the Gods and the
Origins of Agriculture*. Cambridge: Cambridge
University Press.

Chalmers, A. F. 1990. *Science and its Fabrication*.
Minneapolis: University of Minnesota Press.

Chalmers, A. F. 1999. *What is this Thing called
Science?* Milton Keynes: Open University Press.

Chambers, R. 1844. *Vestiges of the Natural History
of Creation*. London: Chambers.

Chauvet, J.-M., Brunel Deschamps, E., & Hillaire,
C. 1996. *Chauvet Cave: The Discovery
of the World's Oldest Paintings*. London:
Thames & Hudson.

Chippindale, C. 2004. *Stonehenge Complete*.
London: Thames & Hudson.

Churchland, P. 2002. *Brain-wise: Studies in
Neurophilosophy*. Cambridge, Mass.: MIT Press.

Clark, J. W., & Hughes, T. M. 1890. *The Life and
Letters of the Reverend Adam Sedgwick*.
Cambridge: Cambridge University Press.

Clottes, J. 1995. Perspectives and traditions in
Palaeolithic rock art research in France. In
Helskog, K., & Olsen, B. (eds). *Perceiving Rock
Art: Social and Political Perspectives*, pp. 35–64.
Oslo: Institute for Comparative Research in
Human Culture.

Clottes, J. 2001. Paleolithic Europe. In Whitley,
D. S. (ed.) *Handbook of Rock Art Research*,
pp. 459–81. Walnut Creek, California: AltaMira
Press.

Clottes, J., Garner, M., & Maury, G. 1994.
Magdalenian bison in the caves of the Ariège.
Rock Art Research 11, 58–70.

Clottes, J., & Lewis-Williams, J. D. 1998. *The
Shamans of Prehistory: Trance and Magic in
the Painted Caves*. New York: Harry Abrams.

Clottes, J., & Lewis-Williams, J. D. 2001. *Les
Chamanes de la Préhistoire: Texte Intégral,
Polémique et Réponses*. Paris: La Maison des
Roches.

Cobb, M., & Coyne, J. 2008. Atheism could be
science's contribution to religion. *Nature* 454,
1049.

Coe, M. D. 1992. *Breaking the Maya Code*.
London: Thames & Hudson.

Cohen, S. 1964. *The Beyond Within: The LSD
Story*. New York: Atheneum.

Conard, N. J. 2003. Palaeolithic ivory sculptures
from southwestern Germany and the origins
of figurative art. *Nature* 426–832.

Conkey, M. W. 1982. Boundedness in art and
society. In Hodder, I. (ed.) *Symbolic and
Structural Archaeology*, pp. 115–28. Cambridge:
Cambridge University Press.

Conkey, M. W. 1987. New approaches in the
search for meaning? A review of research in
'Palaeolithic art'. *Journal of Field Archaeology* 14,
413–30.

Conkey, M. W. 1991. Contexts of action, contexts
for power: material culture and gender in the
Magdalenian. In Gero, J. M., & Conkey M. W.
(eds) *Engendering Archaeology: Women and
Prehistory*, pp. 57–92. Oxford: Basil Blackwell.

Conkey, M. W. 1995. Making things meaningful:
approaches to the interpretation of Ice Age
imagery of Europe. In Lavin, I. (ed.) *Meaning
and the Visual Arts: Views from the Outside*,
pp. 49–64. Princeton: Institute for Advanced
Study.

Copi, I. M. 1982. *Introduction to Logic*. New York:
Macmillan.

Cornwall, J. 2007. *Darwin's Angel: An Angelic
Riposte to the God Delusion*. London: Profile.

Coyne, J. A. 2009. *Why Evolution is True*. Oxford:
Oxford University Press

Crews, F. 1998. *Unauthorized Freud: Doubters
Confront a Legend*. London: Penguin.

Crews, F. 2006. *Follies of the Wise: Dissenting Essays*.
Emeryville, California: Shoemaker Hoard.

Crocker, C. 1985. *Vital Souls: Bororo Cosmology,
Natural Symbolism, and Shamanism*. Tucson:
University of Arizona Press.

D'Aquili, E. G., & Newberg, A. B. 1993. Religious
and mystical states: a neuropsychological
model. *Zygon* 28, 177–99.

D'Aquili, E. G., & Newberg, A. B. 1999. *The Mystical Mind: Probing the Biology of Religious Experience.* Minneapolis: Fortress Press.

Darwin, C. 1839. *Journal of Researches into the Geology and Natural History of the Various Countries Visited by H. M. S. 'Beagle'.* London: Henry Colburn.

Darwin, C. 1934. *Charles Darwin's Diary of the Voyage of H.M.S. 'Beagle'.* Ed. N. Barlow. Cambridge: Cambridge University Press.

Darwin, C. 1968 [1859]. *The Origin of Species by Means of Natural Selection or the Preservation of Favoured Races in the Struggle for Life.* London: John Murray. Penguin edition.

Darwin, C. 1982 [1871]. *The Descent of Man, and Selection in Relation to Sex.* Princeton: Princeton University Press.

Darwin, C. 2002a [1887]. *The Life of Erasmus Darwin.* Cambridge: Cambridge University Press.

Darwin, C. 2002b. *Autobiographies.* Eds M. Neve & S. Messenger. London: Penguin.

Darwin, E. 1800. *Phytologia.* London: Johnson.

Darwin, E. 1818 [1794]. *Zoonomia.* Philadelphia: Edward Earle.

Davies, P. 2006. *The Goldilocks Enigma: Why is the Universe Just Right for Life?* London: Allen Lane.

Davis, R. H. C. 1988. *A History of Medieval Europe from Constantine to Saint Louis.* London: Longman.

Dawes, E. A. 1979. *The Great Illusionists.* Secaucus, New Jersey: Chartwell Books.

Dawkins, R. 1976. *The Selfish Gene.* Oxford: Oxford University Press.

Dawkins, R. 1986. *The Blind Watchmaker.* New York: Norton.

Dawkins, R. 2003. *A Devil's Chaplain: Selected Essays.* Ed. L. Menon. London: Phoenix.

Dawkins, R. 2006a. *The God Delusion.* London: Bantam.

Dawkins, R. 2006b. *The Selfish Gene* (30th anniversary edition). Oxford: Oxford University Press.

Decety, J., & Chaminade, T. 2003. Neural correlates of feeling sympathy. *Neuropsychologia* 41, 127–38.

Dennett, D. C. 1991. *Consciousness Explained.* London: Penguin.

Dennett, D. C. 1995. *Darwin's Dangerous Idea: Evolution and the Meanings of Life.* London: Penguin.

Dennett, D. C. 2006. *Breaking the Spell: Religion as a Natural Phenomenon.* London: Allen Lane.

D'Errico, F., Henshilwood, C., Vanhaeren, M., van Niekerk, K., 2005. *Journal of Human Evolution* 48, 3–24.

Desmond, A., Moore, J., & Browne, J. 2007. *Charles Darwin.* Oxford: Oxford University Press.

De Waal, F. 2005. *Our Inner Ape: The Best and Worst of Human Nature.* London: Granta Books.

De Waal, F. 2006. *Primates and Philosophers: How Morality Evolved.* Princeton: Princeton University Press.

Dobkin de Rios, M. 1974. The influence of psychotropic flora and fauna on Maya religion. *Current Anthropology* 15, 147–64.

Dobkin de Rios, M. 1978. The Maya and the water lily. *The New Scholar* 5, 299–307.

Dobkin de Rios, M. 1990. *Hallucinogens: Cross-Cultural Perspectives.* New York: Prism Press.

Dobres, M.-A. 1995. Gender and prehistoric technology: on the social agency of technical strategies. *World Archaeology* 27(1), 25-49.

Dobres, M.-A. 2000. *Technology and Social Agency: Outlining a Practice Framework for Archaeology.* Oxford: Blackwell.

Dobzhansky, T. 1945. Review of F. L. Marsh 'Evolution, Creation, and Science'. *American Naturalist* 79:45.

Dornan, J. 2004. Beyond belief: religious experience, ritual, and cultural neuro-phenomenology in the interpretation of past religious systems. *Cambridge Archaeological Journal* 14, 25–36.

Douglas-Smith, B. 1971. An empirical study of religious mysticism. *British Journal of Psychiatry* 118, 549–54.

Drab, K. J. 1981. The tunnel experience: reality of hallucination? *Anabiosis* 1, 126–52.

Drake, H. A. 2000. *Constantine and the Bishops: The Politics of Intolerance.* Baltimore: John Hopkins University Press.

Drake, S. 1957. *Discoveries and Opinions of Galileo.* New York: Doubleday.

Dronfield, J. C. 1995. Subjective vision and the source of Irish megalithic art. *Antiquity* 69, 539–49.

Dronfield, J. C. 1996. Entering alternative realities: cognition, art and architecture in Irish passage-tombs. *Cambridge Archaeological Journal* 6, 37–72

Drummond, H. 1894. *The Ascent of Man.* London: Hodder & Staughton.

Dummett, M. 2006. *Thought and Reality.* London: Clarendon Press.

Dunbar, R. 2004. *The Human Story: A New History of Mankind's Evolution.* London: Faber & Faber.

Durkheim, E. 1964 [1912]. *The Elementary Forms of Religious Life.* London: Allen & Unwin.

Durkheim, E. 1972. *Emile Durkheim: Selected Writings*. Ed. A. Giddens. Cambridge: Cambridge University Press.

Eade, J. 2000. Order and power at Lourdes: lay helpers and the organization of a pilgrimage shrine. In Eade J., & Sallnow, M. J. (eds) *Contesting the Sacred: The Anthropology of Christian Pilgrimage*, pp. 51–76. Chicago: University of Illinois Press.

Eade, J., & Sallnow, M. J. 2000. *Contesting the Sacred: The Anthropology of Christian Pilgrimage*. Chicago: University of Illinois Press.

Eldredge, N. 2005. *Darwin: Discovering the Tree of Life*. New York: W. W. Norton.

Eliade, M. 1964. *Shamanism: Archaic Techniques of Ecstasy*. New York: Bollingen Foundation.

Endersby, J. 2007. Creative designs? How Darwin's *Origin* caused the Victorian crisis of faith, and other myths. *Times Literary Supplement* 5424, 3–4.

England, N. M. 1968. Music among the Ju/wa-si of South West Africa and Botswana. Harvard University, PhD thesis.

Enright, D. J., & de Chickera, E. (eds) 1962. *English Critical Texts*. Oxford: Oxford University Press.

Eogan, G. 1986. *Knowth and the Passage Tombs of Ireland*. London: Thames & Hudson.

Evans-Pritchard, E. E. 1937. *Witchcraft, Oracles, and Magic among the Azande*. London: Oxford University Press.

Evans-Pritchard, E. E. 1956. *Nuer Religion*. Oxford: Clarendon Press.

Everitt, N. 2004. *The Non-Existence of God*. London: Routledge.

Fauvel, J., Flood, R., Shortland, M., & Wilson, R. (eds) 1988. *Let Newton Be!* Oxford: Oxford University Press.

Fenwick, P. 1996. The neurology of religious experience. In Bhugra, D. (ed.) *Psychiatry and Religion*, pp. 167–77. London: Routledge

ffytche, D. H. 2002. Cortical bricks and mortar. *Journal of Neurology, Neurosurgery and Psychiatry* 73, 472.

ffytche, D. H., & Howard, R. J. 1999. The perceptual consequences of visual loss: 'Positive pathologies of vision'. *Brain* 122, 1247–60.

ffytche, D. H., Howard, R. J., Brammer, M. J., David, A., Woodruff, P. & Williams, S. 1998. The anatomy of conscious vision: an fMRI study of visual hallucinations. *Nature Neuroscience* 1, 738–42.

Finegan, J. 1954. *Light from the Ancient Past*. Princeton: Princeton University Press.

Finkelstein, I., & Silberman, N. A. 2002. *The Bible Unearthed: Archaeology's New Vision of Ancient Israel and the Origin of its Sacred Texts*. New York: Touchstone.

Finocchiaro, M. A. 2006. *Retrying Galileo, 1633-1992*. Berkeley: University of California Press.

Flanagan, O. 2008. *The Really Hard Problem: Meaning in a Material World*. London: Wiley.

Flanagan, S. 1989. *Hildegard of Bingen, 1098-1179: A Visionary Life*. London: Routledge.

Flanagan, S. 1998. *Secrets of God: Writings of Hildegard of Bingen*. Boston: Shambhala.

Fortes, M. 1966. Religious premises and logical technique in divinatory ritual. In Huxley, J. (ed.). *A Discussion of Ritualization of Behaviour in Animals and Man*. Philosophical Transactions of the Royal Society of London. Series B. 251.

Foucault, M. 1984. What is Englightenment? In Rabinow, P (ed.). *The Foucault Reader*. New York: Pantheon Books.

Fox, M. 2003. *Religion, Spirituality and the Near-Death Experience*. London: Routledge.

Francis, J. E., & Loendorf, L. L. 2002. *Ancient Visions: Petroglyphs and Pictographs of the Wind River and Bighorn Country, Wyoming and Montana*. Salt Lake City: University of Utah Press.

Franklin, M. S., & Zyphur, M. J. 2005. The role of dreams in the evolution of the human mind. *Evolutionary Psychology* 3, 59–78.

Frazer, J. G. 1961 [1890]. *The Golden Bough*. New York: Doubleday.

Freeman, C. 2003. *The Closing of the Western Mind: The Rise of Faith and the Fall of Reason*. London: Pimlico.

Freeman, C. 2008. *AD 381: Heretics, Pagans and the Christian State*. London: Pimlico.

Freidel, D., Schele, L., & Parker, J. 1993. *Maya Cosmos: Three Thousand Years on the Shaman's Path*. New York: William Morrow.

Furst, P. T. 1972. *Flesh of the Gods: The Ritual Use of Hallucinogens*. New York: Praeger.

Furst, P. T. 1982. *Hallucinogens and Culture*. Novato, California: Chandler & Sharp.

Garwood, C. 2008. *The History of an Infamous Idea*. London: Pan.

Gaukroger, S. 2008. *The Emergence of a Scientific Culture*. Oxford: Oxford University Press.

George, W. 1964. *Biologist Philosopher: A Study of the Life and Writings of Alfred Russel Wallace*. London: Abelard-Schuman.

George, W. 1982. *Darwin*. London: Fontana.

Goadsby, P. J. 2007. Recent advances in understanding migraine mechanisms, molecules and therapeutics. Trends in molecular medicine 13(1):39–44.

Goodman, F. D. 1992. *Ecstasy, Ritual, and Alternate Reality*. Bloomington: Indiana University Press.

Gosso, F. & Camilla, G. 2000. *Allucinogeni e Cristianesimo: Evidenze nell'arte sacra*. Paderno Dugnano: Cooperativa Colibri.

Gould, S. J. 1999a. *Rocks of Ages: Science and Religion in the Fullness of Life*. New York: Ballantine.

Gould, S. J. 1999b. A Darwinian gentleman at Marx's funeral. *Scientific American*, September.

Gould, S. J. 1999c. The odd friendship of an evolutionist and a revolutionist. *Natural History*, September.

Grant, E. 2004. *Science and Religion: 400 BC–AD 1550*. Baltimore: John Hopkins University Press.

Gray, J. 2008. *Black Mass: Apocalyptic Religion and the Death of Utopia*. London: Penguin.

Grayling, A. C. 2007a. *Against all Gods: Six Polemics on Religion and an Essay on Kindness*. London: Oberon Books.

Grayling, A. C. 2007b. *Towards the Light: The Story of the Struggles for Liberty and Rights that Made the Modern West*. London: Bloomsbury

Greene, J., & Haidt, J. 2002. How (and where) does moral judgement work? *Trends in Cognitive Science* 16, 517–23.

Greenfield, S. 1997. *The Human Brain: A Guided Tour*. London: Phoenix.

Greenfield, S. 2000. *The Private Life of the Brain*. London: Penguin.

Gregory, R. L. (ed.) 1987. *The Oxford Companion to the Mind*. Oxford: Oxford University Press.

Gribbin, J. 2002. *Science: A History*. London: Penguin.

Guenther, M. 1999. *Tricksters and Trancers: Bushman Religion and Society*. Bloomington: Indiana University Press.

Hachinski, V. C., Porchawka, J, & Steele, J. C. 1973. Visual symptoms in the migraine syndrome. *Neurology* 23, 570–79.

Hacking, I. 1983. *Representing and Intervening: Introductory Topics in the Philosophy of Natural Science*. Cambridge: Cambridge University Press.

Haeckel, E. H. P. 1883. *The History of Creation*. Trans. E. R. Lankester. London: Kegan Paul, Trench.

Halifax, J. 1979. *Shamanistic Voices: A Survey of Visionary Narratives*. Harmondsworth: Penguin.

Hamayon, R. 1998. Ecstasy or the West-dreamt shaman. In Wautischer, H. (ed.) *Tribal Epistemologies*, pp. 175–87. Aldershot: Brookfield.

Hamer, D. 2004. *The God Gene: How Faith is Hardwired into our Genes*. New York: Doubleday.

Harner, M. J. 1972. *The Jívaro: People of the Sacred Waterfalls*. Berkeley: University of California Press.

Harries, R. 2008. *The Re-enchantment of Morality: Wisdom for a Troubled World*. London: SPCK.

Hayden, B. 2003. *Shamans, Sorcerers and Saints: A Prehistory of Religion*. Washington: Smithsonian Institution.

Hedges, K. 1983. The shamanistic origins of rock art. In Van Tilburg, J. A. (ed.) *Ancient Images on Stone: Rock Art of the Californias*, pp. 46–59. Los Angeles: UCLA Institute of Archaeology.

Heinz, R.-I. 1986. More on mental imagery and shamanism. *Current Anthropology* 27, 154.

Hempel, C. G. 1966. *Philosophy of Natural Science*. Englewood Cliffs, NJ: Prentice-Hall.

Henry, J. 2008. *The Scientific Revolution and the Origin of Modern Science*. London: Palgrave Macmillan.

Henshilwood, C. S. 2007. Fully symbolic *sapiens* behaviour: innovation in the Middle Stone Age at Blombos Cave, South Africa. In Mellars, P., Boyle, K., Bar-Yosef, O., & Stringer, C. (eds) *Rethinking the Human Revolution: New Behavioural and Biological Perspectives on the Origin and Dispersal of Modern Humans*, pp. 123–32. Cambridge: McDonald Institute for Archaeological Research.

Henshilwood, C. S., d'Errico, F., Yates, R., Jacobs, Z., Tribolo, C., Duller, G. A. T., Mercier, N., Sealy, J. C., Valladas, H., Watts, I., Wintle, A. G. 2002. Emergence of modern human behaviour: middle stone age engravings from South Africa. *Science* 295, 1278–80. Published online: http://www.sciencemag.org

Herrin, J. 2007. *Byzantium: The Surprising Life of a Medieval Empire*. London: Allen Lane.

Higley, S. L. 2007. *Hildegard of Bingen's Unknown Language*. New York: Palgrave Macmillan.

Hillel, D. 2005. *The Natural History of the Bible: An Environmental Exploration of the Hebrew Scriptures*. New York: Columbia University Press.

Hinde, R A. 1999. *Why Gods Persist: A Scientific Approach to Religion*. London: Routledge.

Hitchens, C. 2007. *God is Not Great: The Case against Religion*. London: Atlantic Books.

Hobson, J. A. 2002. *Dreaming: An Introduction to the Science of Sleep*. Oxford: Oxford University Press.

Hoffmeier, J. K. 2005. *Ancient Israel in Sinai: The Evidence for the Authenticity of the Wilderness Tradition*. Oxford: Oxford University Press.

Horowitz, M. J. 1964. The imagery of visual hallucinations. *Journal of Nervous and Mental Disease* 138, 513–23.

Horowitz, M. J. 1975. Hallucinations: an information-processing approach. In Siegel, R. K., & West, L. J. (eds) *Hallucinations: Behaviour, Experience and Theory*, pp. 163–95. New York: John Wiley.

Horrobin, D. 2001. *The Madness of Adam and Eve: How Schizophrenia Shaped Humanity*. London: Bantam.

Hovers, E., Ilani, S., Bar-Josef, O., & Vandermeersch, B. 2003. An early use of colour symbolism: ochre use by modern humans in Qafzeh Cave. *Current Anthropology* 44, 491–522.

Howard, J. 1982. *Darwin*. Oxford: Oxford University Press.

Hultkranz, Å. 1973. A definition of shamanism. *Tenemos* 9, 25–37.

Hultkranz, Å. 1996. Ecological and phenomenological aspects of shamanism. In Diószegi, V., & Hoppál, M. (eds) *Shamanism in Siberia*, pp. 1–32. Budapest: Akadémiai Kiadó.

Hume, D. 2004 [1748]. Of miracles (part 2). In *An Enquiry Concerning Human Understanding*. New York: Dover.

Humphreys, J. 2007. *In God we Doubt: Confessions of a Failed Atheist*. London: Hodder & Stoughton.

Hutton, R. 2001. *Shamans: Siberian Spirituality and the Western Imagination*. London: Hambledon.

Huxley, A. 1959. *The Doors of Perception*. Harmondsworth: Penguin.

Ings, S. 2007. *The Eye: A Natural History*. London: Bloomsbury.

Insoll, T. 2004. *Archaeology, Ritual, Religion*. London: Routledge.

Israel, J. 2001. *Radical Enlightenment: Philosophy and the Making of Modernity, 1650-1750*. Oxford: Oxford University Press.

Jacobs, Z., Duller, G. A. T., Wintle, A. G., Henshilwood, C. S. 2006. *Journal of Human Evolution* 51, 255–73.

Jakobsen, M. D. 1999. *Shamanism: Traditional and Contemporary Approaches to the Mastery of Spirits and Healing*. New York: Berghahn Books.

James, A. 2001. *The First Bushman's Path: Stories, Songs and Testimonies of the /Xam of the Northern Cape*. Pietermaritzburg: University of Natal Press.

James, W. 1890. *The Principles of Psychology*. New York: Henry Holt.

James, W. 1960 [1902]. *The Varieties of Religious Experience: A Study in Human Nature*. London: Collins.

James, W. 1907. *Pragmatism: A New Way for Some Old Ways of Thinking, Popular Lectures on Philosophy*. New York: Longman Green.

Jay, M. 1999. *Artificial Paradises: A Drugs Reader*. London: Penguin.

Jones, R., & N. White. 1988. Point blank: stone tool manufacture at the Ngilipitji quarry, Arnhem Land, 1981. In Meehan, B., & Jones, R. (eds) *Archaeology with Ethnography: An Australian Perspective*, pp. 51–93. Canberra: Australian National University.

Jordan, P. 2001a. The materiality of shamanism as a 'world-view': praxis, artifacts and landscape. In Price, N. (ed.) *The Archaeology of Shamanism*, pp. 87–104. London: Routledge.

Jordan, W. C. 2001b. *Europe in the High Middle Ages*. London: Allen Lane.

Kasser, R., Meyer, M., & Wurst, G. (eds) 2006. *The Gospel of Judas, from Codex Tchacos*. Washington: National Geographic.

Katz, R. 1976. Education for transcendence. In Lee, R. B., & De Vore, I. (eds) *Kalahari Hunter-Gatherers*, pp. 281–301. Cambridge, Mass: Harvard University Press.

Katz, R. 1982. *Boiling Energy: Community Healing among the Kalahari !Kung*. Cambridge, Massachusetts: Harvard University Press.

Katz, R., Biesele, M., & St Denis, V. 1997. *Healing Makes our Hearts Happy: Spirituality and Cultural Transformation among the Kalahari Ju/'hoansi*. Rochester: Inner Traditions.

Keeney, B. 1999. *Kalahari Bushmen Healers*. Philadelphia: Ringing Rocks.

Keeney, B. 2003. *Ropes to God: Experiencing the Bushman Spiritual Universe*. Philadelphia: Ringing Rocks.

Kehoe, A. B. 2000. *Shamans and Religion: An Anthropological Exploration in Critical Thinking*. Prospect Heights, Illinois: Waveland Press.

Kelly, H. A. 2006. *Satan: A Biography*. Cambridge: Cambridge University Press.

Keyser, J. D., Poetschat, G., & Taylor, M. W. (eds) 2006. *Talking with the Past: The Ethnography of Rock Art*. Portland: Oregon Archaeological Society.

Kingsley, C. n.d. *Letters to the Christian*, no. 2

Kirchner, H. 1952. Ein archäologischer Beitrag zur Urgeschichte des Schamanismus. *Anthropos* 47, 244–86.

Kirk, G. S., & Raven, J. E. 1957. *The Presocratic Philosophers*. Cambridge: Cambridge University Press.

Klüver, H. 1942. Mechanisms of hallucinations. In McNemar, Q., & Merrill, M. A. (eds) *Studies in Personality*, pp. 175–207. New York: McGraw-Hill.

Klüver, H. 1966. *Mescal and Mechanisms of Hallucination*. University of Chicago: Chicago University Press.

Knight, C. 1991. *Blood Relations: Menstruation and the Origins of Culture*. New Haven: Yale University Press.

Knoll, M., J. Kugler, O. Höher, & S. D. Lawder. 1963. Effects of chemical stimulation of electrically induced phosphenes on their bandwidth, shape, number, and intensity. *Confinia Neurologica* 23, 201–226.

Krishna, G. 1997. *Kundalini: The Evolutionary Energy in Man*. Boston: Shambhala.

Kroll, J., & Bachrach, B. 1982a. Visions and psychopathology in the Middle Ages. *Journal of Nervous and Mental Disease* 170, 41–9.

Kroll, J., & Bachrach, B. 1982b. Medieval visions and contemporary hallucinations. *Psychological Medicine* 12, 709–721.

Krosney, H. 2006. *The Lost Gospel: The Quest for the Gospel of Judas Iscariot*. Washington: National Geographic.

Kuhn, T. 1966. *The Copernican Revolution*. Cambridge, Massachusetts: Harvard University Press.

Kuhn, T. S. 1970. *The Structure of Scientific Revolutions*. Chicago: University of Chicago Press.

Kurtz, P. 2003. *Science and Religion: Are they Compatible?* Amherst: Prometheus Books.

La Barre, W. 1970. *The Ghost Dance: Origins of Religion*. New York: Random House.

La Barre, W. 1972. Hallucinogens and the shamanic origins of religion. In Furst, P. (ed.) *Flesh of the Gods: The Ritual Use of Hallucinogens*, pp. 261–78. London: Allen & Unwin.

La Barre, W. 1980. *Culture in Context: Selected Writings of Weston LaBarre*. Durham, North Carolina: Duke University Press.

Lakoff, G., & Johnson, M. 1980. *Metaphors we Live By*. Chicago: University of Chicago Press.

Lara, J. 2008. *Christian Texts for the Aztecs: Art and Liturgy in Colonial Mexico*. Notre Dame: University of Notre Dame Press.

Larson, E. J. 1998. *Summer of the Gods: The Scopes Trial and America's Continuing Debate over Science and Religion*. New York: Basic Books; Cambridge, Massachusetts: Harvard University Press.

Laudan, L. 1996. *Beyond Positivism and Relativism: Theory, Method, and Evidence*. Boulder: Westview Press.

Laughlin, C. D., McManus, J., & d'Aquili, E. G. 1992. *Brain, Symbol and Experience: Towards a Neurophenomenology of Human Consciousness*. New York: Colombia University Press.

Lawson, A. J. 2007. *Chalkland: An Archaeology of Stonehenge and its Region*. Salisbury: Hobnob Press.

Leach, E. R. 1961. Golden bough or gilded twig? *Daedalus, Journal of the American Academy of Arts and Sciences* 90 (2), 371–87.

Leach, E. R. 1969. *Genesis as Myth and Other Essays*. London: Jonathan Cape.

Leach, E. R. 1970. *Lévi-Strauss*. London: Fontana.

Leach, E. R. 1976. *Culture and Communication: The Logic by which Symbols are Connected*. Cambridge: Cambridge University Press.

LeDoux, J. 2002. *Synaptic Self: How our Brains Become Who we Are*. New York: Macmillan.

Lee, R. B. 1967. Trance cure of the !Kung Bushmen. *Natural History* 76(9), 31–7.

Lee, R. B. 1993. *The Dobe Ju/'hoansi*. New York: Harcourt Brace.

Leroi-Gourhan, André. 1968. *The Art of Prehistoric Man in Western Europe*. London: Thames & Hudson.

Leroi-Gourhan, Arlette, & Allain, J. 1979. *Lascaux Inconnu*. Paris: Éditions du CNRS.

LeRon Shults, F. (ed.) 2006. *The Evolution of Rationality: Interdisciplinary Essays in Honor of J. Wentzel van Huyssteen*. Grand Rapids: Erdmans.

Lévi-Strauss, C. 1963. *Structural Anthropology*. London: Penguin.

Lévi-Strauss, C. 1967. The story of Asdiwal. In Leach, E. (ed.) *The Structural Study of Myth and Totemism*, pp. 1–47. London: Tavistock.

Lewis, C. S. 1957. *The Problem of Pain*. London: Fontana.

Lewis, J. E. 2003. *The Mammoth Book of Eyewitness: Ancient Rome*. New York: Carroll & Graf.

Lewis-Williams, J. D. 1981. *Believing and Seeing: Symbolic Meanings in Southern San Rock Paintings*. London: Academic Press.

Lewis-Williams, J. D. 1986. The last testament of the southern San. *South African Archaeological Bulletin* 41, 10–11.

Lewis-Williams, J. D. 1987. A dream of eland: an unexplored component of San shamanism and rock art. *World Archaeology* 19, 165–77.

Lewis-Williams, J. D. 1990. On Palaeolithic art and the neuropsychological model. *Current Anthropology* 31, 407–408.

Lewis-Williams, J. D. 1991a. Wrestling with analogy: a problem in Upper Palaeolithic art research. *Proceedings of the Prehistoric Society* 57(1):149-162.

Lewis-Williams, J. D. 1991b. Wrestling with analogy: a methodological dilemma in Upper Palaeolithic art research. *Proceedings of the*

Prehistoric Society 57, 149–60. (Reprinted in D. S. Whitley (ed.) *Reader in Archaeological Theory*, pp. 157–75. London: Routledge.)

Lewis-Williams, J. D. 1992. Ethnographic evidence relating to 'trance' and 'shamans' among northern and southern Bushmen. *South African Archaeological Bulletin* 47, 56–60.

Lewis-Williams, J. D. 1994. Rock art and ritual: southern Africa and beyond. *Complutum* 5, 277–89.

Lewis-Williams, J. D. 1995a. Perspectives and traditions in southern African rock art research. In Helskog, K., and Olsen B. (eds) Perceiving rock art: social and political perspectives, pp. 65–86. Oslo: Novus forlag.

Lewis-Williams, J. D. 1995b. Some aspects of rock art research in the politics of present-day South Africa. In Helskog, K., and Olsen B. (eds) Perceiving rock art: social and political perspectives, pp. 317–337. Oslo: Novus forlag.

Lewis-Williams, J. D. 1995c. Modelling the production and consumption of rock art. *South African Archaeological Bulletin* 50, 143–54.

Lewis-Williams, J. D. 1995d. Seeing and construing: the making and 'meaning' of a Southern African rock art motif. *Cambridge Archaeological Journal* 5(1), 3–23.

Lewis-Williams, J. D. 1996. A visit to the Lions' house: the structure, metaphors and socio-political significance of a nineteenth-century Bushman myth. In Deacon, J., & Dowson, T. A. (eds) *Voices from the Past: /Xam Bushmen and the Bleek and Lloyd Collection*, pp. 122–41. Johannesburg: Witwatersrand University Press.

Lewis-Williams, J. D. 1997a. Agency, art and altered consciousness: a motif in French (Quercy) Upper Palaeolithic parietal art. *Antiquity* 71, 810–30.

Lewis-Williams, J. D. 1997b. Harnessing the brain: vision and shamanism in Upper Palaeolithic Western Europe. In Conkey, M. W., Soffer, O., Stratmann, D., & Jablonski, N. G. (eds) *Beyond Art: Pleistocene Image and Symbol*, pp. 321–42. San Francisco: Memoirs of the California Academy of Sciences, No. 23.

Lewis-Williams, J. D. 1997c. Prise en compte du relief naturel des surfaces rocheuses dans l'art pariétal sud africain et paléolithique ouest européen: étude culturelle et temporelle croisée de la croyance religieuse. *L'Anthropologie* 101, 220–37.

Lewis-Williams, J. D. 2001. Brainstorming images: neuropsychology and rock art research. In *Handbook of Rock Art Research*, pp. 332–57. Walnut Creek, California: AltaMira Press.

Lewis-Williams, J. D. 2002. *The Mind in the Cave: Consciousness and the Origins of Art*. London: Thames & Hudson.

Lewis-Williams, J. D. 2003a. Putting the record straight: rock art and shamanism. *Antiquity* 77:165–170.

Lewis-Williams, J. D. 2003b. Review feature: 'The mind in the cave: consciousness and the origins of art.' *Cambridge Archaeological Journal* 13, 263–79.

Lewis-Williams, J. D. 2004a. Neuropsychology and Upper Palaeolithic art: observations on the progress of altered states of consciousness. *Cambridge Archaeological Journal* 14, 107–111.

Lewis-Williams, J. D. 2004b. Constructing a cosmos: architecture, power and domestication at Çatalhöyük. *Journal of Social Archaeology* 4, 28–59.

Lewis-Williams, J. D., & Biesele, M. 1978. Eland hunting rituals among northern and southern San groups: striking similarities. *Africa* 48(2), 117–34.

Lewis-Williams, J. D., & Clottes, J. 1998. The mind in the cave – the cave in the mind: altered consciousness in the Upper Palaeolithic. *Anthropology of Consciousness* 9(1), 13–21.

Lewis-Williams, J. D., & Dowson, T. A. 1988. The signs of all times: entoptic phenomena in Upper Palaeolithic art. *Current Anthropology* 29, 201–245.

Lewis-Williams, J. D., & Dowson, T. A. 1990. Through the veil: San rock paintings and the rock face. *South African Archaeological Bulletin* 45, 5–16.

Lewis-Williams, J. D., & Dowson, T. A. 1993. On vision and power in the Neolithic: evidence from the decorated monuments. *Current Anthropology* 34, 55–65.

Lewis-Williams, J. D., & Pearce, D. G. 2004. *San Spirituality: Roots, Expressions and Social Consequences*. Walnut Creek, California: AltaMira Press; Cape Town: Double Storey.

Lewis-Williams, J. D., & Pearce, D. G. 2005. *Inside the Neolithic Mind: Consciousness, Cosmos and the Realm of the Gods*. London: Thames & Hudson.

Lindberg, D. & Westman, R. (eds) 1990. *Reappraisals of the Scientific Revolution*. Cambridge: Cambridge University Press.

Linden, D. J. 2007. *The Accidental Mind: How Brain Evolution has Given us Love, Memory, Dreams, and God*. Cambridge, Massachusetts: Belknap Press.

Liverani, M. 2003. *Israel's History and the History of Israel*. London: Equinox.

Lloyd, G. E. R. 1970. *Early Greek Science: Thales to Aristotle.* New York: W. W. Norton & Co.

Loendorf, L. L., Chippindale, C., & Whitley, D. S. (eds) 2005. *Discovering North American Rock Art.* Tucson: University of Arizona Press.

Lommel, A. 1967. *Shamanism: The Beginnings of Art.* New York: McGraw-Hill.

Losee, J. 1980. *A Historical Introduction to the Philosophy of Science.* Oxford: Oxford University Press.

Lubbock, J. 1865. *Prehistoric Times, as Illustrated by Ancient Remains, and the Manners and Customs of Modern Savages.* London: Williams & Norgate.

Malinowski, B. 1922. *Argonauts of the Western Pacific.* New York: Dutton.

Malinowski, B. 1954. *Magic, Science and Religion.* Garden City, New York: Doubleday.

Marcus Aurelius. 2006. *Meditations.* Trans. M. Hammond. London: Penguin.

Marean, C. W., Bar-Matthews, M., Bernatchez, J., Fisher, E., Goldberg, P., Herries, A. I. R., Jacobs, Z., Jerardino, A., Karkanas, P., Minichillo, T., Nilssen, P. J., Thompson, E., Watts, I., & Williams, H. M. 2007. Early human use of marine resources and pigment in South Africa during the Middle Pleistocene. *Nature* 449, 905–908.

Marshack, A. 1991. *The Roots of Civilization: Revised and Expanded.* Mount Kisco, New York: Moyer Bell.

Marshall, L. 1999. *Nyae Nyae !Kung: Beliefs and Rites.* Cambridge, Massachusetts: Harvard University Press.

Marshall Thomas, E. 2006. *The Old Way: A Story of the First People.* New York: Farrar, Straus & Giroux.

Marx, K. 1843–4. *A Contribution to the Critique of Hegel's Philosophy of Right.* Introduction

McBrearty, S. 2007. Down with the revolution. In Mellars, P., Boyle, K., Bar-Yosef, O., & Stringer, C. (eds) *Rethinking the Human Revolution: New Behavioural and Biological Perspectives on the Origin and Dispersal of Modern Humans*, pp. 133–51. Cambridge: McDonald Institute for Archaeological Research.

McBrearty, S., & Brooks, A. S. 2000. The revolution that wasn't: a new interpretation of the origin of modern human behaviour. *Journal of Human Evolution* 39, 453–63.

McGrath, A. 2004. *The Twilight of Atheism: The Rise and Fall of Disbelief in the Modern World.* London: Rider.

McGrath, A. 2005. *Dawkins' God: Genes, Memes, and the Meaning of Life.* Oxford: Blackwell.

McGrath, A., & McGrath, J. C. 2007. *The Dawkins Delusion? Atheist Fundamentalism and the Denial of the Divine.* London: SPCK

McMullin, E. (ed.) 2006. *The Church and Galileo.* Notre Dame: University of Notre Dame Press.

Mellars, P. 2006. Why did modern human populations disperse from Africa *c.* 60,000 years ago? A new model. *Proceedings of the National Academy of Sciences USA* 103, 9381–6.

Mellars, P., Boyle, K., Bar-Yosef, O., & Stringer, C. (eds) 2007. *Rethinking the Human Revolution: New Behavioural and Biological Perspectives on the Origin and Dispersal of Modern Humans.* Cambridge: McDonald Institute.

Midgley, M. 2007. How myths work. In: Cunningham, M. K. (ed.) *God and Evolution: A Reader*, pp. 28–33. London: Routledge.

Miller, M., & Martin, S. 2004. *Courtly Art of the Ancient Maya.* London: Thames & Hudson.

Millhauser, M. 1959. *Just before Darwin.* Middletown, Connecticut: Wesleyan University Press.

Milner, R. 1996. Charles Darwin and Associates, Ghostbusters. *Scientific American* October

Mithen, S. 1996. *The Prehistory of the Mind: A Search for the Origins of Art, Religion and Science.* London: Thames & Hudson.

Moorehead, A. 1969. *Darwin and the Beagle.* London: Hamish Hamilton.

Müller-Beck, H., & Albrecht, G. (eds). 1987. *Die Anfänge der Kunst vor 30000 Jahren.* Stuttgart: Konrad Theiss Verlag.

Myerhoff, B. G. 1974. *Peyote Hunt: The Sacred Journey of the Huichol Indians.* New York: Cornell University Press.

Nagel, T. 1997. *The Last Word.* Oxford: Oxford University Press.

Narby, J., & Huxley, F. (eds) 2001. *Shamans Through Time: 500 Years on the Path to Knowledge.* London: Thames & Hudson.

Needham, R. 1967. Percussion and transition. *Man* 2, 606–614.

Neve, M. 2002. Introduction. In Darwin, C. *Autobiographies* (ed. Neve, M. & Messenger, S.), pp. ix–xxiii. London: Penguin.

Newberg, A., d'Aquili, E., & Rause, V. 2002. *Why God Won't Go Away: Brain Science and the Biology of Belief.* New York: Ballantine Books.

Newman, B. 1987. *Sister of Wisdom: St Hildegard's Theology of the Feminine.* Berkeley: University of California Press.

Noll, R. 1985. Mental imagery cultivation as a cultural phenomenon: the role of visions in shamanism. *Current Anthropology* 26, 443–61.

Norenzayan, A., & Shariff, A. F. 2008. The origin and evolution of religious prosociality. *Science* 322, 58–62.

Northrop, W. B. 1913. Alfred Russel Wallace: an interview. *The Outlook* 105, 622.

O'Kelly, M. J. 1982. *Newgrange: Archaeology, Art and Legend*. London: Thames & Hudson.

Orpen, J. M. 1874. A glimpse into the mythology of the Maluti Bushmen. *Cape Monthly Magazine* (N.S.) 9(49), 1–13.

Orru, M., & Wang, A. 1992. Durkhein, religion and Buddhism. *Journal of the Scientific Study of Religion* 31, 47–61.

Oster, G. 1970. Phosphenes. *Scientific American* 222(2), 83–7.

Oubré, A. Y. 1997. *Instinct and Revelation: Reflections on the Origins of Numinous Perceptions*. Amsterdam: Gordon & Breach.

Owen, D. 2008. *In Sickness and in Power: Illness in the Heads of Government During the Last 100 Years*. London: Methuen.

Pagels, E. 1980. *The Gnostic Gospels*. London: Weidenfeld & Nicolson.

Paley, W. 1802. *Natural Theology*. New York: American Tract Society.

Pearson, J. L. 2002. *Shamanism and the Ancient Mind: A Cognitive Approach to Archaeology*. Walnut Creek, California: AltaMira Press.

Persinger, M. 1987. *Neurophysiological Bases of God Beliefs*. New York: Praeger.

Pfeiffer, J. E. 1982. *The Creative Explosion: An Enquiry into the Origins of Art and Religion*. New York: Harper & Row.

Pinker, S. 1998. *How the Mind Works*. London: Allen Lane.

Plato. 1935. *The Republic*. Trans. A. D. Lindsay. London: Dent.

Platten, S. 2007. *Rebuilding Jerusalem: the Church's Hold on Hearts and Minds*. London: SPCK.

Popper, K. 1959. *The Logic of Scientific Discovery*. London: Hutchinson.

Porter, J. 2005. *Nature as Reason: A Thomist Theory of the Natural Law*. Grand Rapids, Michigan: Eerdmans.

Price, N. 2001. An archaeology of altered states: shamanism and material culture studies. In Price, N. (ed.) *The Archaeology of Shamanism*, pp. 3–17. London: Routledge.

Pullman, P. 2006. Moral codes. In Bakewell, J. (ed.) *Belief*. London: Duckworth.

Raby, P. 2001. *Alfred Russel Wallace: A Life*. London: Chatto & Windus.

Rahula, W. 1974. *What the Buddha Taught*. New York: Grove Press.

Ramachandran, V. S. 2004. *A Brief Tour of Human Consciousness*. New York: PI Press.

Ramachandran, V. S., & Blakeslee, S. 1998. *Phantoms in the Brain: Probing the Mysteries of the Human Mind*. New York: (Quill) Harper Collins.

Randi, J. 1982. *Flim-Flam*. Amherst, New York: Prometheus Books.

Randi, J. 1986. The Project Alpha experiment: Part 1. The first two years. In Frazier, K. (ed.) *Science Confronts the Paranormal*. Amherst, New York: Prometheus Books.

Rappaport, R. A. 1999. *Ritual and Religion in the Making of Humanity*. Cambridge: Cambridge University Press.

Redondi, P. 1987. *Galileo: Heretic*. Princeton: Princeton University Press.

Reichel-Dolmatoff, G. 1975. *The Shaman and the Jaguar: A Study of Narcotic Drugs among the Indians of Colombia*. Philadelphia: Temple University Press.

Reichel-Dolmatoff, G. 1978. *Beyond the Milky Way: Hallucinatory Imagery of the Tukano Indians*. Los Angeles: UCLA Latin America Center.

Reichel-Dolmatoff, G. 1988. *Goldwork and Shamanism: An Iconographic Study of the Gold Museum*. Medellín: Compañía Litográfica Nacional.

Renfrew, C. 2008. *Prehistory: The Making of the Human Mind*. New York: Random House.

Reznikoff, I., & M. Dauvois. 1988. La dimension sonore des grottes ornées. *Bulletin de la Société Préhistorique Française* 85, 238–46.

Richards, W. 1971. The fortification illusions of migraines. *Scientific American* 225, 89–96.

Ripinsky-Naxon, M. 1989. Hallucinogens, shamans and the cultural process: symbolic archaeology and dialectics. *Anthropos* 84, 219–24.

Robertson, D. 2007. *The Dawkins Letters: Challenging Atheist Myths*. Fearn: Christian Focus Publications.

Robinson, J. A. T. 1963. *Honest to God*. London: SCM Press.

Rock, A. 2004. *The Mind at Night: The New Science of How and Why we Dream*. New York: Basic Books.

Rolston, H. 1999. *Genes, Genesis and God: Values and their Origins in Natural and Human History*. Cambridge: Cambridge University Press.

Rowland, I. D. 2008. *Giordano Bruno: Philosopher/Heretic*. New York: Farrar, Straus & Giroux.

Rudgley, R. 1993. *The Alchemy of Culture: Intoxicants in Society*. London: British Museum Press.

Ruse, M. 2001. *The Evolution Wars: A Guide to the Debates*. New Brunswick, New Jersey: Rutgers University Press.

Sacks, O. W. 1970. *Migraine: The Evolution of a Common Disorder*. London: Faber.

Sacks, O. W. 1993. *Migraine*. Berkeley: University of California Press.

Santhouse, A. M., Howard, R. J. & ffytche, D. H. 2000. Visual hallucinatory syndromes and the anatomy of the visual brain. *Brain* 123, 2055–2064.

Scarre, C. 1989. Painting by resonance. *Nature* 338, 328.

Schama, S. 2008. *The American Future: A History*. London: Bodley Head.

Schele, L., & Freidel, D. 1990. *A Forest of Kings: The Untold Story of the Ancient Maya*. New York: Morrow.

Schele, L., & Miller, M. E. 1986. *The Blood of Kings: Dynasty and Ritual in Maya Art*. Fort Worth: Kimbell Art Museum.

Schmidt, K. 2001. Göbekli Tepe, southeastern Turkey: a preliminary report on the 1995–1999 excavations. *Paléorient* 26, 45–54.

Schmidt, K. 2006. *Sie bauten die ersten Tempel: das rätselhafte Heiligtum der Steinzeitjäger*. Munich: C. H. Beck.

Schrödinger, E. 1996. *Nature and the Greeks and Science and Humanism*. Cambridge: Cambridge University Press.

Searle, J. 2006. *Freedom and Neurobiology*. New York: Columbia University Press.

Secord, J. A. 2000. *Victorian Sensation: The Extraordinary Publication, Reception and Secret Authorship of 'Vestiges of the natural history of creation'*. Chicago: University of Chicago Press.

Selby, G. 1983. *Migraine and its Variants*. Sydney.

Settle, M. L. 1972. *The Scopes Trial: The State of Tennessee v. John Thomas Scopes*. New York: Franklin Watts.

Shanon, B. 2002. *The Antipodes of the Mind: Charting the Phenomenology of the Ayahuasca Experience*. Oxford: Oxford University Press.

Shea, W. R., & Artigas, M. 2003. *Galileo in Rome: The Rise and Fall of a Troublesome Genius*. Oxford: Oxford University Press.

Shermer, M. 2003. *How We Believe: Science, Scepticism and the Search for God*. New York: A. W. Freeman.

Shirokogoroff, S. M. 1935. *Psychomental Complex of the Tungus*. London: Kegan Paul, Trench, Trubner.

Siegel, J. M. 1999. The evolution of REM sleep. In *Handbook of Behavioural State Control*, pp. 87–1000. CRC Press.

Siegel, R. K. 1977. Hallucinations. *Scientific American* 237, 132–40.

Siegel, R. K., & Jarvik, M. E. 1975. Drug-induced hallucinations in animals and man. In Siegel, R. K., & West, L. J. (eds) *Hallucinations, Behaviour, Experience, and Theory*, pp. 81–161. New York: Wiley & Sons.

Silverman, K. 1996. *Houdini!!!: The Career of Ehrich Weiss*. New York: Harper Collins.

Sinclair, A. 1995. The technique as a symbol in Late Glacial Europe. *World Archaeology* 27(1), 50–62.

Singer, C. 1958. *From Magic to Science: Essays on the Scientific Twilight*. New York: Dover.

Skeat. W. W. (ed.) 1912. *The Complete Works of Geoffrey Chaucer*. London: Oxford University Press.

Smith, B., Lewis-Williams, J. D., Blundell, G., & Chippindale, C. 2000. Archaeology and symbolism in the new South African coat of arms. *Antiquity* 74:467–8.

Smith, H. 2003. *Cleansing the Doors of Perception: The Religious Significance of Entheogenic Plants and Chemicals*. Boulder, Colorado: Sentient Publications.

Smith, N. W. 1992. *An Analysis of Ice Age Art: Its Psychology and Belief System*. New York: Peter Lang.

Solso, R. L. (ed.) 1997. *Mind and Brain Sciences in the 21st Century*. Cambridge, Massachusetts: MIT Press.

Spenser, N., & White, R. 2007. *Christianity, Climate Change and Sustainable Living*. London: SPCK.

Spinoza, B. 1883. *Tractatus Politicus*. Trans. R. H. M. Elwes. London: G. Bell & Son.

Stickgold, R., Hobson, J. A., Fosse, R., & Fosse, M. 2001. Sleep, learning and dreams: off-line memory reprocessing. *Science* 294, 1052–7.

Stoichita, V. I. 1995. *Visionary Experience in the Golden Age of Spanish Art*. London: Reaktion Books.

Strong, R. 2007. *A Little History of the English Country Church*. London: Cape.

Stuart, D., & Stuart, G. 2008. *Palenque: Eternal City of the Maya*. London: Thames & Hudson.

Sullivan, L. E. 1988. *Icanchu's Drum: An Orientation to Meaning in South American Religions*. New York: Macmillan.

Sulloway, F. J. 1982. Darwin's conversion: the *Beagle* voyage and its aftermath. *Journal of the History of Biology* 15, 325–96.

Sutherland, P. D. 2001. Shamanism and the iconography of Palaeo-Eskimo art. In Price, N. (ed.) *The Archaeology of Shamanism*, pp. 135–45. London: Routledge.

Taçon, P. S. C. 1991. The power of stone: symbolic aspects of stone use and tool development in Western Arnhem Land, Australia. *Antiquity* 65, 192–207.

Taylor, C. 2007. *A Secular Age*. Cambridge, Massachusetts: Harvard University Press.

Taylor, J. B. 2006. *My Stroke of Insight: A Brain Scientist's Personal Journey*. New York: Viking.

Thompson, D. 2008. *Counter Knowledge: How We Surrender to Conspiracy Theories, Quack Medicine, Bogus Science and Fake History*. London: Atlantic Books.

Thoren, V. 1990. *Lord of Uraniborg: A Biography of Tycho Brahe*. Cambridge: Cambridge University Press.

Tongue, H. 1909. *Bushman Paintings*. London: Clarendon Press.

Townsend, J. B. 1997. Shamanism. In Glazier, S. (ed.) *Anthropology of Religion: A Handbook*, pp. 429–69. Westport, Connecticut: Greenwood.

Turnbull, R. 2008. *Anglican or Evangelical?* London: Continuum.

Turner, V. 1967. *The Forest of Symbols: Aspects of Ndembu Ritual*. Ithaca: Cornell University Press.

Turner, V., & Bruner, E. (eds) 1986. *The Anthropology or Experience*. Urbana, Illinois: University of Illinois Press.

Turner, V., & Turner, E. 1978. *Image and Pilgrimage in Christian Culture*. New York: Columbia University Press.

Van Gennep, A. 1960. *The Rites of Passage*. Chicago: Chicago University Press.

Vanhaeren, M., d'Errico, F., Stringer, C., James, S. L., Todd, J. A., Miemis, H. K., 2006. Middle Paleolithic shell beads in Israel and Algeria. *Science* 312, 1785–8.

Van Huyssteen, J. W. 2006. *Alone in the World? Human Uniqueness in Science and Theology*. Grand Rapids: Erdmans.

Van Wyhe, J. 2006. *Notes and Records of the Royal Society* DOI:101098/rsnr.2006.0171.

Vialou, D. 1991. *La Préhistorire*. Paris: Galllimard.

Vitebsky, P. 1995. *The Shaman*. London: Macmillan.

Wallace, A. R. 1905. *My Life*. London:

Waller, S. J. 1993. Scientific correspondence: sound and rock art. *Nature* 363, 501.

Walter, M. N., & Fridman, E. J. N. (eds) 2004. *Shamanism: An Encyclopedia of World Beliefs, Practices, and Culture*. Santa Barbara: ABC-CLIO.

Ward, K. 2006. *Is Religion Dangerous?* Oxford: Lion Hudson plc.

Watson, P. 2005. *Ideas: A History from Fire to Freud*. London: Weidenfeld & Nicolson.

Westfall, R. 1980. *Never at Rest: A Biography of Isaac Newton*. Cambridge: Cambridge University Press.

Wheen, F. 2004. *How Mumbo-Jumbo Conquered the World: A Short History of Modern Delusions*. London: Harper Perennial.

White, R. 2003. *Prehistoric Art: The Symbolic Journey of Human Kind*. New York: Harry Abrams.

Whitehead, A. N. 1926. *Science and the Modern World*. Cambridge: CambridgeUniversity Press.

Whitehouse, H. 2000. *Arguments and Icons: Divergent Modes of Religiosity*. Oxford: Oxford University Press.

Whitley, D. S. 1992. Shamanism and rock art in far western North America. *Cambridge Archaeological Journal* 2, 89–113.

Whitley, D. S. 1998a. Cognitive neuroscience, shamanism, and the rock art of Native California. *Anthropology of Consciousness* 9(1), 22–37.

Whitley, D. S. 1998b. Finding rain in the desert: landscape, gender and far Western North American rock art. In Chippindale, C., & Taçon, P. S. C. (eds) *The Archaeology of Rock Art*, pp. 11–29. Cambridge: Cambridge University Press.

Whitley, D. S. 2000. *The Art of the Shaman: Rock Art of California*. Salt Lake City: University of Utah Press.

Whitley, D. S., & Hays-Gilpin, K. (eds) 2008. *Belief in the Past: Theoretical Approaches to the Archaeology of Religion*. Walnut Creek, California: Left Coast Press.

Wilbert, J. 1987. *Tobacco and Shamanism in South America*. New Haven: Yale University Press.

Williams, S. 1985. *Diocletian and the Roman Recovery*. London: Routledge.

Wilmsen, E. N. 1989. *Land Filled with Flies: A Political Economy of the Kalahari*. Chicago: University of Chicago Press.

Winkelman, M. 1990. Shamans and other 'magico-religious' healers: a cross-cultural study of their origins, nature, and social transformations. *Ethos* 18, 308–352.

Winkelman, M. 1992. *Shamans, Priests, and Witches: A Cross-Cultural Study of Magico-Religious Practitioners*. Tempe: Anthropology Department, Arizona State University.

Winkelman, M. 2000. *Shamanism: The Neural Ecology of Consciousness and Healing*. Westport: Bergin & Garvey.

Winkelman, M. 2008. Cross-cultural and biogenetic perspectives on the origins of shamanism. In Whitley, D. S., & Hays-Gilpin, K. (eds) *Belief in the Past: Theoretical Approaches to the Archaeology of Religion*, pp. 43–66. Walnut Creek, California: Left Coast Press.

Wiseman, R. 1997. *Deception and Self-Deception: Investigating Psychics*. Amherst, New York: Prometheus Books.

Wolberg, F. L., & Ziegler, D. K. 1982. Olfactory hallucinations in migraine. *Archives of Neurology* 39, 382.

Wolpert, L. 2006. *Six Impossible Things Before Breakfast: The Evolutionary Origins of Belief*. London: Faber & Faber.

Wordsworth, W. 1888. *The Complete Poetical Works of William Wordsworth*, London: Macmillan.

Wright, R. 1996. *The Moral Animal: Why We Are the Way We Are*. London: Abacus.

Wylie, A. 1989. Archaeological cables and tacking: the implications of practice for Bernstein's 'Options beyond objectivism and relativism'. *Philosophy of Science* 19, 1–18.

Wylie, A. 1999. Rethinking unity as a 'working hypothesis' for philosophy of science: how archaeologists exploit the disunities of science. *Perspectives in Science* 7, 293–317.

Wylie, A. 2000. Questions of evidence, legitimacy, and the (dis)unity of science. *American Antiquity* 65, 227–37.

Wylie, A. 2002. *Thinking from Things: Essays in the Philosophy of Archaeology*. Berkeley: University of California Press.

Yonge, C. D. (trans.). 2006. Diogenes Laertius, 'Thales', in *The Lives And Opinions Of Eminent Philosophers*. Whitefish, Montana: Kessinger Publishing.

Zimmerman, L. J. 1996. *Native North America*. London: Macmillan.

Acknowledgments

Many people have contributed, knowingly and unwittingly, to the writing of this book. I am grateful to all of them. The Master and Fellows of St John's College, Cambridge, elected me to a visiting overseas fellowship for the Lent Term, 2008. Many discussions in the welcoming ethos of that college enabled me to explore different understandings of religion. At the same time, discussions with members of the Cambridge Archaeology Department and questions after lectures that I gave in the McDonald Institute sharpened my ideas. The whole Cambridge experience was made possible by the generosity of the Oppenheimer Trust; it was good that Jennifer Oppenheimer and Susan Ward were able to visit Cambridge while I was at St John's. I am also grateful to the University of the Witwatersrand, Johannesburg, for what now amounts to just over thirty years of support. The Rock Art Research Institute is today funded principally by the University of the Witwatersrand and the National Research Foundation. Members of the Rock Art Research Institute, especially David Pearce, my co-author on other projects, William (Sen) Challis, who also assisted greatly with the illustrations, and Ben Smith, were willing to engage in endless debates about science and religion. In France, I am grateful to Jean Clottes, another highly valued co-author, and to Robert Bégouën and his son Eric for permission to visit the Volp Caves and for guiding me through the labyrinth on a number of occasions. Marcus Peters pointed me to many sources. David Wilson drew my attention to many useful points and engaged in stimulating discussions. Elwyn Jenkins read and astutely commented on various chapters. As before, the staff at Thames & Hudson, especially Colin Ridler, was meticulous and encouraging and found some excellent illustrations. Their anonymous external readers saved me from numerous errors. Christopher Dell performed miracles in bringing order to the text and references.

Sources of Illustrations

Unless otherwise indicated, diagrams have been provided by the Rock Art Research Institute, University of the Witwatersrand, Johannesburg.

1, 2 Courtesy Chris Henshilwood; 3 Craig Chiasson/iStockphoto.com; 4 Alinari Archives, Florence; 6 Kunsthistorisches Museum, Vienna; 7 Alinari Archives, Florence; 8 Anderson/Alinari Archives, Florence; 9 Museu Nacional de Arte Antiga, Lisbon; 10 Finsiel/Alinari Archives, Florence; 11 British Library, London; 12 Royal Society, London; 14 Darwin, *Foundations of the Origins of Species, A Sketch Written in 1842*, 1909; 15 Darwin, *A Naturalist's Voyage Round the World*, 1890; 16 Fitzroy, *Narrative of Surveying Voyages of H.M. ships 'Adventure' & 'Beagle'*, 1839; 17 A. J. R. Wallace and R. R. Wallace; 18 Commissioner of Public Works, National Parks & Monuments, Dublin, Ireland; 20 © Enrique Marcarian/ Reuters/Corbis; 21 Rock Art Research Institute, University of the Witwatersrand, Johannesburg; 22 Chris Hurtt/iStockphoto.com; 23 Rock Art Research Institute, University of the Witwatersrand, Johannesburg; 24 Dante, *Vision of Paradise*, 1861; 25 Dorset Fine Arts; 26, 27 British Museum, London; 28 Anderson/Alinari Archives, Florence; 30 TIPS Images; 33, 34 after Breuil; 37–43 Abtei St. Hildegard, Eibingen; 45, 46 Rock Art Research Institute, University of the Witwatersrand, Johannesburg; 47 © Andy Rain/epa/Corbis; 48 Staatliche Museen zu Berlin; 49 © Andy Rain/epa/Corbis.

Index

Numerals in *italics* refer to text illustration numbers.

Abraham 178, 204, 213, 279
Absolute Unitary Being (AUB) 151, 153, 185, 189, 190, 256
Acts of the Apostles (book of the Bible) 33, 36, 38, 47
Adam 241
adaptation 155–6; advantages of 157; mechanism 133
Afghanistan 56
afterlife 123; *see also* Heaven
agnosticism 80, 103, 105, 109, 122, 182, 265, 290
AIDS 121, 155, 271, 275, 282
Albright, William 175
alchemy 72, 73
Alexander the Great 28
altars 196, 249; *30, 43*
Amazon 107, 170; Basin 169
Ananias 35
Anaximander 24
Anaximenes 24
angels 6, 21, 54, 65, 66, 136, 162, 186, 240, 241, 269, 270; *23*; fallen 180
Anglicanism 77, 60, 90, 96, 105, 173, 177, 266, 267, 291; splits in 261
animism 134, 136
antelope 150, 249; *21; 45*
Anne, Queen of England 72
apocalypse 77
Apollo (god) 44
Apostles' Creed 164, 185; *see also* Nicene Creed
Aquinas, Thomas 57, 58, 59, 61, 63, 91, 113, 116, 121, 130, 138, 181, 236, 257, 273; *10; Summa Theologica* 58; proofs 62
Aretaeus 237
Arianism 45, 53, 73
Archimedes 28
Aristotle 27, 28, 29, 30, 31, 32, 37, 38, 39, 48, 49, 57, 59, 64, 66, 75, 95, 108, 180, 198
Arnold, Matthew: *On Dover Beach* 259, 261
asceticism 51, 58, 181, 240
Asensi, Juana 191
astrology 27, 72
astronomy 27, 31, 71, 74
atheism 68, 69, 76, 121, 122, 257ff., 265; *47*
Athens 38, 157, 270
Atlantis 31
Augustine of Hippo, St 8, 50, 52, 54, 55, 58, 63, 64, 66, 87, 91, 97, 116, 117, 118, 121, 147, 173, 180, 181, 195, 273, 283; *9*; conversion of 57;

City of God (*De Civitate Dei*) 50, 54, 147; *Confessions* 52
Aurignacian 221
Auslander, Shalom 204
autism 143
Avalos, Hector 267
Aveling, Edward 6

Babylonia 126
Bacon, Francis 70, 75
baptism 44, 45, 47; failure to be baptized 52
Baptists 266
Basket, Fuegia (person) 98; *16*
Batchelor, Stephen: *Buddhism without Beliefs* 182
Bates, Henry Walter 107
Beagle, HMS 93, 94, 97, 99, 279; *15*
bear (depictions of) 220
Bégouën family 210
Bégouën, Robert 211
Benedictine Order 233
Bermuda 107
Bernard of Clairvaux, St 234
Bernini, Gianlorenzo: *Ecstasy of St Teresa* 190; *28*
Bible 59, 130; divinely inspired 267; King James translation 290; *see also individual books of the Bible*
Bible Belt 164
Big Bang 60, 61, 130
bison (depictions of) 214, 218, 220; *34*
Blair, Tony 283
Blake, William 52, 80, 172
blasphemy 76
bliss 181
Blombos 12, 13, 126, 156, 171, 281; *1*; ochre engravings 14, 19, 155, 171, 196, 207; *2*
blood 15, 16, 18, 41, 67, 69, 178, 203, 245, 249, 268; of Christ 185, 249; *43*
bloodletting 18, 19, 99, 186, 188; *26, 27*
Book of Common Prayer 164, 204
Borneo 170
Bororo (people) 169
Boyle, Robert 70
Brahe, Tycho 70, 71
Brazil 169
Breuil, Henri 217, 219; *33*
brimstone 20, 85
British Empire 136
British Museum 105, 107
Bronze Age 207
Brown, Derren 113
Bruno, Giordano 68
Buddhism 119, 151, 152, 158, 181–2; *22*
Bultmann, Rudolf 179
Bush, George W. 283

Bushmen 232
Button, Jemmy 97, 98; *16*
Byron, George Gordon, Lord 88

Calvin, John 66
Cambridge 92
Canaanites 188
cannibalism 99, 185
Cantor, Norman 58, 185
Carib (people) 169
Carracci, Ludovico: *The Conversion of St Paul* 35; *7*
catastrophism 104
Cathars 234–5
Catholic Church *see* Roman Catholic Church
Catullus 63
caves 11, 16, 17, 165; *see also individual caves*
celestial spheres 65, 71, 72, 85, 287, *11*; music of 65, 66
Chambers, Robert 80, 106, 111; *Vestiges of the Natural History of Creation* 80, 82, 101, 107, 111
chanting 185, 214, 222, 227, 233
Charles, Prince of Wales 280
Chartres Cathedral 198–200; *30*
Chaucer: *Canterbury Tales* 201–203
Chauvet Cave 221
Chesterton, G. K. 273
China: supernatural beliefs in 126
Christ *see* Jesus Christ
Christ's College, Cambridge 91, 92
Christianity 16, 41, 47, 53; *see also* Christ, Jesus; *names of individual communions*
Chukchee (people) 226
Church Fathers 42, 50, 52, 53, 54, 56, 63, 90, 178, 180
Church of England *see* Anglicanism
circumcision 37, 203–205
'City of God' 197ff., 243; *see also* Augustine of Hippo, St
clapping 246–7, 250, 252
Clottes, Jean 211
Clovis I, King of France 45, 63
Coleridge, Samuel Taylor 88, 279
communion 47; *see also* Eucharist; Mass
communism 118
Conan Doyle, Sir Arthur 112, 113, 123
confession (sacrament) 124
consciousness 110, 137, 140, 141, 154, 157; altered states of 36, 46, 137, 144, 209, 229, 247, 251; evolution of 280; spectrum of 144, 155, 158, 172, 182, 186, 210; *19*
Constantine, Emperor 8, 41, 43, 44, 45, 57, 63, 77, 91, 117, 118, 121, 147,

263; vision of 57; 8; see also Nicea, Council of
Constantinople 57
contraception: opposition to 275, 280, 282
conversion 35, 44, 52, 121, 283; by force 55
Cooke, George 112
Copernicus, Nicolas 64–5, 68, 84, 130, 163, 164; De Revolutionibus Orbis Coelestium 64–5; 11
Corinthians (book of the Bible) 35, 39, 242
cosmos and cosmology 16, 17, 18, 162, 186, 192; in religion 162ff.; tiered 27, 170, 173, 197, 210, 216, 226; geocentric structure 63, 71
Cougnac Cave 230
Cosquer Cave 208
Cranmer, Archbishop Thomas 174
Creationism 60, 70, 74, 81, 127, 270, 271; see also Intelligent Design
credo 161–2; see also individual creeds
Crusades 44, 53, 203, 206, 283
crystal spheres see celestial spheres

Dalai Lama 182
Damascus 35, 157; 7
dancing 19, 186, 222, 227, 246, 248, 250, 252, 253, 255; 45
Daniel (Prophet) 45, 46, 73
Dante 58, 104, 164, 225, 226; 23
Darwin, Charles 87ff., 279, 289; 14; on religion 6; beliefs 9; theory of evolution by natural selection 30, 62, 73, 78, 79, 84, 90, 101, 107, 108, 111, 113, 114, 119; burial 290–91; 49; The Descent of Man 101, 281; On the Origin of Species by Means of Natural Selection, or the Preservation of Favoured Races in the Struggle for Life 66, 73, 78, 82, 94, 101, 103, 109, 134, 292; Journal of Researches 106
Darwin, Charles Waring 102
Darwin, Erasmus 78, 79, 80, 88, 100; 13
Darwin, Fox W. 92
Darwin, Robert 88, 90
Davenport Brothers 112
David (Old Testament King) 204
Dawkins, Richard 69, 92, 205, 259, 271; The Blind Watchmaker; The God Delusion 257–8, 278; The Selfish Gene 280
death 18, 81, 131, 132
Declaration of Independence (United States) 75
deduction 30, 75; see also induction
Defoe, Daniel 104
deism 76
de Landa, Bishop 19
Democritus 28

demons 175, 180; see also Devil; Satan
Denis, St 45
Dennett, Daniel 69, 259, 271
Descartes, René 275
Deuteronomy (book of the Bible) 41, 204
Devil 6, 115, 136, 144, 179, 180, 275; see also demons; Satan
De Waal, Frans 281
Diderot, Denis 77
Dieu cornu, Le 218ff.; 33
Diocletian 40, 41
Dobkin de Rios, Marlene 188
Dobzhansky, Theodosius 6
Docherty, Reverend George 260
Dominican Order 58, 68
Donne, John 268, 269
Doré, Gustave 23
Dostoevsky, Fyodor 244
Douglas-Smith, Basil 154
Down House 111
Drakensberg Mountains 255
dreams and dreaming 43, 46, 132, 137, 142, 144, 155, 156, 157, 159, 180, 210, 226; as method of revelation 155, 163, 247; see also consciousness
drumming 19, 169, 214; see also music
Drummond, Henry 85
Dummett, Sir Michael 272
Durkheim, Emile 120, 125, 126, 132, 139, 185

Eastern Orthodox Church see Orthodox Church
Ecclesia 236, 245; 43
eclipse 27, 36
ecstasy 142, 145, 166, 185, 186, 189, 190, 192; 20, 28; through dancing 151, 160
ecumenism 46
Eden 13, 178, 229, 241
education: control by the Church 49, 279
Egypt: ancient 27, 126, 197, 207
Eisenhower, President Dwight 260
Eleusinian cult 47
Eliade, Mircea 170; Shamanism: Archaic Techniques of Ecstasy 166
Eliot, T. S.: The Love Song of J. Alfred Prufrock 272
Elizabeth of Schönau 240
Empyrean 65, 70, 23
Encyclopaedia Britannica 77
L'Encyclopédie 77
endogenous percepts 144
Enlène Cave see Volp Caves, Enlène
Enlightenment 75, 76, 260, 263
entomology 92, 105, 106
entoptic phenomena 144, 148, 239, 248
entheogens see hallucinogens
Ephesians (book of the Bible) 188, 284
Ephesus 36

Epicureans 38
epilepsy 35, 143, 237
epistemological 48
Erasmus 88
eschatology 269
Eucharist 47, 48, 69, 99; see also communion; Mass
Euclid 28
eugenics 122
Eurydice 175
Eusebius 42, 43, 44, 45
Eve 181, 241, 245
evil 52, 62, 105, 125, 180
evolution 40, 62, 78, 84, 101, 105, 106, 132, 277, 280; of the human brain/mind 155; resistance to theories of 114, 260, 286; see also Darwin, Charles
excommunication 47, 81, 259
Exodus (book of the Bible) 41

faith 39, 40, 49, 55, 56, 57, 59, 62, 114, 149, 259, 269, 270, 274, 280, 284; salvation through 37; faith-based education 279
Falkland Islands 97
Fall (Biblical event) 32, 74, 80
fasting 143, 240
Finkelstein, Israel 175
First Cause 60, 130; see also Prime Mover
FitzRoy, Captain Robert 93, 94, 97, 98, 114; 16
flagellation 22, 188; self 143, 190
flight (sensation of) 168, 169, 170, 171, 190, 210; 25, 46
Flood (Biblical event) 85, 177, 260
food taboos 37
form constants 144
Forms (of Plato) 55
fortification scotoma see scotoma, fortification
fossils 105, 260
Foucault, Michel 77
Francis, St 63, 191
Franciscan Order 62
Frazer, James Sir 134, 135; The Golden Bough 134
Freeman, Charles 264
French Revolution 263
Freud, Sigmund 155

Gabillou Cave 210
Galapagos Islands 95
Galatians (book of the Bible) 37
Galilee 176
Galilei, Galileo 8, 31, 61, 64, 66, 67, 163; 12; trial of 68, 69, 72, 78; Dialogue Concerning the Two Chief World Systems 66
Gaudry, Bishop of Laon 49
Genesis (book of the Bible) 20, 21, 175, 178, 241, 242

Gentiles 37, 58; see also heathens
geology 92, 105
Gibbon, Edward: The Decline and Fall of the Roman Empire 77
glossalalia 185, 189
Gnosticism 36
Göbekli Tepe 133, 197, 198, 199, 200
Gomorrah 20, 21, 85, 176–7, 179
Gould, Stephen Jay 286
Gravettian 211
gravity 72, 73, 75
Gray, John 259, 261, 266
Gregory, Bishop of Tours 45
guilt 124; see also confession (sacrament)
Guyana 169

Hades 175; see also Hell
Haeckel, Ernst 40
Halifax, Joan 171
hallucinations 140, 142, 147, 148, 165, 166, 171, 187, 189, 220, 223, 230, 236ff., 247, 265; aural 35, 142, 147, 265; thread 165; olfactory 239, 256; zigzags 240; concentric circles 243; see also flight (sensation of); vortices
hallucinogens 154, 172; hashish 148; mushrooms 189
Handel, G. W.: Messiah 40
Hardy, Thomas 292
harmony of the spheres see celestial spheres
hashish 148
Haydn, Joseph: Creation 76
Hayden, Brian 130
heathens 121
Heaven 47, 49, 123, 163, 164, 183, 241
Hebrews (book of the Bible) 188, 275
Hell 49, 52, 115, 163, 164, 175, 183, 275; see also Hades
Henry VIII, King of England 173, 202
Henslow, John (Reverend) 92, 100
heresy 8, 42, 47, 51–3, 55, 69, 73, 81, 121, 124, 174, 179, 259, 264; 48
Hildegard of Bingen 9, 163, 232–45, 246, 254–5, 275; 37, 39, 40, 41, 42; Scivias 235, 243, 38; Liber Divinorum Operum Simplicis Hominis 235
Hinde Robert 137, 153; Why Gods Persist 137
Hinduism 100, 119
Hitchens, Christopher 271
Hitler, Adolf 32
Hohle-Fels 221
Hohlenstein-Stadel 221
Holy Ghost see Holy Spirit
Holy Land 27, 200
Holy Spirit 144, 145, 164, 178, 236, 240, 275; 20
Homer 88, 89, 104
Homo sapiens 11, 12, 131, 138, 156
homosexuality 53, 121, 177, 262

Hooker, Joseph 100, 102, 104
Horowitz, Mardi 146
horses (depictions of) 220
Houdini, Harry 112, 113
Huichol (people) 171, 172
human sacrifice see sacrifice, human
humanism 182, 278, 282
Humboldt, Alexander von 106
Hume, David 76
humility 62, 159
hunter-gatherers 125, 133
Huxley, Aldous: The Doors of Perception 172
Huxley, Thomas Henry 80, 83, 103, 111
hypnosis 106

icons 196
immanence 173
incense 160
Incubus 180–81
India 166
induction 30; see also deduction
Industrial Revolution 80
infidels 121; see also heathens
Inquisition 68
Intelligent Design 60, 62, 70, 81, 110, 127, 205, 260, 280
Inuit 168, 169, 25
Iraq 56, 282
Irenaeus, Bishop of Lyons 178
Isaac 213
Ishmael 204
Islam 7, 22, 53, 57, 119, 173, 174, 175, 177, 185, 204, 206, 262
Isaiah (Old Testament prophet) 46, 123; vision of 18
Israelites 100, 133, 188, 205

Jacob: vision of 167
James, William 134, 140, 148, 152, 170, 172
Jericho 176
Jerome, St 53
Jerusalem 176
Jesuits see Society of Jesus
Jesus Christ 19, 36, 47, 55, 56, 124, 164, 177, 181, 188, 191, 249, 275, 282; birth of 178, 271; preaching 279; crucifixion of 22, 36, 196, 267; 43; resurrection of 40, 56, 164, 179, 194, 276; ascension of 164; Second Coming of 269
Joan of Arc 147
Joel (book of the Bible) 46
John the Divine, St 188; see also Revelation of St John the Divine
Jonah 178
Joshua 67, 130
Judaism 16, 33, 35, 37, 119, 174, 175, 177, 178, 181, 185, 203–206, 282, 285; Orthodox 138, 204; persecution of 55
Judas: Gospel of 36

Jude (book of the Bible) 177
Judges (book of the Bible) 175–6
Judgment Day 268
Jung, Carl 155
Jutta of Spanheim 233

Kalahari Desert 150, 151, 246, 247, 255; 44
Kepler, Johannes 70, 71; Mysterium Cosmographicum 71
Kingsley, Reverend Charles 82, 123
Klüver, Heinrich 223
Krishna, Gopi 149
Kundalini 149, 151

language 155, 156, 235; secret 255
Lankester, E. Ray 111
Laon Cathedral 49
Lapps (people) 168
Lascaux Cave 208, 210, 220
Lavater, J. C. 93, 94
LeDoux, Joseph 288
Leo XIII, Pope 58
Leonardo da Vinci 168
Les Trois Frères Cave see Volp Caves, Les Trois Frères
Leviticus (book of the Bible) 282
Liberation Theology 124
Linnaean Society 102, 104
lions 151, 217, 254
literacy 47, 263; illiteracy 274
Lloyd, Lucy 165
Locke, John 75
Lodge, Oliver Sir 111
Lot 20, 21, 92, 135, 175, 177, 178, 267; 4
Lourdes (shrine) 147, 200–201
Lubbock, Sir John 104, 134
Luke: Gospel according to 135, 175, 180
Luther, Martin 64, 66, 70, 74, 173
Lyell, Charles 80, 81, 96; Principles of Geology 96, 104; uniformitarianism 100, 102, 107

magic 71, 135, 161, 197
magisteria 286
magnetism 71, 106
Malay Peninsula 107
Malthus, Thomas Robert 108; An Essay on the Principle of Population 108
mammoth (depictions of) 220
Manichaeism 51
mantras 152
Mark: Gospel according to 51
marriage 18, 133
martyrdom 22, 40, 45, 174, 188
Marx, Karl 123, 124
Masada 37
Maskelyne, John Nevil 111, 112
Mass 184, 245; 43; see also communion; Eucharist
Matthew: Gospel according to 51, 124, 178, 180, 243, 264, 275

Matthews, Richard 97
Maxentius, Emperor 42
Maya 16, 17, 27, 124, 126, 133, 165, 186–96, 207, 227, 256; bloodletting 99, 186, 187, 188, 245; 26, 27; human sacrifice by 22, 23, 249; mountain-pyramids 187, 192, 197, 199, 200, 210, 253
McGrath, Alister 261, 266; *The Twilight of Atheism* 260
meditation 142, 143, 151, 152, 160, 186, 189, 227
Memory, Boat 97
Mesmer, Franz Anton 106
Methodism 266
Michelangelo: Sistine Chapel 269, 290
Midgley, Mary 276
migraine 143, 236–40, 256
Miletus 36
Milton, John 104; *Paradise Lost* 180
Milvian Bridge (Battle of the) 42, 43, 44, 117; 8
Minster, York (person) 97, 98; 16
miracles 6, 39, 40, 52, 76–8, 201, 203, 273
monastic orders 159; *see also individual orders*
monotheism: origins of 134ff.
monsters 147, 165, 223
morality 178, 257, 280
More, Thomas Sir: *Utopia* 118
Moses 177, 181
Muhammad, Prophet 277
music 185, 186, 233, 250, 276; mystical 163, 232, 236; of the spheres 65, 66; *see also* chanting; clapping; drumming
mysteries: use of in religion 161, 170
mystery cults 47
mystery plays 161
mysticism 32, 55, 71, 163, 181, 190; *see also names of individual mystics*
myths and mythology 25, 130, 134

natural selection 9, 79, 82, 83, 94, 102, 110; resistance to theory of 114; *see also* evolution; Darwin, Charles
Navajo 226
Neanderthals 11, 210
near-death experiences 143
Nebuchadnezzar 45
Neolithic 128, 197, 207
Neoplatonism 55, 57
neurology 9, 140, 145, 148, 158, 160, 161, 232, 237; experiences generated by 54, 142, 164
New Age 78
New Orleans 53
Newgrange 128; 18
Newton, Isaac 32, 68, 70, 72, 75, 103, 113; *Principia Mathematica* 70, 72
Niger River 104

Nicea, Council of 44, 46, 47, 162, 263, 273
Nicene Creed 46, 121, 162, 164
Nirvana 151, 181
Noah 95, 260
North Africa 53
Numbers (book of the Bible) 143

occult 72
ochre 12, 171; *see also* Blombos, ochre engravings
Old Testament 23, 135; *see also individual books*
Original Sin 52, 161, 178
Orpen, Joseph Millerd 255
Orpheus 175
Orthodox Church 121, 267
Osiander 64, 65, 70
Owen, Richard 105, 283

palaeontology 92, 100, 105
Paley, Reverend William 91, 92, 110, 116; 'watch on the heath' 127, 260
papacy 29, 57, 201, 234, 235, 275; infallibility of 275; 48
Park, Mungo 104
Patagonia 95, 99, 119; natives of 100
Paul, St 8, 35, 36, 40, 42, 43, 45, 47, 48, 51, 55, 58, 81, 87, 91, 117, 118, 121, 147, 157, 188, 204, 233, 242, 262, 270, 274, 275, 284; 7
Pech Merle Cave 220, 230
penance 202
Pérez de Valdivia, Diego 190
persecution 40, 55, 263
Peter, St 177, 262
Peter the Chanter: *Liber de Oratione et Specibus Illius* 190
phosphenes 144, 237–8, 241
physiognomy 93
phrenology 106
piety 56, 234
Piero della Francesca: *St Augustine* 9
pilgrims and pilgrimage 199–203
Pitseciak, Mary: *In the Night Sky* 25
Plato 28, 31, 32, 47, 48, 49, 57, 62, 65, 66, 113; Academy of 38; *Timaeus* 31
Platten, Stephen 266, 267
Pleistocene Ice Age 11
Polynesia 126
polytheism 134, 136
Powell, Evan 112
prayer 122, 143, 160, 180, 189, 192, 278; answered 117, 286
Prime Mover 60, 65, 80; *see also* First Cause
priests and priesthood 17, 25, 26, 28, 42, 76, 128, 130; women as 121; 18
Primum mobile see Prime Mover; First Cause
prophecy 73, 76
Promised Land 67, 123
proselytizing 136

Protestantism 70, 290
Proverbs (book of the Bible) 39
Psalms (book of the Bible) 39, 51, 54, 242, 257; Psalter 59; 10
psychics and psychic phenomena 52, 111
psychotropic drugs and substances 188, 189, 220; *see* hallucinogens
Ptolemaic 64, 65
Pullman, Philip 282
Purgatory 163, 201
Pythagoras 26, 28, 64

rainbow 99; in visions 237, 238
Randi, James 113
Rappaport, Roy 15, 209
Redondi, Pietro 69
Reformation 66, 68, 69, 75, 121, 173, 202, 263, 290
Reichel-Dolmatoff, Geraldo 224
reincarnation 182
reindeer (depictions of) 220
relics 203
religion 8, 23, 64, 114; conflict with science 8, 32, 37, 78, 86, 139, 257ff., 285
resurrection 117; *see also* Jesus Christ
Revelation of St John the Divine (book of the Bible) 54, 118, 175, 178
revelation 36, 45, 51, 110; of knowledge 48, 81, 121, 173, 235, 244; progressive 135
reverie 155, 156, 159
rituals 17, 54, 104, 117, 119, 125, 130, 134, 184, 193, 197, 198, 221, 227, 231, 246, 267; specialists 164
Robertson, David 135, 265
Roman Empire 8, 40, 48, 53, 121
Roman Catholic Church 19, 29, 31, 53, 58, 75, 121, 123, 155, 173, 198; priests in 124; no salvation outside 97
Romans (book of the Bible) 39, 58, 204
Rome 42, 53, 77, 234
Royal Society, London 67, 104; 12

Sacks, Oliver 237, 238, 241, 243, 256
sacraments 234
sacrifice 7, 18, 99, 100, 119, 124, 187, 213, 216; animal 37; human 19, 99 (by Maya) 22, 23, 213, 249
saints 136, 162, 185, 189, 241; *see also names of individual saints*
salvation 48, 51, 121, 183, 185, 189
San 9, 126, 146, 150, 232, 246–53; 21, 44, 45; /Xam 165, 169, 247 (!gi:ten) 149, 150, 165, 247, 249, 251, 254, 255, 256; n/om 150, 151, 198, 247; shamans 151, 168; as egalitarian society 173
Santiago de Compostela 200
Sarai (Sarah) 20

Satan 180, 181, 275; see also Devil; demons
Saul see Paul
scepticism 9, 25, 113, 257ff.
schizophrenia 143, 167, 265
Scholasticism 58, 77, 236, 245
science 23, 29, 32, 40, 48, 56, 57, 58, 64, 122, 125; origin of 27; within its religious milieu 33; conflict with religion 8, 32, 37, 78, 86, 139, 257ff., 285; scientific method 8, 87
Scientific Revolution 73
scotoma 145, 238–40, 248, 253, 256; 35, 36; fortification 144, 145, 147, 238, 243, 248, 250; 40, 41, 42, 45, 46
séances 87, 109, 110, 111, 112
secularism 53, 117, 118, 144, 184, 198, 201
Sedgwick, Adam 81, 92, 100, 101
Shakespeare, William 66, 114
shamans and shamanism 17, 22, 125, 134, 149, 165, 166, 168, 169, 171, 190, 216, 221; training 225
Shanon, Benny 147
Shield Jaguar 187; 26
Siberia 166, 168, 169, 226
Siegel, Ralph 238
Siegel, Ronald 223
Silberman, Neil 175
sin 56, 77, 115, 124, 183, 259; 48; sexual 177; punishment for 177
Sinai 177
Singer, Charles 241
Slade, Henry 111
slavery 45, 56, 97, 262; abolition of 91
sleep 155, 210; see also dreams and dreaming
Smith, Huston: Cleansing the Doors of Perception 172
Smith, Noel 221
Society of Jesus 174
Socrates 31
Sodom 21, 85, 176–7, 179
Sol Invictus 44
solar system 64, 67, 70, 194, 235, 287; heliocentric 64, 66, 68, 71, 73, 74, 84
sorcerers 218–224; 33, 34; see also shamans and shamanism
soul 29, 32, 86, 132, 169, 190, 275, 276; moment of 'ensoulment' 85; immatriality of 288
speaking in tongues see glossalalia
Spencer, Herbert 109; On the Tendency of Varieties to Depart Indefinitely from the Original Type 109
Spenser, Nick 270
spirit realm 7, 19, 117, 149, 151, 159, 165, 171, 187, 228, 248, 251; portals to 248
spiritualism 9, 87, 105, 106, 109, 112, 275; 17
Spottiswode, William 104
Stalin, Josef 32
Stanford, Leland 109

stem cell research 282
Stoichita, Victor 190, 191
Stoicism 38, 47
Stonehenge 128, 195–6
Straits of Magellan 93; 15
Strong, Roy 266
Succubus 180
suicide bombers 22, 123, 206, 280; see also martyrdom
sun 54, 130, 196
Sun-Father 145
supernatural 6, 8, 17, 23, 31, 39, 45, 115, 157, 170, 290; beings 162; interventions 24; realm 49, 62, 161, 179, 180, 184, 203, 271
superstition 76, 181, 236
Sustermans, Justus 67; 12
symbolic thought 15, 155, 156
synesthesia 148

Tapirapé 169
Tarsus 36, 37
teleology 119
televangelists 195, 256, 257, 277, 284
Templeton Foundation 285
Tennyson, Alfred Lord 82
Tertullian 90
Teresa of Avila, St 185–6, 190–91; 28
Thales of Miletus 24, 26, 27, 29, 30, 32, 36, 39, 66
theocracy 32, 38
Theodosius 42
theology 36, 37, 46, 51, 52, 54, 55, 57, 59, 68, 116, 117, 173, 179, 200, 243, 262, 273, 278, 288
therianthropic figures 218ff., 226
Thessalonians (book of the Bible) 275
Thomas Becket, St 202
Thompson, Damian 270
Tierra del Fuego 96, 97
Tillich, Paul 179
torture 45, 81, 124, 235, 263; 48
totemism 134, 136
Tower of Babel 99
trance 35, 132, 151, 186, 209, 225, 240, 246, 249, 251, 256; 45
transmutationism 78, 81, 83
transubstantiation 69, 78, 283
Trinity 45, 46, 53, 54, 56, 136, 161, 174, 183, 243, 273; 39
Tukano (people) 145, 146, 224
tunnel 170, 171; see also vortices
Tupinamba (people) 169
Turkey 133, 197
Turnbull, Richard 267, 270
Turney, Victor and Edith 202
Tylor, Sir Edward 132, 134, 135

underworld 16, 18, 168, 210
uniformitarianism 95, 104
Unitarian 73, 79; 13
United States of America 75, 260–61, 283

Upper Palaeolithic 6, 8, 9, 197; religion in 208ff.; Magdalenian period 126, 210
Ursus, Nicolai Reymers 71
Utopia 118

Valhalla 123
Van Gennep, Arnold 202
Vatican 48, 200
Virgil 88, 89
Virgin Birth 40, 47, 56, 117, 135, 179; see also Jesus Christ
Virgin Mary 40, 136, 164, 178, 185, 198, 200, 202; Immaculate Conception of 40, 271, 283
visions 9, 35, 38, 39, 42–6, 49, 132, 143, 144, 156, 157, 163, 185, 186, 191, 233; 7, 8
Vision Serpent 186, 187, 189; 27
Vitebsky, Piers 170, 171
Volmar 236
Volp Caves (including Enlène Cave and Les Trois Frères Cave) 210ff.; 31
vortices 165–8, 171, 210, 219, 222, 226, 229, 238; 23, 24

Wakefield, Bishop of 270
Wallace, Alfred Russel 9, 87, 102, 104, 105, 275, 280; 17; On the Law which has Regulated the Introduction of New Species 107; 1855 article 108, 114, 117
Walsingham (shrine) 202
Ward, Keith: Is Religion Dangerous? 205, 277
'watch on the heath' 91, 127, 260; see also Paley, Reverend William
Wedgewood, Emma 88, 101, 102
Wedgewood, Josiah 88, 92
Wesley, John 74
Westminster Abbey 113
witches 85, 124, 180
Whitehead, A. N. 10
Whitehouse, Harvey 184
Wilberforce, Bishop Samuel 83
Wilde, Oscar 291
Winkelman, Michael 125
Wolpert, Lewis 271
Wordsworth, William 72, 77, 88, 138, 153, 256, 265
World War I 112
Wright, Tom 268
Wylie, Alison 209

Yakut 166
Yaxchilan 187
Yucatan 19

Xoc, Lady 187; 27

Zen 151
zoology 105